Mitochondria: From Physiology to Pathology

Mitochondria: From Physiology to Pathology

Editor

Francesco Bruni

MDPI • Basel • Beijing • Wuhan • Barcelona • Belgrade • Manchester • Tokyo • Cluj • Tianjin

Editor
Francesco Bruni
Università degli Studi di Bari Aldo Moro
Italy

Editorial Office
MDPI
St. Alban-Anlage 66
4052 Basel, Switzerland

This is a reprint of articles from the Special Issue published online in the open access journal *Life* (ISSN 2075-1729) (available at: https://www.mdpi.com/journal/life/special_issues/mitochondria_life).

For citation purposes, cite each article independently as indicated on the article page online and as indicated below:

LastName, A.A.; LastName, B.B.; LastName, C.C. Article Title. *Journal Name* **Year**, *Volume Number*, Page Range.

ISBN 978-3-0365-2151-0 (Hbk)
ISBN 978-3-0365-2152-7 (PDF)

Contents

About the Editor

Francesco Bruni My main research interests are in the field of biogenesis and mitochondrial function. During my PhD studies, carried out at the University of Bari, I investigated mitochondrial transcription termination mechanisms in Drosophila melanogaster, characterizing two different transcription factors, DmTTF and D-MTERF3. Later, during my post doc scholarship, I studied the function of D-MTERF5 as well as the role of human NRF-2 and Drosophila DREF in mitochondrial biogenesis. In 2010, I began my research activity abroad at the Wellcome Centre for Mitochondrial Research (Newcastle University), a European center of excellence for the study of mitochondrial diseases. I identified and characterized a novel mitochondrial ribonuclease, named REXO2, and participated in numerous clinical projects in collaboration with the diagnostic laboratory. At the WCMR, which possessed two vital resources, the Mitochondrial Disease Patient Cohort and the Mitochondrial BioResource, I also had the opportunity to study mitochondrial translation mechanisms in both physiological and pathological conditions in patients with mitochondrial diseases. In 2017, I came back to Italy and, since then, I have been working at the Department of Biosciences, Biotechnologies and Biopharmaceutics (University of Bari), dividing my time between research and teaching. I investigate the biochemical and clinical features of mitochondrial aa-tRNA synthetases, exploring the therapeutic use of these enzymes. I am also interested in the mechanisms by which mitochondria exercise a level of quality control between mitochondrial transcription and translation processes, as well as the factors responsible for this activity. As a professor, I currently teach molecular biology on the bachelor's degree course in Biological Sciences.

 life

MDPI

Editorial

Mitochondria: From Physiology to Pathology

Francesco Bruni

Department of Biosciences, Biotechnologies and Biopharmaceutics, University of Bari Aldo Moro, Via Orabona 4, 70125 Bari, Italy; francesco.bruni@uniba.it

Citation: Bruni, F. Mitochondria: From Physiology to Pathology. *Life* **2021**, *11*, 991. https://doi.org/10.3390/life11090991

Received: 17 September 2021
Accepted: 19 September 2021
Published: 21 September 2021

Over the past decade, the role of mitochondria has extended beyond those tasks for which these organelles are historically known. Recent proteomics investigations have highlighted the extraordinary complexity of mitochondrial protein organization, which is a reflection of the numerous and disparate mitochondrial functions [1]. These include the synthesis of most of the ATP present in the cell, apoptosis, ion homeostasis, cellular stress response, antioxidant control, redox regulation, mitophagy, involvement in various biosynthetic pathways and many more processes. Furthermore, mitochondria are dynamic organelles, and their morphology can vary significantly within a cell. This is mainly due to the mitochondrial fusion and fission processes, referred as mitochondrial dynamics, which are crucial for the organelles' interactions with other cell compartments and for the cross-talk with the cell cycle and metabolism as well as differentiation and senescence [2].

An intriguing feature of mitochondria is that they possess their own DNA (mtDNA), which is functionally coordinated with the nuclear genome. Indeed, numerous nucleus-encoded proteins are required for complex molecular processes such as replication, transcription, RNA processing and degradation, translation, and assembly of respiratory chain complexes that take place inside the mitochondria [3,4]. On the other hand, mitochondrial gene expression plays an important role in the communication between mitochondria and the nucleus, contributing to the regulation of cell physiology. Recent studies have pointed to novel RNA-driven mechanisms according to which mitochondrial-derived transcripts, in addition to their role in maintaining mitochondrial function, act as signalling molecules modulating the innate immune response [5].

Dysfunctions affecting the mitochondria can lead to various pathological conditions, including aging and neurodegenerative disorders, and are associated with a whole range of complex genetic disorders, known as mitochondrial diseases [6]. In recent years, great advances have been made in the field of mitochondrial diagnostics thanks to the rapid development of next-generation sequencing (NGS) technologies, particularly the whole-genome approach [7]. However, most of the aspects and mechanisms underlying mitochondrial diseases remain unclear and, to date, effective therapies are still lacking. Therefore, a thorough understanding of the molecular processes occurring in mitochondria is essential from a pathological perspective.

The Special Issue "Mitochondria: from Physiology to Pathology" published in *Life* (ISSN 2075-1729) collects a series of research and review articles and aims to provide an updated view of the main topics covering the physiological and the pathological aspects of mitochondrial biology.

Several contributions focus on the mitochondrial genome and its link to the physiopathological aspects of the various mitochondrial processes. Chapman et al. [8] present a comprehensive review on mtDNA metabolism, highlighting the interplay occurring between the organization of the mitochondrial genome, mitochondrial dynamics and cristae structure. After a brief excursus on the organization of mtDNA and on its main processes, such as replication and transcription, the authors deal in depth with the processes of fission and fusion relative to the structure of mtDNA, discussing the known genetic defects that affect mitochondrial dynamics. To complete the picture, they describe the close relationship between the nucleoid, OPA1 and the other "cristae-shaping proteins" constituting the MICOS complex, whose interactions modulate the formation and dynamics of mitochondrial cristae. These are critical process that, if not well regulated, can prompt mitochondrial dysfunction with the onset of neuromuscular diseases. Another important process whose dysregulation can lead to severe pathologies is the elimination of mitochondrial genome. Orishchenko and colleagues [9] review the mechanisms underlying the regulation of heteroplasmy level and mtDNA segregation, illustrating in detail the natural process of mitochondrial genome elimination in both the somatic cells and the germline. The development of methodologies for artificial manipulation of the mtDNA heteroplasmy level that aim either to prevent or potentially cure human diseases associated with mitochondrial dysfunction, is also examined.

Mitochondrial encephalopathy with lactic acidosis and stroke-like episodes (MELAS) is a metabolic disorder caused by point mutations in the mitochondrial tRNALeu$^{(UUR)}$ gene, with a prevalence of A>G substitution in position 3243. Several papers have reported that overexpression of human mitochondrial leucyl-tRNA synthetase (LARS2) or its C-terminal domain (Cterm) has proven effective in rescuing the pathological phenotype in cellular models. In their research article, Capriglia et al. [10] investigated the molecular basis underlying the ability of the Cterm domain in rescuing the MELAS phenotype. The cellular model employed, consisting of a trans-mitochondrial cybrid line with a >95% mutation load, confirmed the therapeutic potential of the Cterm peptide but also showed that its rescuing ability was independent of the mitochondrial bioenergetics, unlike what has previously been observed in other cybrid lines. The authors proposed that the beneficial effect of this peptide could also be mediated by retrograde mitochondrial signals or, alternatively, by its potential ability to bind regulatory RNA in the cytosol. This research indicates that the full understanding of Cterm-rescuing mechanisms imposes the development of tissue-specific cellular models that accurately reproduce the pathological MELAS phenotype.

Another study, based on a cohort of 468 subjects from the Siberian region, was conducted by Kirichenko et al. [11] with the purpose of determining the impact of mitochondrial heteroplasmy measurements in the prognosis of atherosclerosis development. They investigated the association of nine different mtDNA mutations with carotid intima-media thickness (cIMT), a measurement of the artery wall thickness obtained by ultrasound imaging. Several mtDNA variants correlated with the mean cIMT, thus constituting potential prognostic markers. Interestingly, the mutations m.13513G>A and m.14846G>A showed a significant inverse correlation being associated with a low value of cIMT, representing good candidates for the development of anti-atherosclerotic gene therapies.

The dysregulation of mitochondrial functions certainly depends on faulty mitochondrial gene expression at different levels, but also on other factors. One of these, which takes on particular importance in mitochondrial physiology, is the post-translational modification of mitochondrial proteins. Phosphorylation is commonly employed in mitochondria either to modify protein functions or to activate fundamental signalling pathways, inside and outside the mitochondria. Kotrasová et al. [12] center their review on mitochondrial kinases and their role in keeping organelles fully efficient. This function is exerted on various and distinct substrates: mitochondrial import machinery, subunits of respiratory chain complexes and proteins involved in the main steps of mtDNA maintenance and expression. In addition, the activity of mitochondrial kinases is also important for organelle quality control and apoptosis. One of the central key players in mitochondrial quality control pathways is PINK1 (PTEN-induced kinase 1). Barroso Gonçalves and Morais [13] emphasize how valuable the supply of PINK1 is for the mitochondrial clearance through mitophagy and for the maintenance of mitochondrial homeostasis, thereby keeping cells healthy and functional. The precise mechanisms that mediate PINK1 function in the different mitochondrial pathways remain to be elucidated. Nevertheless, the genetic link between PINK1 and Parkinson's Disease (PD) has long been proven, particularly for the familial form of this pathology. Further research needs to be carried out to define whether the mitochondrial homeostasis imbalance, due to the aberrant function of PINK1, is the key element that contributes to the pathological phenotype in both the familiar and sporadic forms of PD.

An important aspect of mitochondrial physiology is its hormone-mediated regulation. Medar et al. [14] analysed mitochondrial response to luteinizing hormone (LH) stimulation in Leydig cells, the major producers of steroid hormones that regulate sexual differentiation and development. In these cells, endocrine function requires the cAMP signalling pathway, whose involvement in modulating respiratory chain complexes activity, ATP production, biogenesis, import, dynamics, and mitochondrial-dependent apoptosis has been widely investigated. The reported results, obtained by different experimental approaches in rat Leydig cells ex vivo and in vivo with a different LH environment and steroidogenic capacity, supported the involvement of LH-cAMP pathway in the regulation of mitochondrial biogenesis and dynamics coupled with mitochondria-mediated steroidogenesis. Additional studies carried out in Leydig and other steroidogenic cells, e.g., derived from adrenal glands and placenta, will shed light on the pathogenic mechanisms triggered by the presence of unhealthy mitochondria in these tissues.

The last research article of this Special Issue focuses on the *Saccharomyces cerevisiae FAD1* gene, encoding the FAD synthase that adenylates the flavin mononucleotide (FMN) to flavin adenine dinucleotide (FAD), an essential coenzyme for various flavoenzymes. Barile and colleagues previously proved that FAD-forming activities, paralleled by FAD precursors uptake in mitochondria and mitochondrial FAD export to cytosol, could be specifically revealed in mitochondria. However, a protein responsible for the synthesis of FAD had never been identified in yeast mitochondria. In this paper [15], the presence of two Fad1p echoforms, dually localised to the cytosol and mitochondria, was reported. Intriguingly, the authors demonstrated the existence of two pools of *FAD1* mRNAs with 3′ untranslated regions (UTRs) of different length and containing a mitochondrial targeting signal. Therefore, the 3′ UTRs would be responsible for the fate of Fad1p echoforms, with the long *FAD1* mRNA generating the mitochondrial Fad1p. In this context, the authors discussed the role of specific RNA binding proteins (e.g., Puf3p) that modulate the import of the mitochondrial Fad1p echoform. Overall, the paper adds a useful piece of knowledge to the post-transcriptional control of genes encoding mitochondrial proteins, proposing the existence of novel regulatory mechanisms in yeast as well as in higher eukaryotes.

Author Contributions: F.B. wrote the manuscript. All authors have read and agreed to the published version of the manuscript.

Funding: Fondi di Ateneo "Contributo Ordinario di Supporto alla Ricerca" 2015/16 and 2017/18 (University of Bari Aldo Moro) to F.B.

Acknowledgments: The guest editor wishes to thank Pasqua Gramegna for the critical reading of the manuscript and Veronica Wang for her precious help with the editorial process. All the contributors and those involved in the peer-review activity of the Special Issue of *Life* 'Mitochondria: from Physiology to Pathology' are gratefully acknowledged.

Conflicts of Interest: The author declares no conflict of interest.

References

1. Pfanner, N.; Warscheid, B.; Wiedemann, N. Mitochondrial Proteins: From Biogenesis to Functional Networks. *Nat. Rev. Mol. Cell Biol.* **2019**, *20*, 267–284. [CrossRef] [PubMed]
2. Giacomello, M.; Pyakurel, A.; Glytsou, C.; Scorrano, L. The Cell Biology of Mitochondrial Membrane Dynamics. *Nat. Rev. Mol. Cell Biol.* **2020**, *21*, 204–224. [CrossRef] [PubMed]
3. Gustafsson, C.M.; Falkenberg, M.; Larsson, N.-G. Maintenance and Expression of Mammalian Mitochondrial DNA. *Ann. Rev. Biochem.* **2016**, *85*, 133–160. [CrossRef] [PubMed]
4. Pearce, S.F.; Rebelo-Guiomar, P.; D'Souza, A.R.; Powell, C.A.; Van Haute, L.; Minczuk, M. Regulation of Mammalian Mitochondrial Gene Expression: Recent Advances. *Trends Biochem. Sci.* **2017**, *42*, 625–639. [CrossRef] [PubMed]
5. Kotrys, A.V.; Szczesny, R.J. Mitochondrial Gene Expression and Beyond—Novel Aspects of Cellular Physiology. *Cells* **2020**, *9*, 17. [CrossRef] [PubMed]
6. Gorman, G.S.; Chinnery, P.F.; DiMauro, S.; Hirano, M.; Koga, Y.; McFarland, R.; Suomalainen, A.; Thorburn, D.R.; Zeviani, M.; Turnbull, D.M. Mitochondrial Diseases. *Nat. Rev. Dis. Primers* **2016**, *2*, 16080. [CrossRef] [PubMed]

7. Schon, K.R.; Ratnaike, T.; van den Ameele, J.; Horvath, R.; Chinnery, P.F. Mitochondrial Diseases: A Diagnostic Revolution. *Trends Genet.* **2020**, *36*, 702–717. [CrossRef] [PubMed]
8. Chapman, J.; Ng, Y.S.; Nicholls, T.J. The Maintenance of Mitochondrial DNA Integrity and Dynamics by Mitochondrial Membranes. *Life* **2020**, *10*, 164. [CrossRef] [PubMed]
9. Zakirova, E.G.; Muzyka, V.V.; Mazunin, I.O.; Orishchenko, K.E. Natural and Artificial Mechanisms of Mitochondrial Genome Elimination. *Life* **2021**, *11*, 76. [CrossRef] [PubMed]
10. Capriglia, F.; Rizzo, F.; Petrosillo, G.; Morea, V.; d'Amati, G.; Cantatore, P.; Roberti, M.; Loguercio Polosa, P.; Bruni, F. Exploring the Ability of LARS2 Carboxy-Terminal Domain in Rescuing the MELAS Phenotype. *Life* **2021**, *11*, 674. [CrossRef] [PubMed]
11. Kirichenko, T.V.; Ryzhkova, A.I.; Sinyov, V.V.; Sazonova, M.D.; Orekhova, V.A.; Karagodin, V.P.; Gerasimova, E.V.; Voevoda, M.I.; Orekhov, A.N.; Ragino, Y.I.; et al. Impact of Mitochondrial DNA Mutations on Carotid Intima-Media Thickness in the Novosibirsk Region. *Life* **2020**, *10*, 160. [CrossRef] [PubMed]
12. Kotrasová, V.; Keresztesová, B.; Ondrovičová, G.; Bauer, J.A.; Havalová, H.; Pevala, V.; Kutejová, E.; Kunová, N. Mitochondrial Kinases and the Role of Mitochondrial Protein Phosphorylation in Health and Disease. *Life* **2021**, *11*, 82. [CrossRef] [PubMed]
13. Gonçalves, F.B.; Morais, V.A. PINK1: A Bridge between Mitochondria and Parkinson's Disease. *Life* **2021**, *11*, 371. [CrossRef] [PubMed]
14. Medar, M.L.J.; Marinkovic, D.Z.; Kojic, Z.; Becin, A.P.; Starovlah, I.M.; Kravic-Stevovic, T.; Andric, S.A.; Kostic, T.S. Dependence of Leydig Cell's Mitochondrial Physiology on Luteinizing Hormone Signaling. *Life* **2020**, *11*, 19. [CrossRef] [PubMed]
15. Bruni, F.; Giancaspero, T.A.; Oreb, M.; Tolomeo, M.; Leone, P.; Boles, E.; Roberti, M.; Caselle, M.; Barile, M. Subcellular Localization of Fad1p in Saccharomyces Cerevisiae: A Choice at Post-Transcriptional Level? *Life* **2021**, *11*, 967. [CrossRef]

Article

Impact of Mitochondrial DNA Mutations on Carotid Intima-Media Thickness in the Novosibirsk Region

Tatiana V. Kirichenko [1,*], Anastasia I. Ryzhkova [2], Vasily V. Sinyov [2,3], Marina D. Sazonova [2], Varvara A. Orekhova [2,3,4], Vasily P. Karagodin [2,5], Elena V. Gerasimova [6], Mikhail I. Voevoda [7], Alexander N. Orekhov [1,2,4], Yulia I. Ragino [7], Igor A. Sobenin [1,2,3] and Margarita A. Sazonova [2,3]

1 Laboratory of Cellular and Molecular Pathology of Cardiovascular System, Research Institute of Human Morphology, 3 Tsyurupy Str., 117418 Moscow, Russia; a.h.opexob@gmail.com (A.N.O.); igor.sobenin@gmail.com (I.A.S.)

2 Laboratory of Angiopathology, Institute of General Pathology and Pathophysiology, 8 Baltiyskaya Str., 125315 Moscow, Russia; ryzhkovaai@gmail.com (A.I.R.); centaureaceanus@mail.ru (V.V.S.); marinasazon1990@gmail.com (M.D.S.); varvaraao@gmail.com (V.A.O.); vpka@mail.ru (V.P.K.); daisy29@mail.ru (M.A.S.)

3 Laboratory of Medical Genetics, National Medical Research Center of Cardiology, 15A 3 Cherepkovskaya Str., 121552 Moscow, Russia

4 Institute for Atherosclerosis Research, Skolkovo Innovative Center, 143025 Moscow, Russia

5 Department of Commodity Science and Expertise, Plekhanov Russian University of Economics, 125993 Moscow, Russia

6 V.A.Nasonova Research Institute of Rheumatology, 34A Kashirskoe sh., 115522 Moscow, Russia; gerasimovaev@list.ru

7 Research Institute of Internal and Preventive Medicine–Branch of the Institute of Cytology and Genetics, Siberian Branch of Russian Academy of Sciences, 630089 Novosibirsk, Russia; mvoevoda@ya.ru (M.I.V.); ragino@mail.ru (Y.I.R.)

* Correspondence: t-gorchakova@mail.ru; Tel.: +7-910-461-5845

Received: 18 July 2020; Accepted: 20 August 2020; Published: 22 August 2020

Abstract: The search for markers of predisposition to atherosclerosis development is very important for early identification of individuals with a high risk of cardiovascular disease. The aim of the present study was to investigate the association of mitochondrial DNA mutations with carotid intima-media thickness and to determine the impact of mitochondrial heteroplasmy measurements in the prognosis of atherosclerosis development. This cross-sectional, population-based study was conducted in 468 subjects from the Novosibirsk region. It was shown that the mean (carotid intima-media thickness) cIMT correlated with the following mtDNA mutations: m.15059G>A ($r = 0.159$, $p = 0.001$), m.12315G>A ($r = 0.119$; $p = 0.011$), m.5178C>A ($r = 0.114$, $p = 0.014$), and m.3256C>T ($r = 0.130$, $p = 0.011$); a negative correlation with mtDNA mutations m.14846G>A ($r = -0.111$, $p = 0.042$) and m.13513G>A ($r = -0.133$, $p = 0.004$) was observed. In the linear regression analysis, the addition of the set of mtDNA mutations to the conventional cardiovascular risk factors increased the ability to predict the cIMT variability from 17 to 27%. Multi-step linear regression analysis revealed the most important predictors of mean cIMT variability: age, systolic blood pressure, blood levels of total cholesterol, LDL and triglycerides, as well as the mtDNA mutations m.13513G>A, m.15059G>A, m.12315G>A, and m.3256C>T. Thus, a high predictive value of mtDNA mutations for cIMT variability was demonstrated. The association of mutation m.13513G>A and m.14846G>A with a low value of cIMT, demonstrated in several studies, represents a potential for the development of anti-atherosclerotic gene therapy.

Keywords: atherosclerosis; carotid intima-media thickness; mitochondrial mutations; cardiovascular risk factors

1. Introduction

Current knowledge about the mechanisms of atherosclerosis development has significantly increased in recent decades, but atherosclerotic-based diseases still occupy a leading position in the structure of mortality in developed countries [1]. In this regard, the early detection and treatment of patients with a high risk of atherosclerosis development is a primary medical and social problem. Timely prevention of subclinical atherosclerosis has a potential to decrease cardiovascular morbidity and mortality. The search for markers of subclinical atherosclerosis is one of the fundamental factors in the identification of individuals with a high risk of cardiovascular disease (CVD). The thickness of the intimal-medial layer of the carotid arteries (cIMT) is considered to be a direct non-invasive marker of subclinical atherosclerosis, used in clinical and epidemiological studies to assess the effect of conventional and new cardiovascular risk factors (CVRFs) on the progression of atherosclerotic lesions [2,3]. However, conventional CVRFs are poorly associated with cIMT, which suggests the presence of other factors determining the risk atherosclerosis development [4,5].

Currently, the existence of a genetic predisposition to the development of atherosclerosis is not in doubt. In particular, a family history of CVD (coronary artery disease, acute myocardial infarction, and arterial hypertension in first-degree relatives under 60 years of age) is an independent predictor of the risk of these diseases and is taken into account in many modern algorithms for cardiovascular risk prediction [6]. The association of numerous polymorphisms of the nuclear genome with risk of CVD has been revealed [7]. At the same time, the effect of mutations of the mitochondrial genome is more significant for the formation of atherosclerotic lesions since mitochondrial dysfunction leads to activation of the key factors of atherogenesis, such as oxidative stress and inflammation [8]. Nowadays, the role of mitochondrial genome mutations in the predisposition to atherosclerosis development and cardiovascular disease is being investigated [9].

In our previous studies, variants of mitochondrial heteroplasmy were identified in samples of the human aorta with atherosclerotic lesions [10]. In several populations, the association of mitochondrial mutations in blood leukocytes with coronary artery disease and carotid atherosclerosis has been studied [11–13]. It was revealed in the IMPROVE study, a European multicenter study aimed to investigate the prognostic value of cIMT for future cardiovascular events, that latitude is an important determinant of cIMT in Europe and traditional CVRFs do not fully explain cIMT variability in European population, which suggests the presence of other factors, including hereditary ones, in formation of a predisposition to atherosclerosis development [14]. In this regard, replicative studies on the association of mitochondrial heteroplasmy and carotid atherosclerosis need to be carried out in other regions that differ in ethnic composition, as well as in socio-economic and geographical conditions. We have previously investigated the association between mitochondrial heteroplasmy and carotid atherosclerosis in a population of the European part of Russia [15]. The objective of the present research is to study the possible relationship of mitochondrial heteroplasmy and cIMT variability in the Siberian region.

2. Results

In total, 500 participants were included in the study; 32 of them were excluded from the analysis due to unreadable ultrasound images (13) and the insufficient quality of their DNA samples (19). Therefore, 134 male and 334 female samples were analyzed. The clinical and laboratory characteristics of the study participants are presented in Table 1. Female and male groups were different by the following parameters: there were significantly more smokers in the male group ($p = 0.038$); men had higher systolic blood pressure ($p = 0.001$) and a lower level of high-density lipoproteins ($p = 0.032$). The additional estimated parameter in the female group was the period after menopause, which averaged 12.4 (7.4) years.

Table 1. Clinical and laboratory characteristics of the study participants.

Characteristic	Total Group	Men	Women	Difference, *p*-Value
Age, years	62.3 (5.7)	62.9 (5.4)	62.1 (5.8)	0.141
BMI, kg/m^2	28.6 (4.9)	28.4 (4.9)	28.7 (4.9)	0.515
SBP, mm Hg	130 (16)	134 (18)	127 (15)	0.001
DBP, mm Hg	81 (10)	82 (9)	80 (10)	0.121
Smoking, %	9	14	7	0.038
Diabetes, %	6	3	8	0.190
Family history of CVD, %	34	32	35	0.398
Hypotensive therapy, %	55	51	57	0.442
Total Cholesterol, mg/dL	231.7 (42.3)	227.4 (40.4)	233.5 (42.9)	0.131
TG, mg/dL	119.1 (73.3)	120.9 (74.6)	118.4 (72.8)	0.732
HDL, mg/dL	49.8 (14.2)	47.8 (12.8)	50.6 (14.7)	0.032
LDL, mg/dL	158.5 (38.3)	155.2 (37.0)	159.8 (38.8)	0.213
cIMT, mm	0.778 (0.170)	0.772 (0.149)	0.777 (0.150)	0.731

Mitochondrial DNA (mtDNA) mutations m.13513G>A, m.3336T>C, m.15059G>A, m.12315G>A, m.1555A>G, m.5178C>A, m.14459G>A, m.14846G>A, and m.3256C>T were estimated in all the study participants. Figure 1 represents an example of the results of the pyrosequencing for mutation m.5178.

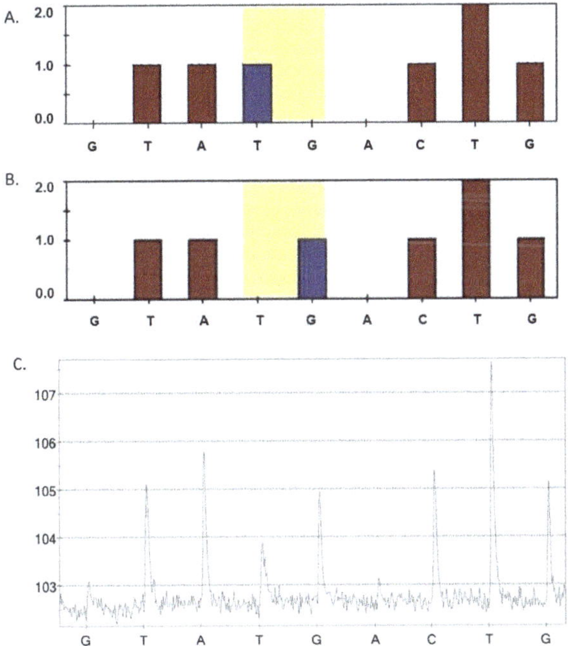

Figure 1. Sample of the sequencing results for mutation m.5178C>A. For the detection of mutation m.5178C>A, a reverse primer for sequencing was used, so the replacement of G by T was analyzed. (**A**) The theoretical height of the nucleotide peaks in homoplasmy for the mutant allele 5178A in the mitochondrial genome; (**B**) the theoretical height of the nucleotide peaks in homoplasmy for the normal allele 5178C in the mitochondrial genome; and (**C**) a pyrogram of a study participant's DNA sample revealed the heteroplasmy level for mutation m.5178C>A is 48%.

Table 2 demonstrates the levels of mtDNA mutations in total and in men and women separately. The difference between the male and female group was observed only for mutation m.15059G>A, which was significantly higher in women ($p = 0.009$).

Table 2. The MtDNA mutation levels of the study participants.

MtDNA Mutations, %	Total Group	Men	Women	Difference, *p*-Value
m.13513G>A	24.3 (13.8)	23.8 (13.5)	24.5 (13.9)	0.635
m.3336T>C	3.9 (0.4)	3.1 (4.7)	4.2 (9.5)	0.086
m.15059G>A	11.7 (14.3)	9.5 (9.3)	12.6 (15.8)	0.009
m.12315G>A	36.0 (24.1)	35.4 (23.3)	36.2 (24.4)	0.714
m.1555A>G	22.8 (12.5)	24.2 (12.8)	22.2 (12.4)	0.070
m.5178C>A	5.8 (12.3)	4.8 (11.3)	6.3 (12.7)	0.216
m.14459G>A	4.1 (5.5)	4.4 (5.4)	4.1 (5.6)	0.540
m.14846G>A	17.1 (16.6)	17.7 (18.5)	16.9 (15.8)	0.654
m.3256C>T	15.0 (16.3)	15.4 (15.5)	14.9 (16.7)	0.772

Further, the association of mtDNA with mean cIMT was estimated and demonstrated in Table 3. In the total group, the mean cIMT correlated significantly with the following mtDNA mutations: m.15059G>A, m.12315G>A, m.5178C>A, and m.3256C>T, as well as inversely correlated with m.13513G>A and m.14846G>A. While the total group was divided by sex, the negative correlation of m.13513G>A with mean cIMT was observed in both men and women, while the correlation of m.14846G>A with cIMT reached statistical significance neither in men nor in women. A positive correlation of mutations m.15059G>A, m.12315G>A, and m.5178C>A was found only in women and m.3256C>T only in men.

Table 3. Correlation of mtDNA mutations with mean cIMT.

MtDNA Mutations (*r*; *p*)	Total Group	Men	Women
m.13513G>A	−0.133 *; 0.004	−0.173 *; 0.048	−0.119 *; 0.030
m.3336T>C	0.064; 0.176	−0.101; 0.251	0.097; 0.079
m.15059G>A	0.159 *; 0.001	0.006; 0.942	0.208 *; <0.001
m.12315G>A	0.119 *; 0.011	−0.040; 0.655	0.179 *; 0.001
m.1555A>G	−0.030; 0.525	−0.120; 0.193	0.011; 0.843
m.5178C>A	0.114 *; 0.014	−0.078; 0.375	0.181 *; 0.001
m.14459G>A	0.058; 0.220	−0.152; 0.085	0.138 *; 0.012
m.14846G>A	−0.111 *; 0.042	−0.144; 0.100	−0.071; 0.195
m.3256C>T	0.130 *; 0.011	0.297 *; 0.002	0.069; 0.257

r, Pearson's correlation coefficient, with the significance of the correlation shown. *, statistical significance at $p < 0.05$.

The next stage of the study was a multi-step regression analysis that was performed to identify the factors that may affect cIMT progression in a population from the Novosibirsk region. Mean cIMT was used as a dependent variable, while variants of mitochondrial heteroplasmy as well as conventional cardiovascular risk factors were used as independent variables. The initial independent variables were age, sex, systolic and diastolic blood pressure, BMI, status of smoking, diabetes, a family history of CVD, parameters of the lipid profile—total cholesterol, triglycerides, and low- and high-density lipoproteins—as well as the levels of all the determined mtDNA mutations. The initial linear regression model, including the 20 variables listed above, explained the variability of mean cIMT with $r = 0.521$, $r^2 = 0.274$, $p < 0.001$. The least significant independent variables were sequentially excluded from the analysis to obtain the most predictive linear regression model. In the resulting linear regression model ($r = 0.462$, $r^2 = 0.218$, $p < 0.001$) the following parameters determined the variability of the dependent variable, mean cIMT: age, systolic blood pressure, blood levels of total cholesterol, LDL and triglycerides, and the mtDNA mutations m.13513G>A, m.15059G>A, m.12315G>A, and m.3256C>T. The exclusion of all mtDNA mutations from the initial analysis led to a decrease in the determination coefficient, i.e., deteriorate the explanatory capacity of the model to $r = 0.413$, $r^2 = 0.171$, $p < 0.001$. A set of levels of mitochondrial mutations without other factors explained the variability of mean cIMT with $r = 0.333$, $r^2 = 0.111$, $p < 0.001$.

In women, the initial set of predictors determined a cIMT value with $r = 0.583$, $r^2 = 0.340$, $p = 0.005$; the most predictive variables were age, period after menopause, diabetes, blood levels of total cholesterol, LDL and triglycerides, and mtDNA mutations m.13513G>A, m.14459G>A, and m.5178C>A ($r = 0.540$, $r^2 = 0.292$, $p < 0.001$). For men, the initial linear regression model was not significant ($r = 0.657$, $r^2 = 0.431$, $p = 0.227$); the most important predictors of mean cIMT variability in the male group were age, blood levels of total cholesterol, HDL and triglycerides, smoking, diabetes, and mitochondrial heteroplasmy m.3256C>T ($r = 0.600$, $r^2 = 0.360$, $p = 0.003$).

3. Discussion

In present study, the relationship between mutations of the mitochondrial genome with carotid atherosclerosis in the Novosibirsk region was studied for the first time. It was demonstrated that variants of mitochondrial heteroplasmy m.15059G>A, m.12315G>A, m.5178C>A, and m.3256C>T significantly correlated with the intima-media thickness of common carotid arteries; a negative correlation of m.13513G>A and m.14846G>A with mean cIMT was revealed. When a correlation analysis was carried out in men and women separately, it was shown that the mean cIMT correlated with m.13513G>A, m.15059G>A, m.12315G>A, m.5178C>A, and m.14459G>A in the female group and only with m.13513G>A and m.3256C>T in the male group. Probably, these results are associated with one of the main limitations in this study—the unequal volume of the male and female groups, so most correlations did not reach statistical significance in male group due to the small sample size. In previous studies, the association of variants of mitochondrial heteroplasmy was demonstrated in other populations. The association of mitochondrial heteroplasmy in the European part of Russia partially coincides with the results of the present study in the Siberian region. A previous study in the Moscow region revealed the positive correlation between carotid atherosclerosis and the following mitochondrial mutations: m.652delG, m.3256C>T, m.3336T>C, m.5178C>A, m.12315G>A, m.14459G>A, and m.15059G>A; and an inverse correlation for m.652insG, m.13513G>A, and m.14846G>A. [15]. In subjects derived from a Kazakhstan population, m.12315G>A was associated with cIMT only in women, but a negative correlation of m.13513G>A with cIMT was confirmed in the total group [16].

The present study also included an estimation of the predictive value of mitochondrial heteroplasmy for cIMT variability along with conventional CVRFs using a linear regression analysis. The linear regression model, including a set of conventional cardiovascular risk factors together with the investigated mitochondrial mutations, predicted the cIMT variability with $r^2 = 0.274$. The coefficient of determination r^2 shows that this set of factors explains only 27% of variability of the mean cIMT in the presented linear regression model; however, the exclusion of the levels of mitochondrial heteroplasmy from the analysis led to a decrease in the predictive ability of this model by 10% ($r^2 = 0.171$), which indicates a high contribution of mitochondrial mutations in cIMT variability in participants of the present study. Exclusion of the non-significant variables during the multi-step regression analysis resulted in a decrease in the determination coefficient to 0.218, i.e., reduced the explanting ability of the model to 22%, but allowed identifying the most predictive variables—age, systolic blood pressure, parameters of the lipid profile, and the mtDNA mutations m.13513G>A, m.15059G>A, m.12315G>A, m.3256C>T. In our previous study, the factors affecting the cIMT value in subjects from Moscow were determined. It was shown that the same set of conventional cardiovascular risk factors explained 21% of the cIMT variability. The inclusion of the m.652delG, m.3256C>T, m.13513G>A, m.14459G>A, and m.15059G>A heteroplasmy levels in the linear regression model provided a significantly better explanatory level of 36% [17].

The association of mitochondrial genome mutations, occurring during ontogenesis or inherited through the maternal line, with the development of a number of pathological conditions in humans is widely known [18,19]. One more limitation of the present study was the fact that detected levels of mitochondrial heteroplasmy did not reach 50%, while it is known that the level of mitochondrial heteroplasmy should exceed 50% for development of clinical manifestations [20]. However, a number of other studies have demonstrated a significant relationship between cardiovascular diseases associated with atherosclerosis and some mutations in the mitochondrial genome, determined at a level comparable to the results of the current study. For example, in a study of 482 patients with coronary heart disease, the variant of mitochondrial heteroplasmy m.16189T>C, at a level of 21.6%, was associated with an increased risk of coronary heart disease compared to healthy participants who had a 4.5% prevalence of m.16189T>C [21]. In another study it was shown that mtDNA with a 4977 bp deletion in blood cells was at significantly higher level in patients with coronary artery disease in comparison with healthy subjects (26.2% vs. 4.5%) [22].

4. Materials and Methods

4.1. Study Design

The present study aimed to investigate the association of mitochondrial mutations with cIMT and to determine the most important factors affecting the development of carotid atherosclerosis in a Siberian population with a special focus on genetic predisposition, namely, mtDNA mutations. This cross-sectional, population-based trial was conducted in men and women aged 55–79 years from the Novosibirsk region. Conventional CVRFs, such as body mass index (BMI), systolic and diastolic blood pressure, smoking status, history diabetes mellitus, family history of CVD, treatment of arterial hypertension, blood lipids parameters (total cholesterol, triglycerides, and low- and high-density lipoproteins) were assessed. All study participants were free of CVD. The recruitment of patients was carried out at the Research Institute of Internal and Preventive Medicine, Novosibirsk, Russia. The additional inclusion criterion in the female group was time after menopause (>5 years). History of angina pectoris, myocardial infarction, intermittent claudication, stroke or transient ischemic attacks, as well as arterial revascularization was exclusion criteria. The thickness of the intima-media layer of the common carotid arteries was determined as a direct quantitative characteristic of carotid atherosclerosis. Blood samples of the study participants were obtained to measure the lipid profile and the levels of mtDNA mutations previously identified as related to atherosclerosis and cardiovascular disease. The study was performed in accordance with the Declaration of Helsinki of 1975 and its revised version of 2013. The study protocol was approved by the Institute for Atherosclerosis Research Committee on Human Research (Moscow, Russia). All study participants provided written informed consent prior the inclusion in the study.

4.2. Measurements of cIMT

B-mode high-resolution ultrasonography with a linear array vascular probe 7.5 MHz on ultrasonic scanner SonoScape SSI-1000 (SonoScape, Shenzhen, China) was used for examination of the carotid arteries. The left and right common carotid arteries, the carotid bifurcation area, as well as the external and internal carotid arteries were scanned [14]. The measurements of cIMT were performed at the far wall of the common carotid artery 10 mm opposite the top of the carotid bifurcation in three fixed projections—lateral, anterolateral, and posterolateral. The carotid ultrasound was carried out by one researcher throughout the study. Frozen images of the far wall of right and left carotid arteries in three projections (6 images for each person) were saved for subsequent analysis using the dedicated software package M'Ath (Metris, SRL, Argenteuil, France). The cIMT was measured as the distance from the leading edge of the first echogenic area to the leading edge of the second echogenic area. The average of six measurements was considered as an integral indicator of the intima-media thickness.

4.3. DNA Isolation, PCR and Pyrosequencing

To perform pyrosequencing, total mtDNA was isolated from whole venous blood by phenol-chloroform extraction using the method previously developed by our group on the basis of the method by Maniatis et al. [23]. Polymerase chain reaction (PCR) was performed to obtain fragments of mtDNA containing the region of the investigated mutations [24]. MtDNA samples were kept in TE buffer at a concentration of 0.03 μg/μL. The analysis of the heteroplasmy level in the mutations was carried out using the quantitative method, developed by M. A. Sazonova et al. Pyrosequencing was performed by using the PSQ HS96MA (Biotage, Uppsala, Sweden) device, to determine the defective allele quantified by analyzing the peak heights in the pyrogram of the one-chained PCR-fragments of a mitochondrial genome [25]. The heteroplasmy level was measured as a percent of the mtDNA mutant copies. In our previous study we investigated 40 mitochondrial mutations in samples of the aorta's intima and revealed a high level of heteroplasmy for some mutations in the atherosclerotic lesions [24]. Further, these mutations were selected to assess the relationship between mitochondrial heteroplasmy in the blood leukocytes and ultrasound indicators of carotid atherosclerosis. The levels of the following mtDNA mutations were measured: m.13513G>A, m.3336T>C, m.15059G>A, m.12315G>A, m.1555A>G, m.5178C>A, m.14459G>A, m.14846G>A, and m.3256C>T.

4.4. Statistical Analysis

Statistical analysis was performed using SPSS 27.0 (IBM SPSS Statistics, IBM Corp., Armonk, NY, USA,) [26]. A Kolmogorov–Smirnov test with Lilliefors's correction was performed to estimate the data distribution. The comparison of mean values for continuous variables was performed by a Mann–Whitney test, and for categorical variables by a chi-square Pearson's test. Results were expressed in terms of the mean and standard deviation. Significance was defined at the 0.05 level of confidence. Pearson's correlation coefficient was used for the correlation analysis between the cIMT and mtDNA mutations. Multi-step linear regression analysis was performed to identify the factors affecting the cIMT.

5. Conclusions

Association of the mtDNA mutations m.12315G>A, m.15059G>A, m.3256C>T, and m.5178C>A, and an inverse correlation of m.13513G>A and m.14846G>A, with thickening of the intimal-medial layer of carotid arteries was found in the study participants from the Novosibirsk region. The set of investigated mtDNA mutations had a high predictive value for cIMT variability and increased the explanatory ability of the linear regression models by 10% when added to conventional cardiovascular risk factors. The strong association of the mutation m.13513G>A and m.14846G>A with a low value of cIMT, demonstrated in several studies, represents a potential for the development of anti-atherosclerotic gene therapy, but further replication studies in other populations and in larger samples are required.

Author Contributions: Conceptualization, M.A.S. and A.N.O.; methodology, M.A.S. and I.A.S.; validation, T.V.K., V.V.S. and A.I.R.; formal analysis, T.V.K., M.A.S., V.V.S. and I.A.S.; investigation, A.I.R., M.D.S., V.A.O., Y.I.R. and V.V.S.; resources, E.V.G., Y.I.R. and M.I.V.; data curation, V.P.K., V.A.O. and T.V.K.; writing—original draft preparation, T.V.K.; writing—review and editing, I.A.S., T.V.K. and V.P.K.; visualization, E.V.G., M.D.S. and A.I.R.; supervision, I.A.S. and M.I.V.; project administration, I.A.S., A.N.O. and M.I.V.; funding acquisition, M.A.S., I.A.S., A.N.O. All authors have read and agreed to the published version of the manuscript.

Funding: This study was supported by Russian Science Foundation (Grant # 20-15-00364).

Conflicts of Interest: The authors declare no conflict of interest. The funders had no role in the design of the study; in the collection, analyses, or interpretation of data; in the writing of the manuscript, or in the decision to publish the results.

Abbreviations

BMI	Body mass index
cIMT	Carotid intima-media thickness
CVD	Cardiovascular disease
CVRFs	Conventional cardiovascular risk factors
DBP	Diastolic blood pressure
HDL	High-density lipoproteins
LDL	Low-density lipoproteins
mtDNA	Mitochondrial DNA
SBP	Systolic blood pressure

References

1. Benjamin, E.J.; Muntner, P.; Alonso, A.; Bittencourt, M.S.; Callaway, C.W.; Carson, A.P.; Chamberlain, A.M.; Chang, A.R.; Cheng, S.; Das, S.R.; et al. Heart disease and stroke statistics-2019 update: A report from the American Heart Association. *Circulation* **2019**, *139*, e56–e528. [CrossRef] [PubMed]
2. Bots, M.L.; Evans, G.W.; Tegeler, C.H.; Meijer, R. Carotid intima-media thickness measurements: Relations with atherosclerosis, risk of cardiovascular disease and application in randomized controlled trials. *Chin. Med. J.* **2016**, *129*, 215–226. [CrossRef] [PubMed]
3. Kirichenko, T.V.; Myasoedova, V.A.; Ravani, A.L.; Sobenin, I.A.; Orekhova, V.A.; Romanenko, E.B.; Poggio, P.; Wu, W.-K.; Orekhov, A.N. Carotid atherosclerosis progression in postmenopausal women receiving a mixed phytoestrogen regimen: Plausible parallels with kronos early estrogen replacement study. *Biology* **2020**, *9*, 48. [CrossRef] [PubMed]
4. Adams, M.R.; Nakagomi, A.; Keech, A.; Robinson, J.; McCredie, R.; Bailey, B.P.; Freedman, S.B.; Celermajer, D.S. Carotid intima-media thickness is only weakly correlated with the extent and severity of coronary artery disease. *Circulation* **1995**, *92*, 2127–2134. [CrossRef]
5. Lucaroni, F.; CicciarellaModica, D.; Macino, M.; Palombi, L.; Abbondanzieri, A.; Agosti, G.; Biondi, G.; Morciano, L.; Vinci, A. Can risk be predicted? An umbrella systematic review of current risk prediction models for cardiovascular diseases, diabetes and hypertension. *BMJ Open* **2019**, *9*, e030234. [CrossRef]
6. Hindieh, W.; Pilote, L.; Cheema, A.; Al-Lawati, H.; Labos, C.; Dufresne, L.; Engert, J.C.; Thanassoulis, G. Association between family history, a genetic risk score, and severity of coronary artery disease in patients with premature acute coronary syndromes. *Arterioscler. Thromb. Vasc. Biol.* **2016**, *36*, 1286–1292. [CrossRef]
7. Rankinen, T.; Sarzynski, M.A.; Ghosh, S.; Bouchard, C. Are there genetic paths common to obesity, cardiovascular disease outcomes, and cardiovascular risk factors? *Circ. Res.* **2015**, *116*, 909–922. [CrossRef]
8. Geto, Z.; Molla, M.D.; Challa, F.; Belay, Y.; Getahun, T. Mitochondrial dynamic dysfunction as a main triggering factor for inflammation associated chronic non-communicable diseases. *J. Inflamm. Res.* **2020**, *13*, 97–107. [CrossRef]
9. Finsterer, J. Atherosclerosis can be mitochondrial: A review. *Cureus* **2020**, *12*, e6987. [CrossRef]
10. Sobenin, I.A.; Sazonova, M.A.; Postnov, A.Y.; Bobryshev, Y.V.; Orekhov, A.N. Mitochondrial mutations are associated with atherosclerotic lesions in the human aorta. *Clin. Dev. Immunol.* **2012**, *2012*, 832464. [CrossRef]
11. Kirichenko, T.V.; Sobenin, I.A.; Khasanova, Z.B.; Orekhova, V.A.; Melnichenko, A.A.; Demakova, N.A.; Grechko, A.V.; Orekhov, A.N.; Ble Castillo, J.L.; Shkurat, T.P. Data on association of mitochondrial heteroplasmy and cardiovascular risk factors: Comparison of samples from Russian and Mexican populations. *Data Brief* **2018**, *12*, 16–21. [CrossRef] [PubMed]
12. Mitrofanov, K.Y.; Zhelankin, A.V.; Shiganova, G.M.; Sazonova, M.A.; Bobryshev, Y.V.; Postnov, A.Y.; Sobenin, I.A.; Orekhov, A.N. Analysis of mitochondrial DNA Heteroplasmic mutations A1555G, C3256T, T3336C, C5178A, G12315A, G13513A, G14459A, G14846A and G15059A in CHD patients with the history of myocardial infarction. *Exp. Mol. Pathol.* **2016**, *100*, 87–91. [CrossRef] [PubMed]
13. Sazonova, M.A.; Zhelankin, A.V.; Barinova, V.A.; Sinyov, V.V.; Khasanova, Z.B.; Postnov, A.Y.; Sobenin, I.A.; Bobryshev, Y.V.; Orekhov, A.N. Dataset of mitochondrial genome variants associated with asymptomatic atherosclerosis. *Data Brief* **2016**, *7*, 1570–1575. [CrossRef] [PubMed]

14. Baldassarre, D.; Nyyssönen, K.; Rauramaa, R.; de Faire, U.; Hamsten, A.; Smit, A.J.; Mannarino, E.; Humphries, S.E.; Giral, P.; Grossi, E.; et al. Cross-sectional analysis of baseline data to identify the major determinants of carotid intima-media thickness in a European population: The IMPROVE study. *Eur. Heart J.* **2010**, *31*, 614–622. [CrossRef]
15. Sazonova, M.A.; Sinyov, V.V.; Ryzhkova, A.I.; Galitsyna, E.V.; Khasanova, Z.B.; Postnov, A.Y.; Yarygina, E.I.; Orekhov, A.N.; Sobenin, I.A. Role of mitochondrial genome mutations in pathogenesis of carotid atherosclerosis. *Oxid. Med. Cell. Longev.* **2017**, *2017*, 6934394. [CrossRef] [PubMed]
16. Kirichenko, T.V.; Ragino, Y.I.; Voevoda, M.I.; Urazalina, S.J.; Khasanova, Z.B.; Orekhova, V.A.; Sinyov, V.V.; Sazonova, M.A.; Orekhov, A.N.; Sobenin, I.A. Data on association of mitochondrial heteroplasmy with carotid intima-media thickness in subjects from Russian and Kazakh populations. *Data Brief* **2020**, *29*, 105136. [CrossRef] [PubMed]
17. Sobenin, I.A.; Myasoedova, V.A.; Kirichenko, T.V.; Orekhova, V.A.; Khasanova, Z.B.; Sinyov, V.V.; Melnichenko, A.A.; Grechko, A.V.; Orekhov, A.N. Profiling of risk of subclinical atherosclerosis: Possible interplay of genetic and environmental factors as the update of conventional approach. *Vessel Plus* **2019**, *3*, 15. [CrossRef]
18. Wallace, D.C. A mitochondrial paradigm of metabolic and degenerative diseases, aging, and cancer: A dawn for evolutionary medicine. *Annu. Rev. Genet.* **2005**, *39*, 359. [CrossRef]
19. Finsterer, J. Secondary manifestations of mitochondrial disorders. *J. Zhejiang Univ. Sci. B* **2020**, *21*, 590–592. [CrossRef]
20. Smith, P.M.; Lightowlers, R.N. Altering the balance between healthy and mutated mitochondrial DNA. *J. Inherit. Metab. Dis.* **2011**, *34*, 309–313. [CrossRef]
21. Mueller, E.E.; Eder, W.; Ebner, S.; Schwaiger, E.; Santic, D.; Kreindl, T.; Stanger, O.; Paulweber, B.; Iglseder, B.; Oberkofler, H.; et al. The mitochondrial T16189C polymorphism is associated with coronary artery disease in Middle European populations. *PLoS ONE* **2011**, *6*, 16455. [CrossRef] [PubMed]
22. Botto, N.; Berti, S.; Manfredi, S.; Al-Jabri, A.; Federici, C.; Clerico, A.; Ciofini, E.; Biagini, A.; Andreassi, M.G. Detection of mtDNA with 4977 bp deletion in blood cells and atherosclerotic lesions of patients with coronary artery disease. *Mutat. Res.* **2005**, *570*, 81–88. [CrossRef] [PubMed]
23. Little, P.F.; Treisman, R.; Bierut, L.; Seed, B.; Maniatis, T. Plasmid vectors for the rapid isolation and transcriptional analysis of human beta-globin gene alleles. *Mol. Biol. Med.* **1983**, *1*, 473–488. [PubMed]
24. Sazonova, M.; Budnikov, E.; Khasanova, Z.; Sobenin, I.; Postnov, A.; Orekhov, A. Studies of the human aortic intima by a direct quantitative assay of mutant alleles in the mitochondrial genome. *Atherosclerosis* **2009**, *204*, 184–190. [CrossRef]
25. Sazonova, M.A.; Postnov, A.I.; Orekhov, A.N.; Sobenin, I.A. A new method of quantitative estimation of mutant allele in mitochondrial genome. *Patol. Fiziol. Èksperimental'naia Ter.* **2011**, *4*, 81–84.
26. IBM SPSS Statistics 27 Documentation. Available online: https://www.ibm.com/support/pages/downloading-ibm-spss-statistics-27/ (accessed on 15 July 2020).

Review

The Maintenance of Mitochondrial DNA Integrity and Dynamics by Mitochondrial Membranes

James Chapman [1,2,*], Yi Shiau Ng [1,3] and Thomas J. Nicholls [1,2,*]

1 Wellcome Centre for Mitochondrial Research, Faculty of Medical Sciences, Newcastle University, Newcastle upon Tyne NE2 4HH, UK; Yi.Ng@newcastle.ac.uk
2 Biosciences Institute, Faculty of Medical Sciences, Newcastle University, Newcastle upon Tyne NE2 4HH, UK
3 Translational and Clinical Research Institute, Faculty of Medical Sciences, Newcastle University, Newcastle upon Tyne NE2 4HH, UK
* Correspondence: james.chapman@newcastle.ac.uk (J.C.); thomas.nicholls@newcastle.ac.uk (T.J.N.)

Received: 27 July 2020; Accepted: 23 August 2020; Published: 26 August 2020

Abstract: Mitochondria are complex organelles that harbour their own genome. Mitochondrial DNA (mtDNA) exists in the form of a circular double-stranded DNA molecule that must be replicated, segregated and distributed around the mitochondrial network. Human cells typically possess between a few hundred and several thousand copies of the mitochondrial genome, located within the mitochondrial matrix in close association with the cristae ultrastructure. The organisation of mtDNA around the mitochondrial network requires mitochondria to be dynamic and undergo both fission and fusion events in coordination with the modulation of cristae architecture. The dysregulation of these processes has profound effects upon mtDNA replication, manifesting as a loss of mtDNA integrity and copy number, and upon the subsequent distribution of mtDNA around the mitochondrial network. Mutations within genes involved in mitochondrial dynamics or cristae modulation cause a wide range of neurological disorders frequently associated with defects in mtDNA maintenance. This review aims to provide an understanding of the biological mechanisms that link mitochondrial dynamics and mtDNA integrity, as well as examine the interplay that occurs between mtDNA, mitochondrial dynamics and cristae structure.

Keywords: mitochondria; mtDNA; cristae; mitochondrial fission; mitochondrial fusion; mitochondrial diseas

1. Introduction

Mitochondria act as metabolic hubs within the cell to facilitate a myriad of essential cellular processes such as energy production, the regulation of apoptosis and cellular signalling pathways, amongst others. They are unique organelles in the fact that they harbour their own genome that is distinct from the nuclear genome. In human cells, this consists of a circular double-stranded DNA molecule, referred to as mitochondrial DNA (mtDNA). A typical cell possesses between a few hundred and several thousand copies of mtDNA that are replicated independently of the cell cycle within the mitochondrial matrix and segregated between mitochondria. These genomes are closely interlinked with the cristae ultrastructure of the mitochondrion. Once they have been replicated, mtDNA molecules are subsequently distributed around the mitochondrial network by processes that rely on the plastic nature of mitochondria to undertake fission and fusion events, as well as the modulation of cristae structure. Defects in the fission and fusion machinery, or in proteins associated with modulating cristae structure, disrupt the even allocation of mtDNA throughout the network and subsequently to daughter organelles and cells. Defects in mtDNA metabolism typically manifest as the accumulation of mtDNA molecules harbouring deletions and/or as a depletion in the number of copies of mtDNA per cell. The close association between mtDNA and cellular energy production means that the loss of mtDNA number and integrity limits the capacity for the mitochondria to meet the energy demands of an organism. From a clinical perspective, patients carrying mutations within dynamics or cristae-associated genes display heterogeneous neurological phenotypes. There is a clear requirement to understand the basic biological processes linking mitochondrial structure with mtDNA maintenance and its links to mitochondrial disease.

This review aims to disentangle the relationships that exist between mitochondrial dynamics, cristae structure and the organisation of mtDNA. In addition, the biological mechanisms that may prompt the disruption of mtDNA integrity following the impairment of mitochondrial dynamics are assessed. Finally, these mechanisms are discussed in context with observations from the clinic.

2. Mitochondrial DNA

2.1. The Origins of mtDNA

mtDNA is present in the mitochondria of almost all eukaryotic organisms, and the advent of genome sequencing has allowed its evolutionary origins to be dissected. It is now understood that mtDNA derives from an event whereby a host cell engulfed an alphaproteobacterium [1]. This endosymbiotic occurrence fostered a relationship in which the bacterium was utilised for its energy-producing capacity by the host cell. This gave rise to the first complex eukaryotic cells, and since then the course of evolution and the transfer of mitochondrial genes to the nucleus has led to mitochondrial genomes that vary significantly in both structure and size between modern eukaryotic organisms. For example, higher plants harbour genomes that are typically 200–300 kb in size made up of linear and small circular regions of DNA, whereas, algae and fungi have much smaller linear genomes in the region of 30–90 kb [2,3].

Human mtDNA is 16,569 base pairs in length and is organised into a double-stranded circular structure containing 37 genes which encodes for 13 mitochondrial proteins (Figure 1a) [4]. These proteins are all essential components of the oxidative phosphorylation (OXPHOS) system. The OXPHOS machinery is made up of four respiratory chain complexes and the ATP (adenosine triphosphate) synthase which are responsible for energy production in the currency of ATP [5]. In addition, mtDNA encodes 22 transfer RNA (tRNA) and 2 ribosomal RNA (rRNA) molecules which are components of the mitochondrial translation system [4]. The rest of the mitochondrial proteome, currently estimated at 1158 proteins [6], is encoded by the nucleus as the result of the lateral transfer of mitochondrial genes [1]. This evolutionary pressure towards mtDNA reduction means that human mtDNA possesses very few noncoding regions and contains areas of overlapping genes [4]. The OXPHOS system is made up of approximately 90 proteins, and as such is comprised of the products of both mitochondrial and nuclear genes [7]. This dual-genetic origin requires that nuclear-encoded subunits be translated in the cytosol prior to being imported into the mitochondria. Conversely, subunits encoded by mtDNA are synthesised and assembled within the mitochondria by a dedicated mitochondrial translation machinery. Once the nuclear subunits are imported, they are assembled alongside the mtDNA-encoded subunits to form the respiratory complexes that make up the electron transport chain.

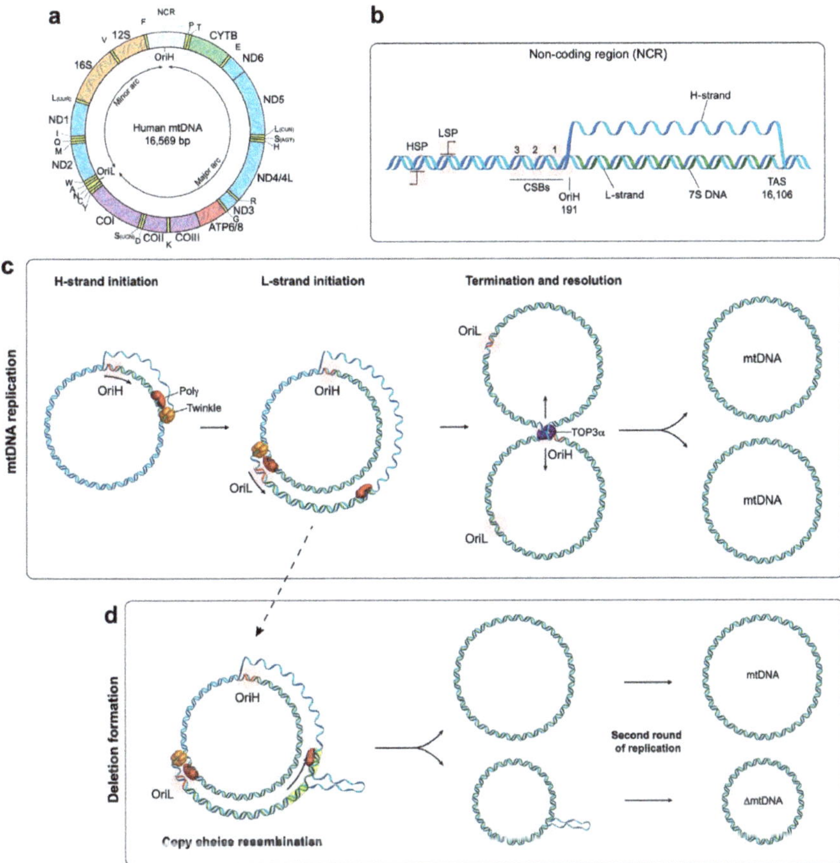

Figure 1. Schematic overview of the human mtDNA genome, its replication and the formation of deletions. (**a**) The structure of mtDNA highlighting the arrangement of protein coding genes, rRNAs (orange) and tRNAs (yellow). The replication origins of the heavy and light strand (OriH and OriL, respectively) are highlighted. (**b**) Enlargement of the mtDNA non-coding region (NCR) depicting the arrangement of the heavy strand promoter (HSP) and light strand promoter (LSP), the three conserved sequence boxes (CSB), OriH and the termination-associated sequence (TAS). The premature termination of the DNA synthesis of the H-strand at TAS results in the formation of a triple-stranded displacement-loop structure termed the D-loop. The short double stranded product formed within the D-loop is termed 7S DNA. (**c**) mtDNA replication is initiated at OriH and proceeds unidirectionally until OriL is reached. At this point, DNA synthesis of the light strand is initiated, and both strands are synthesised simultaneously until two completely replicated genomes are produced. The two replicated genomes are physically interlinked by a single-stranded overlap structure, termed a hemicatenane. This structure is resolved by topoisomerase 3α (TOP3α) to produce two separate mtDNA molecules. (**d**) Copy choice recombination model for the formation of mtDNA deletions. mtDNA deletions generally occur in the major arc. The replication of a repeat sequence in the template heavy strand (yellow boxes) can lead to stalling of POLγ which results in its dissociation from the newly synthesised DNA-end. When Polγ reanneals, it may associate at another repeat sequence further along the template. Following the completion of replication, this slippage event produces two mtDNA genomes; one full length molecule and a second heteroduplex molecule (which has a full-length heavy strand alongside a deletion-containing light strand). The subsequent replication of the heteroduplex molecule culminates in the formation of mtDNA harbouring the deletion.

2.2. mtDNA Replication and Segregation

Unlike nuclear DNA, the replication of mtDNA occurs throughout the cell cycle. The process of mtDNA replication utilises a different set of proteins to those which carry out replication of the nuclear genome [8]. These proteins are all nuclear-encoded and are imported into the mitochondrial matrix from the cytosol. The mitochondrial transcription machinery generates the primers for mtDNA synthesis and can therefore be considered essential components of the mtDNA replication machinery. The mechanisms of replication will be outlined in brief here; for further details there are a number of excellent reviews on the topic [9,10].

2.2.1. mtDNA Transcription

Human mtDNA consists of a 'heavy' (guanine rich) and a 'light' (cytosine rich) strand, with each strand having a single promoter region located close together within the noncoding region (NCR) of the genome (Figure 1b) [11]. These are referred to as the light strand promoter (LSP) and heavy strand promoter (HSP) respectively and facilitate polycistronic transcription of both strands of mtDNA almost in their entirety. Transcription initiation minimally requires the mitochondrial RNA polymerase (POLRMT) with mitochondrial transcription factor A (TFAM) and mitochondrial transcription factor B2 (TFB2M) [12,13]. Following initiation, the mitochondrial transcription elongation factor TEFM promotes POLRMT processivity [14,15]. RNA primers responsible for the initiation of DNA synthesis are generated when transcription from the LSP is terminated prematurely around a series of conserved sequence blocks (CSBs) downstream of LSP, forming an R-loop [16]. This allows the DNA synthesis machinery to assemble and leads to the initiation of DNA synthesis from OriH [17,18].

2.2.2. mtDNA Replication

The protein complex responsible for DNA synthesis is termed the replisome. At the core of the replisome is DNA polymerase-γ (POLγ) which is responsible for synthesising the DNA [19]. POLγ is a heterotrimer made up of a catalytic subunit (POLγA) which functions as a highly accurate proofreader of the newly synthesised DNA [20]. The processivity of POLγA is increased by two copies of the accessory subunit (POLγB) which interact with the DNA substrate [21,22]. In order for POLγ to access and replicate the DNA, it is necessary for the dsDNA to first be unwound. This is performed by the DNA helicase TWINKLE which travels in front of the replication fork and unwinds DNA in the 5′ to 3′ direction [23]. The activity of both TWINKLE and POLγ is enhanced by the mitochondrial single stranded DNA-binding protein (mtSSB) [24]. mtSSB has also been demonstrated to bind to and stabilise the unwound single stranded DNA behind the replication fork [23,24].

During the initial phase of DNA synthesis only the H-strand is replicated. Once the replisome has travelled approximately 12,000 bp it reaches the origin of L-strand replication (OriL), which becomes single stranded and folds into a stem-loop structure [25]. This structure prevents mtSSB from binding and provides a stretch of poly(T) ssDNA that is accessible to POLRMT [26,27]. POLRMT transcribes a short 25–75 bp primer on the single-stranded template, from which synthesis of the L-strand is initiated [27]. See Figure 1c for a schematic overview of mtDNA replication. The strand displacement model proposes that the long tract of exposed H-strand ssDNA is coated and protected by mtSSB until DNA synthesis is initiated from OriL. This model has been reviewed in depth elsewhere [9]. In addition to this model, the bootlace model proposes that the displaced single-stranded lagging-strand template DNA is instead coated by RNA transcripts [28]. Fully double-stranded replication intermediates reminiscent of coupled leading and lagging-strand DNA replication have also been observed and characterised using two-dimensional agarose gel electrophoresis [29–31].

2.2.3. Termination of mtDNA Replication

Once DNA synthesis is complete, the primers at OriL and OriH must be removed to allow for ligation of the DNA ends. Studies have revealed that RNase H1 is involved in the process of primer removal as the loss of RNase H1 in mouse embryonic fibroblasts (MEF) leads to the retention of the primers at sites including both origins of replication; OriL and OriH [32]. The role of RNase H1 in removing primers has been reconstituted in vitro at OriL [33]. Furthermore, the loss of RNase H1 in vivo in mice is embryonically lethal as a consequence of significant mtDNA loss [34]. Following primer removal, the DNA ends need to be prepared to allow ligation by DNA ligase III to occur [35,36]. In a reconstituted system, the mitochondrial genome maintenance exonuclease 1 (MGME1) protein facilitates efficient ligation by modifying single stranded DNA overhangs that occur on the 5′ end of the newly synthesised DNA following primer removal [36]. The loss of MGME1 in vivo has been shown to result in diminished ligation at OriH [37].

Once replication is complete, the two genomes remain connected at the OriH region by a single stranded linkage (termed a hemicatenane) and must be separated (see Figure 1c for a schematic overview). Recent work has revealed that topoisomerase 3α, a type 1A topoisomerase, cleaves the single strand linkage to allow passage of one of the strands to occur, resulting in the separation of the two mtDNA molecules [38].

2.2.4. Formation of mtDNA Deletions

Common manifestations of defects in mtDNA maintenance are deletions and rearrangements of mtDNA. Single, large-scale deletions can cause mitochondrial DNA diseases if they undergo clonal expansion to accumulate beyond a biochemical threshold, typically 60–90% [39]. Such deletions are sporadic and believed to be formed during embryonic development. The most well-studied single deletion is the 4977 bp "common deletion", underlying Pearson's syndrome and Kearns–Sayre syndrome in early life, and chronic progressive external ophthalmoplegia (CPEO) in later life [40]. Alternatively, multiple mtDNA deletions can occur secondarily to disease-causing mutations in nuclear genes that encode factors involved in mtDNA replication, nucleotide metabolism and mitochondrial dynamics [41]. Multiple deletions are also observed in post-mitotic tissues during normal ageing [42,43]. Both the mechanism of deletion formation and the mechanism of clonal expansion of deletions have been the subject of debate. The clonal expansion of mtDNA deletions has been recently reviewed extensively elsewhere [44].

The formation of mtDNA deletions has been proposed to occur either during mtDNA replication or as the result of double-strand breaks [45]. An early model proposed a slip-replication mechanism for the formation of the common deletion [46]. This model involves the annealing of the displaced H-strand to a downstream repeat sequence in the leading-strand template, leading to the removal of the sequence between the two repeats [46]. A more recent model that is supported by in vitro reconstitution experiments suggests that deletions are the result of copy-choice recombination [45,47]. Deletions in mtDNA predominantly form in the major arc between OriH and OriL in the direction of replication. During the synthesis of the L-strand, replication slippage can occur. Specifically, the 3′ end of the nascent L-strand becomes dissociated from the H-strand template at one repeat sequence, and subsequently reanneals to another repeat sequence further along the template. This produces a heteroduplex molecule consisting of a complete H-strand and a shortened L-strand harbouring the deletion. Subsequent rounds of replication will produce shortened mtDNA molecules containing the deletion, see Figure 1d [47].

It has also been observed that inducing high levels of double-strand breaks in mtDNA, which are normally rapidly degraded [48], can result in the formation of deletions [49,50]. Therefore, it has been suggested that limited nucleolytic processing of double-strand breaks could lead to the annealing of repeat sequences and the generation of deletion-containing mtDNA molecules [45,51]. As mitochondria have not been found to possess repair pathways for double-strand breaks comparable to those in the nucleus, this mechanism would presumably operate by a distinct mechanism.

2.3. mtDNA Packaging

A typical mammalian cell harbours between 1000 and 10,000 copies of the mtDNA genome that exist in a DNA-protein complex, termed a nucleoid [52]. Studies using super-resolution microscopy have revealed that nucleoids generally contain a single copy of mtDNA and have a diameter of approximately 100 nm [53,54]. This small size is attributed to the compaction of mtDNA by TFAM. Specifically, TFAM binds along the length of mtDNA at a ratio of 1 subunit per 16–17 bp, instigating bending and loop formation along the DNA backbone that results in compaction [55–58]. The stable protein filaments that are formed prevent POLRMT and TWINKLE from accessing the DNA [52,55]. Reconstituted nucleoids in vitro were demonstrated to become progressively more compact with increasing levels of TFAM, and a wide range of packaging densities were observed at typical physiological concentrations of TFAM [52]. It has therefore been speculated that TFAM may regulate the overall transcription and replication rate of mtDNA by controlling its accessibility to relevant proteins. TFAM is the most abundant protein found associated with nucleoids, however, a number of proteins related to replication, transcription and translation are also commonly found in association with nucleoids [59–61]. It has been proposed that these proteins are located within the "inner core" of the nucleoid, whereas other proteins such as mitochondrial chaperones and membrane binding proteins constitute the peripheral layer [62].

3. Membrane Dynamics and the Organisation of mtDNA

Once the mitochondrial genomes have been separated, they need to be segregated and dispersed around the mitochondrial network. Because the OXPHOS complexes are composed of both mtDNA-encoded and nuclear-encoded subunits, the complexes are assembled in situ proximal to the nucleoid [63,64]. The capacity for mtDNA to spatially diffuse by itself is limited and so nucleoids require mitochondrial membrane structure and dynamics to aid in their distribution around the mitochondrial network [65]. The disruption of nucleoid distribution can lead to a mosaic pattern of respiratory activity, in which only regions of the mitochondrial network that contain nucleoids are capable of assembling OXPHOS complexes and are therefore capable of oxidative ATP production [66]. Structurally, mitochondria are composed of two phospholipid membranes arranged as an outer mitochondrial membrane (OMM) and inner mitochondrial membrane (IMM). The space between these two membranes is referred to as the intermembrane space (IMS). The IMM is intricately folded to form the cristae structures that harbour the respiratory chain complexes. The interior space enclosed by the IMM is referred to as the matrix and contains the mitochondrial genome. mtDNA is closely associated with the IMM and is suggested to be physically attached, a relationship that likely aids in its distribution [67–69]. Early studies identified a protein complex bound to the OriH region of mtDNA and to the IMM, although the factors responsible were not identified [67,70]. It has since been found visually using electron microscopy and immuno-gold labelling that mtDNA is found in close apposition to the IMM [69]. A number of plausible candidates for mtDNA tethering are discussed later in this review. This section aims to evaluate how the dynamic nature of mitochondrial membranes facilitate the distribution of the replicated genomes around the network.

3.1. Mitochondrial Fission and Its Role in mtDNA Distribution

Mitochondria cannot be synthesised de novo; they must have the capacity to grow and divide in order to distribute the replicated genomes to daughter mitochondria. The appearance of mitochondria within a cell can vary significantly; they can exist as isolated entities or be fused together in vast sprawling networks [71]. This versatile nature allows them to sustain energy production as well as act as signaling platforms in complex cellular processes such as apoptosis, autophagy and senescence [72–74]. There are two main mechanisms that underlie mitochondrial dynamics: fission and fusion. First, the relationship between fission and mtDNA distribution and maintenance will be discussed.

3.1.1. Mitochondrial Fission

Mitochondrial fission is the division of a mitochondrion into multiple distinct mitochondria and has been implicated in the distribution of nucleoids around the mitochondrial network [75–77]. Fission requires the progressive constriction, and eventual scission, of the IMM and OMM. The initial constriction occurs at contact sites between the OMM and the endoplasmic reticulum (ER), where actin polymerisation provides the force required to contract the mitochondrial membrane [78–80]. Further constriction is primarily fulfilled by dynamin-related protein 1 (DRP1), a cytosolic protein that translocates to the OMM and interacts with several adaptor proteins [81]. DRP1 binds with mitochondrial fission factor 1 (MFF), mitochondrial dynamics protein of 49kDa (MID49), MID51 and mitochondrial fission 1 protein (FIS1) [82–85]. Utilising GTP (guanine triphosphate), polymerisation of DRP1 with MID49 and MID51 occurs leading to the formation of linear filaments. GTP hydrolysis induces the oligomerisation of DRP1 and filament shortening to create rings which constrict around the mitochondria [79,86]. It has now been established that DRP1-mediated constriction is sufficient for the final scission step to separate mitochondria [87,88]. Recent studies have also highlighted the importance of additional interorganelle contacts for mitochondrial fission, with roles for lysosomes in fission regulation and with Golgi-derived vesicles during final scission [89,90].

It has been observed that ER-OMM contact sites that mark sites of mitochondrial division are often also spatially located adjacent to replicated nucleoids [75–77], suggesting a role for fission in mtDNA segregation. Furthermore, the visualisation of DRP1 and MFF using confocal microscopy has demonstrated colocalisation at sites adjacent to the nucleoid [69,77,91]. Where division occurs between replicated nucleoids, the daughter mtDNA molecules are subsequently observed to be located at the tips of the separated mitochondria [75,76]. This mechanism ensures that following division each mitochondrion receives a copy of the genome, and secondly functions to disperse nucleoids throughout the mitochondrial network. At this stage, it is unclear what signalling takes place to ensure that division occurs between the two replicated nucleoids.

3.1.2. Defects in Mitochondrial Fission and Human Disease

At this point in time, the prevalence of human diseases secondary to defects in mitochondrial fission is not known. However, they would appear to be much less common than diseases that are related to mitochondrial fusion. The most common clinical manifestations that occur in relation to disruption of mitochondrial fission genes are subtypes of Charcot–Marie–Tooth (CMT) neuropathy and optic neuropathy, as identified following mutations that occur in *DRP1*, *GDAP1*, *INF2*, *MFF* and *SLC25A46* [92–96]. Severe neurological presentations, such as neurodevelopmental delay and epilepsy, are observed following mutations of the *DRP1* gene [97]. Leigh-like syndrome has been observed in patients harbouring mutations in *MFF* and *SLC25A46* [98,99]. Mutations in *DNM2* have been linked to CPEO and central core myopathy [100,101], a form of myotubular myopathy. Extra-neurological involvements are relatively uncommon, except cardiac arrhythmia and neutropenia, which have been associated with mutations in *DNM2* [101,102], and glomerular disease, described in INF2-related disease (Appendix A Table A1).

Mitochondrial respiratory chain deficiencies and multiple mtDNA deletions have been demonstrated in muscle biopsies taken from patients harbouring *DNM2* mutations [103]. Normal qualitative and quantitative assessments of mtDNA were reported in the skeletal muscle or fibroblasts that contain mutant forms of other mitochondrial fission-associated genes such as *GDAP1*, *INF2* and *MFF* (Table A1) [93,94,98,104–109]. These observations would suggest that the biological consequences of these genetic defects are tissue specific, given that only *DNM2* mutations primarily manifest with a myopathic phenotype [100,101]. Whilst other fission-associated genes predominantly cause peripheral neuropathy including optic neuropathy or CNS involvement, it is important to note that these affected tissues are far less accessible for further characterisation compared to muscle biopsy.

3.1.3. Mitochondrial Fission and mtDNA Integrity

The disruption of mitochondrial fission by genetic knockdown of the fission factors *DRP1* or *MFF* results in an elongated and fused mitochondrial network that prevents the even distribution of nucleoids around the network, resulting in the clustering of nucleoids [66,77]. Reports in the literature differ on its effects on mtDNA copy number and respiratory activity. In human cells, one study reported that the loss of *DRP1* induced mtDNA loss and a reduction in respiratory activity as measured by ATP production [110], while a separate study found that the knockdown of *DRP1* or *MFF* using siRNA did not affect the overall mtDNA copy number [77]. In mice, mtDNA copy number and respiratory function were shown to be unaffected in *Drp1* knockout MEFs [111], but a whole-body knockout of *Drp1* is embryonically lethal [111,112]. In a tissue-specific knockout of *Drp1* in the heart, mice survived for 11 days and displayed exacerbated mitochondrial fusion, nucleoid clustering, reduced mtDNA and respiratory defects [66]. Interestingly, in this model immunofluorescent staining of the mtDNA-encoded cytochrome c oxidase 1 subunit (COX1) revealed that there was an increased staining intensity in regions where nucleoids were clustered together, and a decreased intensity in areas with a sparse presence of nucleoids [66]. These data suggest that nucleoid clustering leads to a mosaic pattern of respiratory subunit distribution in the mitochondrial network. Similarly, the knockout of *MFF* is associated with premature death and defective mitochondrial respiratory activity [113]. In some cases, it was reported that *DRP1* knockout was associated with enlarged nucleoid size [77]. However, this may be a limitation associated with the microscope resolution used, as it would not be expected that fission would affect the decatenation of the DNA molecules. As such, these apparently enlarged nucleoids may represent decatenated mtDNA molecules located in close proximity, which are beyond the limits of detection, although this remains to be confirmed.

Cardiolipin is an integral structural component of mitochondrial membranes that is synthesised from phosphatidic acid (PA). The IMM contains approximately 20% cardiolipin, the presence of which is considered a signature of the IMM. Both cardiolipin and PA have been implicated in facilitating the fission and fusion processes (this topic has been reviewed in depth elsewhere [114]). Briefly, DRP1 binds to both cardiolipin and PA, cardiolipin at the OMM can stimulate oligomerisation of DRP1 and subsequent GTP hydrolysis to induce fission [115–117]. It has also been shown that DRP1 binding to cardiolipin induces reorganisation of the membrane to an inverted hexagonal, nonbilayer configuration that promotes membrane constriction [118]. Conversely, the activity of DRP1 can be restrained at the OMM by the reversion of cardiolipin to PA by MitoPLD (mitochondria-localised phospholipase D), as the enhanced level of PA inhibits oligomerisation-stimulated GTP hydrolysis that is responsible for membrane constriction [119]. As such, the loss of cardiolipin is associated with a reduced capacity to correctly segregate and guide nucleoids to the daughter mitochondria; this leads to a lack of mtDNA inheritance between replicating cells, resulting in a dysfunctional respiratory phenotype in daughter cells [120–122]. Modulating the level of cardiolipin has also been associated with loss of mtDNA and subsequent mitochondrial dysfunction [121,123]. Clearly, fission plays an important role in facilitating the dissemination of mtDNA around the mitochondrial network and to subsequent daughter organelles (see Figure 2a for a schematic overview). Furthermore, these observations also highlight how mitochondrial fission maintains respiratory function independently of mtDNA copy number or integrity. It is clear that if the fission process is not tightly regulated then cellular respiration will be affected and can contribute to the mitochondrial disease phenotypes discussed earlier.

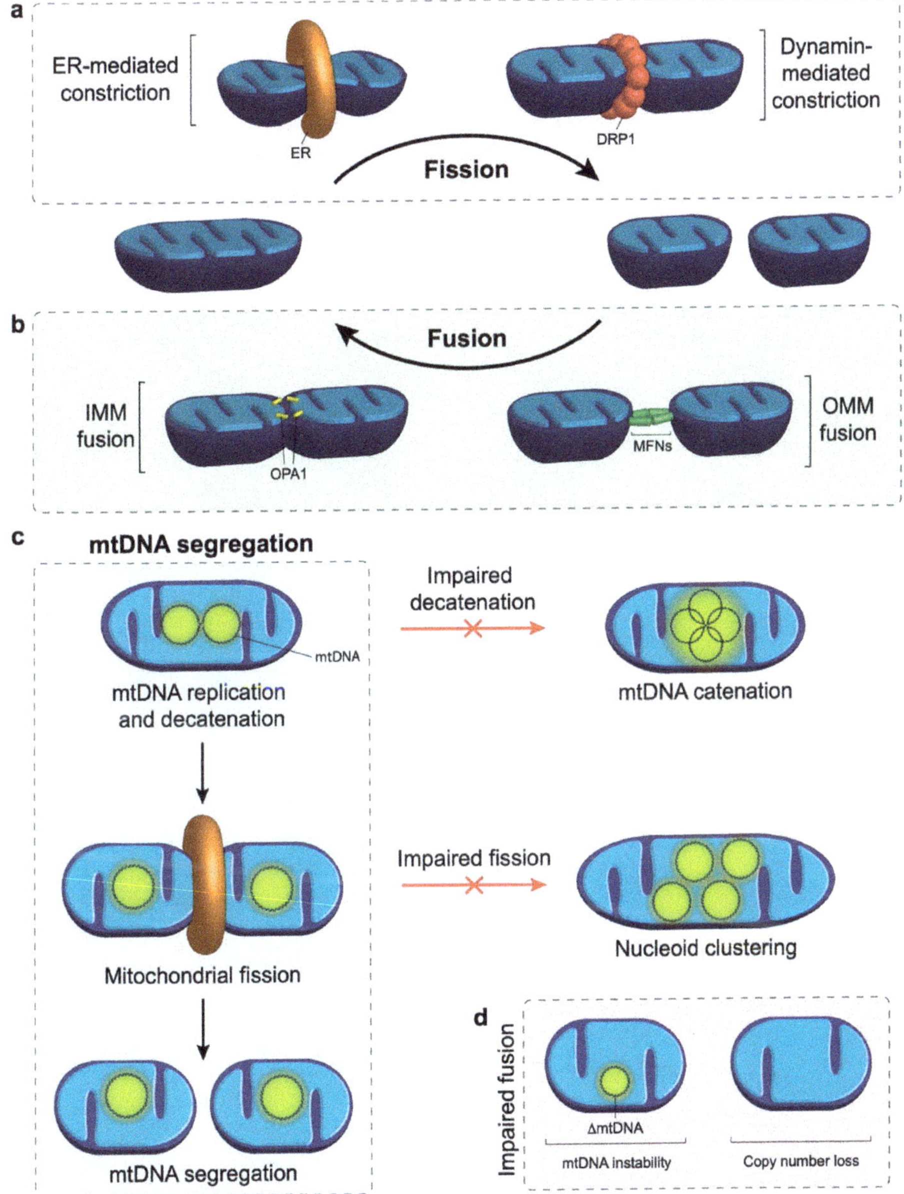

Figure 2. Mitochondrial dynamics and its role in mtDNA distribution. (**a**) Schematic overview of mitochondrial fission. Constriction of the mitochondrion occurs by the endoplasmic reticulum (ER) and dynamin related protein 1 (DRP1). The final scission step is performed by DRP1 to produce multiple independent mitochondria. (**b**) Mitochondrial fusion outline. The mitofusins (MFNs) are responsible for tethering neighbouring mitochondria and fusing the outer membranes together. Fusion of the inner mitochondrial membrane is mediated by optic atrophy 1 (OPA1). (**c**) Summary of mtDNA segregation

and distribution. First, newly replicated mtDNA molecules that are joined by a hemicatenane must be physically separated by topoisomerase 3α. Failure of this process results in the formation of multiple physically joined mtDNA genomes, termed mtDNA catenanes. Following separation, mitochondrial fission is required to distribute the replicated nucleoids into separate mitochondria and facilitate distribution of the genomes around the mitochondrial network. An impairment of fission results in a clustered phenotype whereby many replicated nucleoids are observed in close proximity to each other but are not physically linked together. (**d**) A lack of mitochondrial fusion is associated with mtDNA instability as observed by the progressive onset of mtDNA deletions and point mutations. In addition, the lack of fusion leads to a depletion in mtDNA copy number.

3.2. *Mitochondrial Fusion and Its Role in mtDNA Maintenance and Distribution*

Together with mitochondrial fission, mitochondrial fusion has been implicated in the dissemination of nucleoids around the network. Furthermore, fusion is recognized as an important mediator of mtDNA maintenance. Defects in mtDNA maintenance may manifest as either a quantitative reduction in mtDNA copy number (depletion) or as an accumulation of rearranged mtDNA molecules (deletions and/or duplications), all of which may be indicative of impaired mtDNA replication. Defects in the proteins that regulate mitochondrial fusion dynamics have been implicated in a range of genetic diseases that will be discussed here and are summarised in Table A1. Their contribution to the development of disease will be described in relation to their role in mtDNA maintenance.

3.2.1. Mitochondrial Fusion

Mitochondrial fusion is the joining of two separate mitochondria and is important to enable the sharing of contents between neighbouring mitochondria [124]. The fusion of the OMM and IMM occur sequentially, therefore the outer membranes are joined first. This is regulated by two outer membrane proteins, the dynamin related GTPases mitofusin 1 & 2 (MFN1 and MFN2). MFN1 is responsible for tethering the adjoining mitochondria together in a GTP-dependent dimerization process, prompting a conformational change which in turn mediates GTP hydrolysis by MFN1 to pull the membranes together, resulting in mitochondrial fusion [125–127].The role of MFN2 is less clear, although it has been implicated in interactions between mitochondria, as well as between mitochondria and the ER [128–130]. Following OMM fusion the IMM can be fused. IMM fusion is controlled by optic atrophy 1 (OPA1). OPA1 exists in multiple forms of different sizes which are regulated by proteolytic processing; long OPA1 is IMM-anchored while the short OPA1 is soluble. The presence of OPA1 in its long isoform has been associated with promoting IMM fusion [131]. Processing of OPA1 to its short isoform either by YME1L1 or OMA1 is associated with mitochondrial fission [132,133].

There is evidence that OMM fusion proteins also play a role in the dissemination of nucleoids. It has been reported that the knockout of OMM fusion proteins *Mfn1 & 2* in MEFs leads to the clustering of nucleoids [134]. Using super-resolution microscopy these clusters were confirmed to be separate individual nucleoids in close proximity rather than a cluster of interlinked DNA molecules. The role of IMM fusion in nucleoid distribution is less clear. It has been reported by some groups that the loss of *OPA1* is associated with a reduction in the number of nucleoids per cell, and in some cases a clustered phenotype [77,135,136]. In contrast, a recent study found that in MEFs the removal of *Opa1* was not associated with any alterations in mtDNA distribution, although there was a reduction in copy number [134]. This normal nucleoid distribution following *Opa1* knockout may be explained by the role of MFN2 in forming a tether between the OMM and ER [137,138]. As discussed earlier, it is recognized that the ER localises at positions of nucleoid replication and is implicated in coordinating mitochondrial constriction at these sites [75]. Taken together, this highlights the importance of the OMM acting as a signalling platform for coordinating the circulation of nucleoids around the mitochondrial network.

Interestingly, the knockdown of *MFN1 & 2* in conjunction with *DRP1* has been found to prevent the clustering of nucleoids in human cells [77]. Mechanistically, how the impairment of both fission and fusion would restore nucleoid distribution is not clear as it may be expected that this would render the mitochondria unable to be dynamic, and thus unable to evenly distribute nucleoids. Another study in vivo demonstrated that the dual knockout of *Mff* and *Mfn1* in mice could completely rescue the heart dysfunction, shortened life span, and respiratory chain dysfunction associated with *Mff* knockout [113], although this rescue was tissue specific. It was suggested that this dual knockout reinstates a balance of fission and fusion, rather than hyperfusion or fragmentation. The relationship between mitochondrial dynamics, nucleoid segregation and maintenance of the respiratory chain is complex and therefore further study is still required to disentangle its intricacies. See Figure 2 for an overview of the interplay between fission and fusion in mtDNA distribution.

3.2.2. Defects in Mitochondrial Fusion and Human Disease

In recent years, genetic defects involved in mitochondrial fusion have emerged to be the common cause of several neurological and ophthalmological disorders. *OPA1* mutations account for 60% of dominant optic atrophy (DOA) cases [139], and the prevalence of the disease has recently been revised to 1 in 34,000 [140]. More complex neurological phenotypes such as cerebellar ataxia, spasticity, CPEO [141,142] and more recently, Parkinsonism and dementia, have been identified in patients with *OPA1* mutations [143]. *MFN2* mutations are the fourth most common cause of CMT neuropathy [144,145], accounting for 20% of CMT2 [146], a form of dominant axonal neuropathy. Mutations in *SPG7* were initially described in hereditary spastic paraplegia associated with mitochondrial OXPHOS defects [147]. Independent cohort studies have subsequently shown that cerebellar ataxia could be the most prominent clinical feature without evidence of upper motor signs in cases of *SPG7* mutations, and are the most common or second most frequent cause of recessive genetic ataxia in European populations [148,149]. Interestingly, a recent Spanish study suggested that around one-fifth of SPG7 cases exhibited Parkinsonism [150].

AFG3L2 and SPG7 together form the subunits of the m-AAA metalloprotease complex, which is crucial for the maturation, maintenance and quality control of the mitochondrial proteome [151]. Heterozygous mutations in *AFG3L2* cause spinocerebellar ataxia type 28 (SCA28) and mitochondrial respiratory chain deficiency [151]. Given the close interaction between AFG3L2 and paraplegin (SPG7), there is little surprise that genetic defects result in many overlapping clinical features of neurodegeneration and the classic phenotypes of mitochondrial dysfunction such as CPEO and multiple mtDNA deletions in the muscle [152,153].

Severe, childhood-onset encephalopathy has been observed in mutations in *OPA1* (recessive), *FBXL4* and *YME1L1* (Table A1). On the other hand, mitochondrial DNA depletion is rare in defects of mitochondrial fusion and has only been identified in several cases of severe childhood disease secondary to mutations in *MFN2* and *FBXL4* [154,155]. Intriguingly, multiple lipomatosis, as previously observed in myoclonic epilepsy and ragged-red fibres (MERRF) syndrome but no other forms of primary mtDNA mutations, has been identified in several families of *MFN2* mutations [156–158].

3.2.3. Mitochondrial Fusion and mtDNA Copy Number

A number of studies have highlighted that mitochondrial fusion is critical for the maintenance of mtDNA copy number. In yeast, the loss of fusion activity leads to a loss of mtDNA copy number [159,160]. Consistent with this observation, mtDNA content and respiratory activity is reduced in MEFs following the knockout of OMM fusion factors *Mfn1 &2* either alone or together or following knockout of the IMM fusion factor *Opa1* [134,135]. Mice carrying a mutation in *Opa1* display mtDNA loss and reduced mitochondrial function in the heart [161], while whole-body knockouts of MFN1 & 2 are embryonically lethal [162]. The roles of both MFN1 & 2 have therefore been studied in a tissue-specific context. Work in mice has shown that the knockout of *Mfn1 & 2* in heart and skeletal muscle leads to a drop in overall mtDNA copy number and OXPHOS deficiency [134,136]. Cardiolipin also has a role in regulating fusion as it is necessary for the biogenesis of OPA1 and for the formation of higher order OPA1 oligomers which are required for fusion [163–165]. The cardiolipin precursor PA has also been associated with the induction of fusion in an MFN dependent manner [166].

The disruption of cardiolipin levels is linked with a reduction in mtDNA copy number [121,123]. ATAD3 (ATPase family AAA-domain containing protein 3 A) has also been implicated in maintaining mitochondrial fusion as its manipulation either by knockdown or overexpression leads to mitochondrial fragmentation [167–170]. This fragmented phenotype is mediated by the recruitment of DRP1 to the OMM via oligomerisation of the ATAD3 coiled-coil domain [170]. Furthermore, it was demonstrated that in addition to activating DRP1, the dimerisation of ATAD3 provokes mtDNA instability by disrupting the binding of TFAM and mtDNA. Indeed, the knockdown of ATAD3 has been associated with a reduction in mtDNA content [170]. In human fibroblasts that are either deficient for ATAD3 or harbour a duplication of the ATAD3 gene cluster, nucleoids were found to be enlarged and clustered together suggesting a role of ATAD3 and fusion in mtDNA distribution [171,172]. The knockout of *ATAD3* in skeletal muscle has been associated with the progressive formation of mtDNA deletions and copy number depletion [173].

The mechanisms underlying the loss of mtDNA when fusion is impaired have long been unclear, although recent work has shed light on potential links between the two. In general, the loss of copy number indicates that there is a defect with the process of mtDNA replication. It has been demonstrated that mitochondrial fusion is necessary to facilitate high levels of replication through content mixing to ensure a proper stoichiometry of replisome components. The loss of OPA1 alone or the collective loss of MFN1 & 2 together prompts an imbalance of replisome factors and thus leads to a reduced rate of mtDNA replication, leading to mtDNA depletion [134].

3.2.4. Mitochondrial Fusion and mtDNA Integrity

It has been observed that aside from being required to maintain copy number, fusion appears to play an important role in maintaining the integrity of the mitochondrial genome. For example, mutations in the *OPA1* gene have been associated with the accumulation of mtDNA deletions and OXPHOS defects in the skeletal muscle of patients [141,174,175]. Similarly, patients with mutations in the OMM fusion protein MFN2 display evidence of deletion-containing mtDNA genomes [154,176]. In animal models, the knockout of *Mfn1 & 2* in the skeletal muscle of mice is associated with an increased occurrence of point mutations and deletions of mtDNA [136]. On the other hand, another group studying the cardiac tissue of mice with MFN1 & 2 genetically removed found no differences in levels of mtDNA mutations or deletions [134]. These studies highlight that effects may be tissue-specific, and that care must be taken when comparing effects in patients carrying missense mutations with effects seen in knockout animal models. However, from a clinical perspective, mutations in the fusion machinery have a clear association with the progressive onset of mtDNA mutations and deletions (see Table A1).

Multiple lines of evidence indicate that fusion plays a protective role against mutations. Work using fibroblasts derived from *MFN2* patients found that these cells display a reduced capacity to repair mtDNA damage [176]. This finding potentially suggests that fusion may preserve mtDNA integrity by enabling the repair of DNA damage or by facilitating the clearance of mitochondria harbouring damaged DNA. In support of this notion, excessive mitochondrial fragmentation (which would occur with impaired fusion) has been associated with an increased production of reactive oxygen species [177]. Indeed, in an *Opa1* mutant mouse it was observed that there was an increased level of ROS coupled with a reduced antioxidant capacity [161]. A separate study used the mutator mouse, which has a mutation in the proofreading domain of polymerase *PolgA*, and consequently rapidly accumulates mtDNA mutations, crossed with a knockout of MFN1 [178]. The mutator mouse or the MFN1 knockout mouse individually survive into adulthood, however crosses between the two resulted in embryonic lethality [136]. In this case it is plausible that fusion exerts a protective effect through its ability to "dilute" mutation-containing mtDNA via content mixing and therefore preserve mitochondrial function.

It remains unclear how defects in fusion contribute to the molecular mechanism of deletion formation. Deletions may conceivably accumulate at an increased rate either because of impaired mtDNA replication or because of increased mtDNA damage. It has been speculated that the primary purpose of mitochondrial fusion is to enable the sharing of contents between two mitochondria. Indeed, it has been published that matrix proteins and mtDNA are transferred between fused mitochondria [179]. Conversely, mitochondria from dual MFN knockout cells were found to have increased protein heterogeneity compared to their wild-type counterparts [136]. It may therefore be that the loss of content mixing associated with defective fusion impairs mtDNA replication through an imbalance of replisome components [134]. This could then promote replication stalling and copy-choice recombination, leading to increased deletion formation. Alternatively, it is possible that increased mtDNA damage in the absence of fusion promotes the formation of deletion-containing mtDNA molecules via double-strand break formation. However, a key limitation in the idea of mtDNA repair underlying the formation of deletions is that at this stage the proteins that are responsible for DNA repair have not been identified as being present in mitochondria.

It is also conceivable that deletions accumulate in the absence of fusion because their removal is impaired. However, the observation that defects in fusion are associated with an accumulation of mtDNA mutations and deletions appears counterintuitive, as it would be expected that a fragmented network would be optimal for selective mitophagy to clear dysfunctional mitochondria away. Indeed, in a *Drosophila melanogaster* model of mtDNA heteroplasmy, the knockdown of MFN promoted the mitophagy of fragmented mitochondria and a reduction in the mutant mtDNA load [180]. Similarly, a perpetually fused network due to the inhibition of fission by DRP1 or FIS1 in human cells was associated with a shift in heteroplasmy towards mutant mtDNA, possibly because of reduced mitophagy, although this was not addressed [181]. As such, it appears that the accumulation of mtDNA mutations and deletions in the absence of fusion is not related to a reduction in mitophagy.

Collectively, these studies demonstrate that IMM and OMM fusion have an important role in maintaining the integrity of mtDNA and coordinating its replication. However, the relationship is complex and further study is certainly required. It seems likely that the role of fusion in facilitating content mixing between mitochondria is important to maintain the balance of proteins required for mtDNA replication [134,136]. There is also evidence to suggest that fusion also has a protective role against the effects of ROS and DNA damage [161,177]. This, coupled with an imbalance of replisome proteins leading to replication stalling/slippage, could underlie the onset of mutations and deletions associated with fusion defects, as well as contribute to mtDNA copy number loss.

4. The Relationship between mtDNA and Mitochondrial Cristae Structure

The IMM is a complex structure that can be subdivided into two further regions: the inner boundary membrane (IBM) and the cristae (Figure 3b). The IBM runs parallel to the OMM and houses proteins responsible for localising and importing proteins into the matrix, as well as inserting and assembling proteins into the IMM [182,183]. The IBM is divided at regions where the IMM is folded inwards to form cristae structures. Cristae are functionally active structures that contain the respiratory chain complexes and the ATP synthase [184]. At points where the cristae joins the IBM, the cristae narrows to form a cristae junction (approximately 20–40 nm in diameter) which allows the separation of the intracristal space from the intermembrane space between the IMM and OMM [185–187]. The mitochondrial contact site and cristae organising system (MICOS) complex is located at these junctions and plays a key role in maintaining their structure.

It has long been appreciated that mtDNA is located close to the IMM and to the cristae structure. It has been recently demonstrated using correlative 3D super-resolution fluorescence and electron microscopy that nucleoids are found within cristae regions of the mitochondria [68]. Sophisticated super-resolution imaging has now revealed that nucleoids are typically located in the voids that form between groups of cristae rather than being embedded within the cristae structure itself [188]. This is in line with the observation that the size of the nucleoid, at approximately 100 nm, is greater than the gaps between cristae, which are found tightly packed together [53,54,188]. It is postulated that these voids where nucleoids are located may function to provide space for transcription, replication and segregation to occur [68,188]. In this section we will summarise current data on cristae structure and modulation, and its relationship with mtDNA maintenance and segregation.

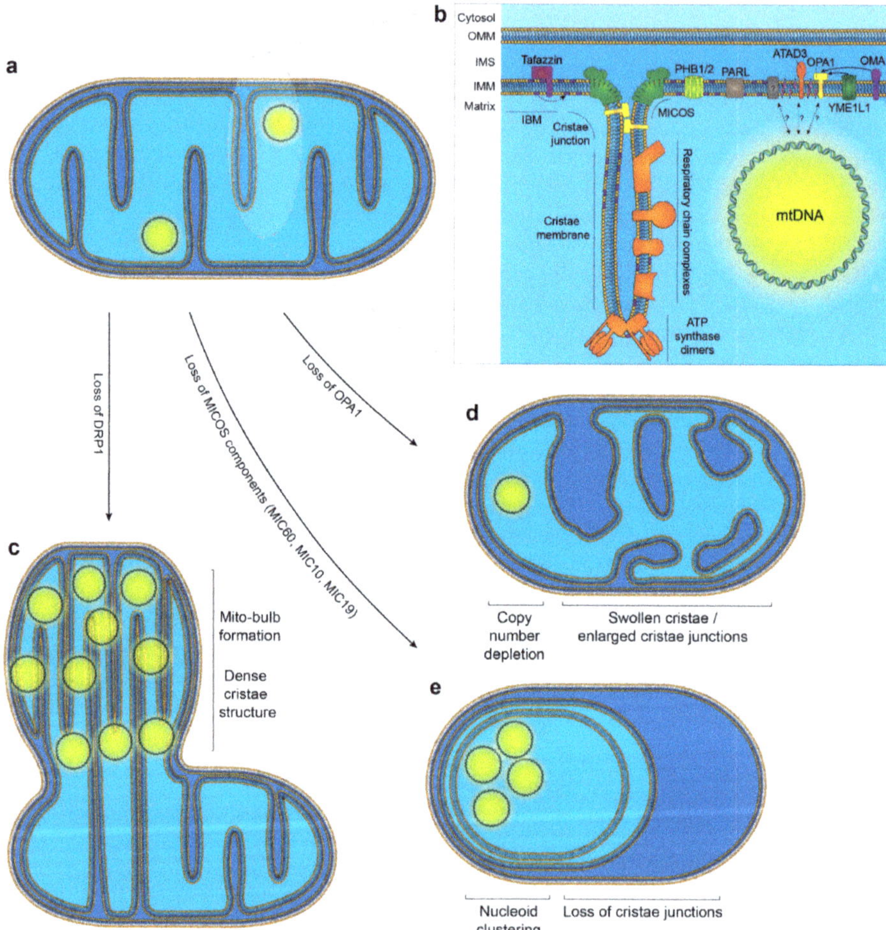

Figure 3. The relationship between cristae modulators and mtDNA organisation. (**a**) Schematic overview of a normal mitochondrion. (**b**) Enlarged cristae region depicting the spatial arrangement of key proteins that are necessary for dictating cristae structure and shape, as well as the organisation of the respiratory chain complexes. Proteins that are potentially responsible for the tethering of mtDNA to the IBM are also highlighted. (**c**) Loss of the fission factor DRP1 has been associated with the formation of mito-bulb structures. These are regions of dense cristae structure which harbour a number of clustered mtDNA molecules. (**d**) The loss of IMM fusion and cristae shaping protein OPA1 results in mitochondria that display enlarged cristae junctions and a perturbed cristae structure. In addition, mtDNA copy number is reduced. (**e**) The loss of MICOS components (MIC60, MIC10 and MIC19) has been associated with a complete loss of cristae junctions and the formation of cristae in concentric circles. Nucleoid clustering is evident following the loss of MICOS components.

4.1. The Importance of Lipids in the IMM and Their Role in mtDNA Tethering

The capacity for mtDNA to freely diffuse around the mitochondrial network is limited, due in part to the relatively large size of the nucleoid and to the high density of proteins located within the matrix and cristae regions [53,69,189]. There is also evidence that has led to the suggestion that mtDNA is tethered to the IMM [67,69,70,190,191]. However, at this stage the factors responsible remain poorly understood.

Several studies have highlighted the importance of the lipid composition of the IMM to mtDNA attachment, replication and organisation. The composition of phospholipids in the IMM is tightly regulated. Phospholipids have a hydrophobic lipid domain tail which is larger than their charged hydrophilic head groups; this configuration forms a rough conical shape which pulls head groups together, inducing membrane tension and facilitating the bending of the IMM to form cristae invaginations [192,193]. Cardiolipin has an important role in stabilising IMM proteins such as the respiratory complexes [194]. Alterations in the level of cardiolipin have been associated with mitochondrial dysfunction, mtDNA loss and the formation of abnormal cristae structures [121,123]. Similarly, in mammalian cells the removal of the IMM phospholipid phosphatidylethanolamine has comparable effects; mitochondrial function is reduced coupled with obvious swelling of mitochondria which lack distinct cristae structures [195].

It has been demonstrated in both yeast and mammalian cells that cardiolipin is physically associated with the nucleoid [120]. Furthermore, as discussed earlier cardiolipin is a key constituent of the IMM and has a role in mitochondrial fission and fusion. As such, the role of cardiolipin in mtDNA maintenance is multi-faceted. The loss of cardiolipin is associated with a reduced capacity to correctly segregate and guide nucleoids to daughter mitochondria. In addition, this defective IMM-nucleoid association coupled with defects in mitochondrial dynamics leads to a lack of mtDNA inheritance between replicating cells resulting in a dysfunctional respiratory phenotype in daughter cells [120–122].

The composition of the mitochondrial IMM is subject to a relatively low ratio of cholesterol to phospholipids [196]. Cholesterol has been implicated in nucleoid organisation as its modulation either through pharmacological inhibition or supplementation prompts the formation of aggregated nucleoids in human fibroblasts [171]. The majority of mitochondrial cholesterol is clustered in specialised structures that span the IMM and OMM [197]. TWINKLE-containing nucleoids were shown to be associated with these cholesterol structures in a complex that also contains ATAD3. ATAD3 is an essential protein that is anchored into the IMM, with its C-terminal AAA ATPase domain located in the mitochondrial matrix while its N-terminus interacts with the OMM where it is present at ER-mitochondria contact sites [62,169,197–199]. It has been proposed that these cholesterol-rich sites provide a platform for mtDNA replication, and act as an attachment site at ER-mitochondria junctions to allow for the coordinated distribution of mtDNA. As discussed earlier, ATAD3 has been demonstrated to have a regulatory role in mitochondrial morphology and dynamics and has been implicated in nucleoid organisation [167–170]. ATAD3 also has an important role in maintaining IMM architecture as the disruption of ATAD3 leads to impaired transfer of cholesterol from the ER into the mitochondria and lipid metabolism [171,200]. The knockout of *ATAD3* in skeletal muscle has been associated with alterations in cholesterol metabolism and the progressive formation of mtDNA deletions and copy number depletion [173]. As discussed earlier, the formation of these mtDNA rearrangements has been associated with stalling of the replication machinery and a loss of cristae structure. It has been suggested that ATAD3 may interact with OPA1 or the MICOS complex to stabilise cristae structure [173]. Collectively, it is clear that ATAD3 is an important regulator of mitochondrial dynamics, mtDNA attachment, maintenance and organisation. However, at this stage it is not entirely clear whether these effects are directly associated with the loss of ATAD3 or due to cholesterol-mediated alterations in the IMM structure as a consequence of the role of ATAD3 in the uptake of cholesterol to the mitochondria [200].

In addition to ATAD3, OPA1 has also been implicated as having a potential role in the attachment of nucleoids to the IMM [135]. OPA1 is embedded in the IMM, co-immunoprecipitates with nucleoids and as discussed earlier, its deletion has profound effects on nucleoid organisation [77,135,136]. Thus, it has been suggested that OPA1 may have a role in membrane attachment of the nucleoid (Figure 3b). Loss of either ATAD3 or OPA1 has been shown to cause significant aberrations in cristae structure [161,197,201].

The structure of proteins and lipids within the IMM is also supported by a group of evolutionarily conserved scaffolding proteins termed prohibitins. Prohibitin 1 and 2 are essential proteins that form high molecular weight ring-shaped heterooligomers in the IMM. They have also been found to associate with nucleoids which prompts speculation that they may be involved in tethering of nucleoids to the IMM [57]. It has been demonstrated that the prohibitins can influence mtDNA copy number through their interactions with TFAM [202]. Cultured cells lacking the prohibitins display impaired cardiolipin maturation, loss of cristae structure and disorganisation of nucleoids [203–205]. It has also been demonstrated that the effects of prohibitin 2 depletion upon cristae structure are dependent upon OPA1, a known modulator of cristae structure [204].

4.2. Modulators of Cristae Structure

Mitochondrial cristae are dynamic structures that can modulate their shape in response to various physiological conditions. As they are the primary site of the OXPHOS machinery, it is important that the folding of the IMM into cristae structures facilitates the most efficient manner of producing ATP. Recent work has found that individual cristae act as autonomous bioenergetic units, highlighting the importance of their structural integrity [206]. Cristae structure is dictated by a variety of "cristae-shaping proteins" (Figure 3b). A key mediator of cristae architecture is the MICOS complex. The MICOS complex consists of at least the subunits MIC10, MIC12, MIC13, MIC19, MIC25, MIC26, MIC27 and MIC60 [207]. Each of these proteins has a specific role in shaping the cristae structure. MIC60 is a core component of the complex and is associated with the formation of cristae junctions and contact sites with the OMM [208–210]. MIC10 can also bend membranes and is known to be responsible for forming hairpin structures in the IMM [211,212]. MIC13 has also been recently recognised as being essential for maintaining the stability of the MICOS complex and cristae junction formation [212]. MIC26 has a role in cristae junction formation, and both MIC26 and MIC27 are necessary to maintain normal cristae architecture [213,214].

The F_1F_0 ATP synthase is embedded in the IMM and is localised to the cristae tips, where it exists in a dimer confirmation. These dimers are also assembled into oligomers. The presence of these dimers prompts a bending of the surrounding IMM lipid bilayer, highlighting their importance as mediators of cristae structure [215,216]. As such, the loss of the F_1F_0 ATP synthase is associated with a loss of cristae invaginations [216]. However, the F_1F_0 ATP synthase may also act indirectly to maintain cristae morphology, as it has also been recognized that the F_1F_0 ATP synthase interacts with the MICOS complex [217]. It has also been reported that oligomerisation of F_1F_0 ATP synthase dimers is promoted by OPA1, which also has a role in cristae shaping [218].

The role of OPA1 in cristae shaping is reflected in its ability to define the diameter and width of cristae junctions [209]. It has been demonstrated in yeast that the OPA1 homologue MGM1 is required to maintain cristae structure by tethering to other OPA1 molecules on the opposing IMM [219]. Indeed, it has been observed that high molecular weight multimers of OPA1 stabilise and induce the formation of tight cristae junctions, whereas lower-order OPA1 oligomers are associated with increased cristae junction and lumen width [209]. The remodeling of cristae is mediated by the integral membrane protease PARL. Processing by PARL produces a short, intramembrane-soluble form of OPA1 that binds with the membrane-bound form to maintain tight cristae junctions [220]. There is some evidence that OPA1 has a broader regulatory role and can adjust mitochondrial respiration by modulating cristae shape. It has been reported that in response to hypoxia there is an increase in the abundance of OPA1 oligomers that prevent cristae remodeling, thereby acting to enhance mitochondrial respiration [221]. It has also been shown that OPA1 can interact with metabolic sensors in response to starvation, acting to adjust the cristae shape to maintain ATP production [222]. High molecular weight OPA1 multimers have also found to be associated with MIC60 and F_1F_0 ATP synthase [209,218,223]. Further work will be required to understand the extent to which OPA1 dictates cristae shape alone, and how it functions as part of an interactive network with other cristae shaping proteins.

A number of other proteins have been observed to have an impact on cristae structure, but for which the precise mechanisms are less well understood. MCL1 has been found to have a role in preserving cristae ultrastructure and the maintenance of oligomeric ATP synthase [224]. The loss of proteins that interact with cardiolipin have also been found to disrupt cristae architecture. Specifically, the prohibitins and UQCC3 have all been demonstrated to bind to cardiolipin, and their loss is associated with alterations in cristae structure [204,225,226].

4.3. The Relationship between Cristae Structure and mtDNA

A number of studies indicate that there is an interesting but poorly-understood link between mtDNA and cristae ultrastructure. Dissecting this relationship is problematic because of the role that mtDNA-encoded proteins, such as components of the ATP synthase, play in maintaining cristae structure. For example Rho-0 (ρ0) fibroblasts, which are devoid of mtDNA, display sparse cristae structures and swollen mitochondria compared to mtDNA-containing cells [227–229]. Embryos lacking TFAM also lack mtDNA and display abnormal cristae structure [230]. Similarly, tissue-specific knockout of TFAM in developed animals has also been shown to ablate cristae structure [231,232]. In cultured cells, the transient depletion of mtDNA using the replication inhibitor 2′-3′-dideoxycytidine also leads to a loss of cristae structure [77]. Other data has suggested a direct role for mtDNA in maintaining cristae structure. The loss of DRP1 is associated with the formation of structures filled with very densely packed cristae and clustered nucleoids termed "mito-bulbs" (Figure 3c) [77]. The formation of these cristae-rich mito-bulbs in DRP1 knockdown cells was not impacted when mitochondrial translation was inhibited using chloramphenicol. However, the depletion of mtDNA prior to DRP1 knockdown prevented the formation of mito-bulbs. This suggests that mtDNA itself may be important for determining cristae architecture rather than its gene products. These data highlight the need for further study to understand how mtDNA contributes to the maintenance of normal cristae structure.

The loss of *OPA1* has been associated with a reduced mtDNA copy number, both in model systems and in patients harbouring *OPA1* mutations [135,161,233]. OPA1-deficient cultured cells also show a distinct loss of cristae structure (Figure 3d) [201,234]. Similarly, in mice, both the mutation of *Opa1* in the heart and deletion of *Opa1* in the skeletal muscle was found to cause the loss of cristae structures [161,235,236]. Other in vivo models in which mtDNA content was reduced, such as the deletion of *Mfn2* in Purkinje cells or the dual knockout of *Mfn1 & 2* in skeletal muscle, have also been shown to be associated with sparse cristae structures. Whilst these observations support the idea that the loss of mtDNA is linked with the disruption of cristae structure, it is likely that the relationship is not this simple. For example, one study noted that the loss of cristae structure in *Opa1* mutant mice occurred prior to the reduction in mtDNA content [235], although this is not surprising because of the key role of OPA1 in maintaining cristae junctions [209].

Analysis of the relationship between mtDNA and cristae structure is also complicated by the fact that a number of interventions which alter mtDNA content also impact on the balance of fission and fusion rates of mitochondria, and therefore the proper organisation of nucleoids around the network. There is also evidence to suggest that normal cristae structure is required to facilitate fission and fusion [210]. The MICOS complex can be destabilized by the loss of the core protein MIC60. This results in a striking phenotype of enlarged mitochondria that display abnormal circular cristae, commonly described as "concentric rings" or an "onion-like structure", alongside an almost complete loss of cristae junctions (Figure 3e) [210,237]. It has also been reported that the loss of MIC60 causes a reduction in the copy number of mtDNA (139). The downregulation of other MICOS subunits such as MIC19 and MIC10 also leads to a disruption of cristae structure [210]. The silencing of *MIC19* and *MIC60* led to the formation of enlarged and disorganised nucleoids in both mammalian cells and yeast, accompanied by a reduction in mtDNA transcription [210,238]. Similarly, it has been noted that mutations in *CHCHD10*, a constituent of the MICOS complex, leads to a loss of cristae junctions and structure. Furthermore, this was also shown to be linked with the formation of mtDNA deletions, as well as a reduction in nucleoid number without an overall effect on copy number [239].

It is worth noting that it is often difficult to determine whether apparently enlarged nucleoids represent one single enlarged nucleoid consisting of multiple physically joined (catenated) mtDNA molecules, or a group of nucleoids clustered in close proximity. An interesting aspect of the enlargement or clustering of nucleoids associated with the loss of *MIC60* is that it was not reported to lead to the formation of cristae-enriched mito-bulb structures as are seen following *DRP1* knockdown [77]. However, it is likely that these cells are unable to form dense cristae regions as a consequence of the fundamental role of MIC60 and the MICOS complex in shaping cristae [207]. Interestingly, *MIC60* knockdown cells showed a normal balance of fission and fusion compared to wild-type cells, but this occurred at a reduced rate with DRP1 being one of the mitochondrial dynamics proteins observed to be downregulated. When DRP1 was overexpressed to promote mitochondrial fragmentation, the presence of enlarged nucleoids associated with the loss of MIC60 was partially reverted [210]. Furthermore, work in yeast has found that when *MIC60 (FCJ1)* is knocked out in conjunction with *DRP1 (DNM1)* the nucleoids are further enlarged compared to the loss of *MIC60* alone [238].

Cristae junctions have also been implicated in enabling nucleoid distribution. Unlike *MIC60* deletion, the knockout of F_1F_0 ATP synthase dimerization factors does not promote nucleoid aggregation or affect the number of cristae junctions. However, its knockout does reduce the number of cristae tips. The dual knockout of *MIC60 (FCJ1)* and dimerization partners of F_1F_0 ATP synthase results in an increased number of ring-like cristae structures that rescues the nucleoid aggregation phenotype associated with *MIC60* deletion alone [238]. This finding indicates that cristae structure is likely important for partitioning nucleoids to prevent their aggregation.

Addressed collectively, these studies support the idea that normal cristae structure is required to support the even distribution of mtDNA nucleoids around the mitochondrial network (Figure 3). Firstly, dynamic cristae structure is required to facilitate normal rates of fission and fusion, which as discussed earlier are key mediators of nucleoid dispersal. Secondly, cristae can also prevent the aggregation of nucleoids by forming partitioning structures. This is in logical agreement with the observation that nucleoids are generally located in voids between groups of cristae structures [188]. At this stage, it is unclear whether nucleoid partitioning is a precursor to the distribution of nucleoids around the network via fission. Therefore, further study will be necessary to understand the intricate interplay between these two events.

4.4. Cristae Remodelling by Independent Fission and Fusion of Cristae Membranes

It has been speculated for some time that cristae are dynamic structures that can undergo remodelling of their membranes in response to various physiological conditions. However, studying these remodelling events in real-time has been particularly challenging until recently. A number of groups have developed methods that allow for the visualisation of cristae structures in live cells using super-resolution microscopy [188,206,240]. Recent work has established that cristae within the same mitochondrion can have different membrane potentials, thus demonstrating that cristae can behave as independent bioenergetic units [206]. This work suggests that within a mitochondrion cristae can functionally isolate themselves from one another, and it is postulated that this mechanism can prevent individual dysfunctional cristae from disrupting membrane potential in the broader mitochondrial network. Further work has revealed that cristae membranes within a mitochondrion undergo fission and fusion events in a MICOS-dependent manner [241]. Specifically, this study provides evidence for a model in which MIC60 is evenly distributed along the IBM, acting as a docking or scaffolding platform to facilitate the formation of cristae junctions following the recruitment of MIC10. The recruitment of other subunits to complete the MICOS complex allows cristae junctions to fully form. Both cristae junctions and cristae membranes are dynamic, and it was observed that cristae junctions have the capacity to split and merge. Cristae membranes can detach from one cristae junction, and fuse with either the same cristae junction or another cristae junction, such as one on the opposing IBM. It was demonstrated that the merging of cristae membranes is associated with changes in membrane potential at that region, and that these events can facilitate content mixing between distinct cristae compartments [241].

The fact that cristae continuously undergo remodelling and have the capacity to function independently raises the question of how these processes are involved in maintaining the integrity of mtDNA, as well as its segregation and normal distribution. This novel ability to study cristae dynamics in live cells allows an unprecedented opportunity for the field to address some of these experimentally difficult questions. For example, what is the importance of dynamic cristae membranes in maintaining mtDNA integrity by content mixing? Other directions may include understanding the interplay between overall mitochondrial dynamics and cristae dynamics, as well as addressing how the modelling of cristae membranes impacts the segregation and movement of mtDNA around the mitochondrial network.

4.5. Genetic Defects Associated with Perturbed Mitochondrial Cristae Structure and Diseases

Mammalian MICOS comprises seven protein subunits of which defects in two encoding genes have been linked to human disease so far. Mutations in *MICOS13* (aka *QIL1*) cause severe infantile mitochondrial disease characterised by failure to thrive, microcephaly, truncal hypotonia, spasticity and cerebellar atrophy, lactic acidosis and 3-methylglutoconic aciduria (3-MGA) [242–244]. Subclinical hepatic involvement manifesting with persistently abnormal liver function tests was also reported. Altered mitochondrial cristae morphology was evident in liver tissue and fibroblasts, but the mitochondrial ultrastructure in muscle appeared normal. Multiple mitochondrial respiratory chain deficiencies were identified in skeletal muscle and liver biopsies, and isolated complex IV deficiency was present in fibroblasts. However, there was no evidence of mtDNA deletions or depletion in muscles and fibroblasts [242–244]. More recently, a mutation in *APOO*, encoding MIC26, has been reported to cause an X-linked disease associated with developmental delay, hypotonia, autistic spectrum disorder, gastrointestinal symptoms, lactic acidosis and abnormal carnitine profile [245]. The pathogenicity of the *APOO* variant was supported by work with fibroblasts and a fly model.

Homozygous mutations in the PINK1 gene, encoding for PTEN-induced kinase 1, were first identified in three consanguineous families with early-onset Parkinsonism through linkage analysis in 2004 [246]. Overall, PINK1 mutations account for less than 10% of autosomal recessive Parkinson's disease (PD) [247]. Both Parkin and PINK1 have been shown to play crucial roles in mediating the mitophagy process [248]. More recently, PINK1 was found to maintain cristae junctions by the phosphorylation of the MICOS subunit MIC60 in both *Drosophila melanogaster* and human neurons [249]. Furthermore, mutations within the mitochondrial targeting sequence of MIC60 were evident in a subset of PD patients. The introduction of these patient mutations into a *Drosophila melanogaster* model disrupted the ability for MIC60 to localise to mitochondria and prompted the formation of abnormal cristae junctions. Furthermore, the overexpression of MIC60 in a $PINK1^{-/-}$ model of PD compensated for the loss of PINK1-mediated phosphorylation and rescued both cristae defects and mitochondrial function [249]. In addition, mutations in the mitochondrial targeting sequence of CHCHD2 have been identified as a risk factor for PD and Lewy body disease [250].

CHCHD10 is a protein located in the mitochondrial inter-membrane space and it has been demonstrated to play a role in the maintenance of cristae integrity [239]. Heterozygous mutations in *CHCHD10* were first linked to motor neuron disease and frontotemporal dementia, myopathy and hyperkalemia, impaired respiratory chain function and multiple mtDNA deletions in the skeletal muscle in a large family of French origin, and in a Spanish family (none of the family members underwent muscle or skin biopsy) in 2014 [251]. Whilst mutations of CHCHD10 display a mitochondrial disease that resembles motor neuron disease, a large consortium of motor neuron disease (amyotrophic lateral sclerosis) patients (n = 4365) and healthy controls (n = 1832) from seven countries subsequently demonstrated that pathogenic *CHCHD10* variants are exceptionally rare [252]. Therefore, in pure forms of amyotrophic lateral sclerosis it is not necessarily associated with *CHCHD10* mutations.

Prominent extra-neurological disease has been observed following the disruption of several proteins implicated in the maintenance of cristae structure, for example, hypertrophic cardiomyopathy occurs in patients harbouring mutated forms of *ATAD3A* and *TAZ* (Barth syndrome) [253,254], and 3-MGA in *ATAD3A*, *TAZ*, *MICOS13* and *ATP5F1E* [243,253,255,256].

5. Conclusions

The structure and dynamics of the mitochondrial membranes are required to maintain both the integrity of mtDNA and well as its distribution within the mitochondrial network. Mitochondrial membranes make a poorly understood contribution to mtDNA replication, the impairment of which manifests either as an inability to maintain a sufficient number of copies of mtDNA, or as rearrangements of the genome. Following replication, the dynamics of mitochondria are also required for the segregation of mtDNA, and the disruption of mitochondrial dynamics can lead to the clustering of nucleoids within cells. The direct association between mtDNA and the respiratory chain means that an uneven distribution of nucleoids can lead to a mosaic pattern of respiratory activity within cells, which may represent an under-appreciated molecular contributor to mitochondrial pathologies. Observations from the clinic highlight the importance of these mechanisms, with mutations in genes associated with fusion, fission and cristae structure manifesting mainly in severe neurological disorders frequently associated with mtDNA depletion or rearrangements.

Work from the laboratory puts forward the notion that mitochondrial fusion is necessary to maintain adequate levels of mtDNA replication, likely highlighting the importance of content mixing for maintaining a proper stoichiometry of replisome components between mitochondria. Indeed, it is plausible that this protein heterogeneity may prompt replication stalling resulting in the formation of deletions and rearrangements of mtDNA. The reduction in mitochondrial DNA copy number observed in conjunction with the disruption of fusion therefore likely occurs due to reduced rates of replication.

Defects in mitochondrial fission on the other hand have less of an association with the onset of mtDNA rearrangements or a reduction in mtDNA copy number in the laboratory. However, mutations of fission genes still result in severe neurological disorders in the clinic. As mitochondrial function is often observed to be normal when fission is impaired, the mechanism underlying these clinical phenotypes is not entirely clear at this stage. It is understood that mitochondrial fission is necessary for mitochondrial quality control which can facilitate the accumulation of dysfunctional mitochondria. Furthermore, fission is essential for the even allocation of nucleoids around the mitochondrial network and to subsequent daughter cells. Disruption of fission leads to the clustering of nucleoids. Indeed, the presence of mtDNA is intricately linked to the formation of normal cristae architecture. The clustering of nucleoids is associated with the formation of dense cristae regions in the form of mito-bulb structures, and the loss of mtDNA is often correlated with the loss of cristae structure. The data available suggests that dynamic cristae are necessary to facilitate normal rates of mitochondrial fission and fusion, as well as allow for the partitioning of mtDNA. As such the interactions between dynamics, cristae and mtDNA organisation are tightly interwoven and interdependent. It is therefore likely that if each of these processes is not tightly regulated then there are subsequent downstream effects that disrupt the delicate balance and prompt the onset of mitochondrial dysfunction.

Author Contributions: Conceptualization, J.C. and T.J.N.; writing—original draft preparation, J.C. and Y.S.N.; writing—review and editing, T.J.N.; supervision, T.J.N.; funding acquisition, T.J.N. All authors have read and agreed to the published version of the manuscript.

Funding: T.J.N. is the recipient of a Sir Henry Dale Fellowship (213464/Z/18/Z), jointly funded by the Wellcome Trust and Royal Society, and a Rosetrees and Stoneygate Trust fellowship (M811). Y.S.N. holds an NIHR Clinical Lectureship in Neurology (CL-2016-01-003).

Conflicts of Interest: The authors declare no conflict of interest. The funders had no role in the design of the study; in the collection, analyses, or interpretation of data; in the writing of the manuscript, or in the decision to publish the results.

Abbreviations

3-MGA	3-methylglutoconic aciduria
AD	autosomal dominant
AR	autosomal recessive
ATAD3	ATPase family AAA domain-containing protein 3
ATP	adenosine triphosphate
CK	creatine kinase
CMT	Charcot-Marie-Tooth
COX1	cytochrome c oxidase 1 subunit
CPEO	chronic progressive external ophthalmoplegia
CSF	cerebrospinal fluid
DOA	dominant optic atrophy
DD	developmental delay
DRP1	dynamin-related protein 1
ER	endoplasmic reticulum
ESRF	end-stage renal failure
FIS1	mitochondrial fission 1 protein
FSGS	focal segmental glomerulosclerosis
GTP	guanine triphosphate
HSP	heavy strand promoter
IBM	inner boundary membrane
IMM	inner mitochondrial membrane
IMS	intermembrane space
LA	lactic acidosis
LS	Leigh syndrome
LSP	light strand promoter
MCL1	induced myeloid leukemia cell differentiation protein
MEF	mouse embryonic fibroblast
MERRF	myoclonic epilepsy and ragged-red fibres
MFN	Mitofusin
MICOS	mitochondrial contact site and cristae organising system
MID49	mitochondrial dynamics protein of 49kDa
MitoPLD	mitochondria-localised phospholipase D
MRC	mitochondrial respiratory chain
mtDNA	mitochondrial DNA
MTERF1	mitochondrial transcription termination factor 1
mtSSB	mitochondrial single stranded DNA-binding protein
NCR	noncoding region
OA	optic atrophy
OMA1	overlapping with the M-AAA protease 1 homolog
OMM	outer mitochondrial membrane
OPA1	optic atrophy 1
OriH	origin of H-strand replication
OriL	origin of L-strand replication
OXPHOS	oxidative phosphorylation
PA	phosphatidic acid
PARL	presenilin-associated rhomboid-like
PD	Parkinson's disease
POLγ	DNA polymerase-γ
POLRMT	mitochondrial RNA polymerase
rRNA	ribosomal RNA
RRFs	ragged-red fibres

RTA	renal tubular acidosis
SCA28	spinocerebellar ataxia type 28
SNHL	sensorineural hearing loss
TEFM	mitochondrial transcription elongation factor
TFAM	mitochondrial transcription factor A
TFB2M	mitochondrial transcription factor B2
tRNA	transfer RNA
VLCFA	very long chain fatty acid
WM	white matter
YME1L1	YME1 like 1 ATPase

Appendix A

Table A1. Summary of clinical phenotype, extra-neurological involvement and integrity of mitochondrial DNA associated with the genetic defects of mitochondrial dynamics.

	Gene	Protein	Protein Function	Inheritance	Clinical Phenotype	Extra-Neurological	mtDNA Integrity	References
Regulators of Mitochondrial Dynamics	DRP1	Dynamin-related protein 1	Mitochondrial fission	AD	Microcephaly, abnormal brain development, OA, LA, elevated VLCFA; DD, refractory epilepsy; isolated DOA (OPA5)	None	EM & confocal microscopy study showing concentric cristae structure in patient-derived fibroblasts [97]	[92,97,257,258]
							CIV deficiency in muscle [257]	
							Elongated mitochondria in optic nerve and RGC layer without axonal degeneration in Dnm1l$^{+/-}$ mice [258]	
							No reports of mtDNA deletions or depletion	
	DNM2	Dynamin 2	Mitochondrial fission	AD	Centronuclear myopathy; dominant intermediate CMT neuropathy type B associated with neutropenia and cataract; CPEO, facial weakness, neck flexor weakness; severe cardiomyopathy and centronuclear myopathy	Cardiac, neutropenia	COX negative fibres and multiple mtDNA deletions in skeletal muscle [103]	[100–103,259,260]
	GDAP1	Ganglioside Induced Differentiation Associated Protein 1	Mitochondrial fission	AR	CMT4A (early onset with rapid progression demyelinating neuropathy)	None	No reports of mtDNA deletions or depletion	[93,106,108,109]
				AD	CMT2RV (axonal neuropathy with hoarse voice and diaphragmatic weakness)			
	INF2	Inverted Formin, FH2 And WH2 Domain Containing	Mitochondrial fission	AD	CMT, dominant, intermediate type, E (CMTDIE) with SNHL and renal phenotype (FSGS, proteinuria and ESRF)	Renal	No reports of mtDNA deletions or depletion	[94,104,105,261]
	MFF	Mitochondrial Fission Factor	Mitochondrial fission	AR	DD, pyramidal signs and neuropathy; Leigh-like disease with infantile spasm, OA and neuropathy	None	Normal mitochondrial respiratory chain activities in muscle; significant branching of mitochondria in patient-derived fibroblasts [98]	[95,98,107]
	SLC25A46	Solute Carrier Family 25 Member 46	Mitochondrial fission	AD	DOA and CMT2; hereditary motor and sensory neuropathy, Type VIB (HMSN6B)	None	Abnormal mitochondrial cristae following knockdown in fibroblasts [99]	[96,99,262,263]
							Mitochondrial elongation following knockdown in Zebrafish, mtDNA integrity not assessed [262]	
				AR	Progressive myoclonic ataxia, OA and neuropathy; LS; pontocerebellar hypoplasia and apnoea		No reports of mtDNA deletions or depletion	

Table A1. *Cont.*

Gene	Protein	Protein Function	Inheritance	Clinical Phenotype	Extra-Neurological	mtDNA Integrity	References
OPA1	OPA1 Mitochondrial Dynamin Like GTPase	IM fusion/cristae shaping	AD	DOA (most common genetic cause); DOA plus syndrome (CPEO, ataxia, axonal neuropathy, deafness); syndromic Parkinsonism and dementia	Cardiac	Abnormal cristae in OPA1$^{+/-}$ mouse model [264]	[141,143,174,265–270]
			AR	Ataxia, early-onset OA, gastrointestinal dysmotility; Behr syndrome (OA and ataxia); Infantile encephalomyopathy, HCM and OA; Leigh-like imaging changes		COX negative fibres and multiple mtDNA deletions in cases of heterozygous mutation [141,174]	
						Multiple respiratory chain deficiencies and mtDNA depletion in muscle with recessive inheritance [265]	
MFN2	Mitofusin 2	OM fusion	AD	CMT2A; DOA plus; DD, progressive weakness, failure to thrive, axonal sensorimotor neuropathy, chorea and elevated CSF lactate	Lipoma	Multiple mtDNA deletions and mtDNA depletion detected in muscles and fibroblasts [154]	[154,157,176,271–275]
			AR	Severe, early-onset axonal neuropathy, OA; multiple symmetric lipomatosis, LA and axonal neuropathy		RRFs, COX deficient fibres, altered mitochondrial morphology and multiple mtDNA deletion detected in muscle [176]	
						COX deficient fibres, ultrastructural changes and mtDNA depletion	
FBLX4	F-Box And Leucine Rich Repeat Protein 4	OM fusion	AR	Encephalopathy, microcephaly and persistent LA; encephalomyopathy, abnormal WM changes, LA, dysmorphism, RTA, seizures	Renal (RTA)	Multiple respiratory chain deficiencies and mtDNA depletion in patient-derived fibroblasts [155]	[155,276–279]
						Normal MRC enzymes but mtDNA depletion in muscle [276]	
						mtDNA depletion, reduced respiratory function and ultrastructure abnormalities in muscle, reduced respiratory function in fibroblasts [277]	
MSTO1	Misato Mitochondrial Distribution And Morphology Regulator 1	Interacts with mitochondrial fusion machinery	AR	Myopathy, elevated CK (1200–4500 IU/L), pigmentary retinopathy, scoliosis and cerebellar ataxia and atrophy	None	Normal MRC enzymes, reduced mtDNA copy number in muscle; mitochondrial fragmentation in patient-derived fibroblasts [280]	[280–282]
			AD	Myopathy, cerebellar ataxia and non-specific neuropsychiatric symptoms			
YME1L1 (YME1)	YME1 like 1 ATPase	OPA1 processing	AR	Intellectual disability, DD, OA, hearing impairment, ataxia, LA, leukoencephalopathic changes on MRI,	None	Altered cristae ultrastructure in muscle, mitochondrial fragmentation in patient-derived fibroblasts [283]	[283]
SPG7	SPG7 Matrix AAA Peptidase Subunit, Paraplegin	OPA1 processing	AR	Spastic paraplegia 7; spastic ataxia; cerebellar ataxia; CPEO and myopathy;	None	Altered mitochondrial ultrastructure; multiple respiratory chain deficiencies in muscle [147,284]	[147,148,153,284,285]
						RRFs, COX-deficient fibres and multiple mtDNA deletions; normal mtDNA copy number in muscle; reduced network size in patient-derived fibroblasts [153]	
AFG3L2	AFG3 Like Matrix AAA Peptidase Subunit 2	OPA1 processing	AD	Spinocerebellar ataxia 28 (SCA28); CPEO	None	RRFs, COX deficient fibres and multiple mtDNA deletions in muscle [152]	[152,286–288]

Table A1. *Cont.*

	Gene	Protein	Protein Function	Inheritance	Clinical Phenotype	Extra-Neurological	mtDNA Integrity	References
Regulators of Cristae Structure	MICOS13 (MIC13; QIL1)	Mitochondrial Contact Site And Cristae Organizing System Subunit 13	Stabilises MIC60 and MIC10 in the MICOS Complex	AR	Failure to thrive, acquired microcephaly, truncal hypotonia, limb spasticity cerebellar atrophy, LA, 3-MGA, elevated liver transaminases and myoclonic seizures	Liver	Normal histology and MRC activities in muscle; low MRC enzymatic activities and abnormal cristae in liver [243]	[243,244]
							Multiple respiratory chain deficiencies in muscle and liver [244]	
	APOO (MIC26)	Apolipoprotein O	Regulatory role in MICOS complex	X-linked	DD, myopathy, LA, cognitive impairment, autistic features, abnormal acylcarnitidine profile	None	Abnormal ultrastructure in fibroblasts and a Drosophila Melanogaster model [245]	[245]
	CHCHD10	Coiled-Coil-Helix-Coiled-Coil-Helix Domain Containing 10	Stabilises MICOS complex	AD	Frontotemporal dementia and amyotrophic lateral sclerosis; late onset spinal motor neuronopathy and elevated CK; myopathy, elevated CK and lactate; CMT2	None	RRFs, COX deficient fibres, multiple respiratory chain deficiencies, abnormal assembly of CV, multiple mtDNA deletions in muscle, reduced fusion and loss of cristae structure in patient-derived fibroblast [251]	[239,251,252,289–292]
							Mildly increased COX deficient fibres and occasional RRFs [290] and multiple mtDNA deletions [290]	
							Multiple respiratory chain deficiencies (worst with CIV), RRFs, lipid accumulation and abnormal circular cristae ultrastructure [291]	
							Loss of mitochondrial cristae junctions with impaired mitochondrial genome maintenance and inhibition of apoptosis in patient-derived fibroblasts [239]	
	PINK1	PTEN Induced Kinase 1	Mitophagy/Interacts with MIC60 to regulate CJs	AR	Parkinson disease 6 (PARK6)	None	Complex I deficiency in Drosophila Melanogaster model [293]	[246,293–297]
	ATP5F1E F_1F_o-ATP synthase (sub unit E)	ATP Synthase F1 Subunit Epsilon	Cristae shaping/ATP Production	AR	LA, 3-MGA, mild mental retardation and development of peripheral neuropathy	None	Complex V deficiency in patient-derived fibroblasts [256]	[256]
	ATP5F1A F_1F_o-ATP synthase (sub unit a)	ATP Synthase F1 Subunit Alpha	Cristae shaping/ATP production	AR	Intrauterine growth retardation, microcephaly, hypotonia, pulmonary hypertension, failure to thrive, encephalopathy, and heart failure; neonatal-onset encephalopathy with intractable seizures, extensive signal changes involving WM, basal ganglia and thalamus	Cardiac	Multiple respiratory chain deficiency and mtDNA depletion in muscle [298]	[298,299]
							Complex V deficiency in muscle [299]	
IMM Structure	ATAD3A	ATPase Family AAA Domain Containing 3A	mtDNA and cristae organisation	De novo & AR	Global DD, hypotonia, OA, axonal neuropathy, mildly elevated lactate, seizures, 3-MGA and hypertrophic cardiomyopathy; fatal congenital pontocerebellar hypoplasia with a simplified gyral pattern; cerebellar atrophy with dystonia and ataxia	Cardiac	Abnormal cristae in Drosophila Melanogaster and patient-derived fibroblasts [253]	[171,253]
							mtDNA abnormalities in patient-derived fibroblasts [171]	
	TAZ	Tafazzin	Organisation of cardiolipin in the IM	X-linked	Severe infantile cardiomyopathy (left ventricular noncompaction, Barth syndrome with dilated cardiomyopathy), sudden cardiac death, myopathy, neutropenia, growth failure, 3-MGA, facial dysmorphism	Cardiac, neutropenia	Altered mitochondrial ultrastructure, stacked and circular cristae, in patient-derived lymphoblasts [300]	[254,255,300–304]

References

1. Gray, M.W. Mitochondrial Evolution. *Cold Spring Harb. Perspect. Biol.* **2012**, *4*, a011403. [CrossRef]
2. Gualberto, J.; Mileshina, D.; Wallet, C.; Niazi, A.K.; Weber-Lotfi, F.; Dietrich, A. The plant mitochondrial genome: Dynamics and maintenance. *Biochimie* **2014**, *100*, 107–120. [CrossRef]
3. Foury, F.; Roganti, T.; Lecrenier, N.; Purnelle, B. The complete sequence of the mitochondrial genome of Saccharomyces cerevisiae. *FEBS Lett.* **1998**, *440*, 325–331. [CrossRef]
4. Andrews, R.M.; Kubacka, I.; Chinnery, P.F.; Chrzanowska-Lightowlers, Z.M.; Turnbull, D.M.; Howell, N. Reanalysis and revision of the Cambridge reference sequence for human mitochondrial DNA. *Nat. Genet.* **1999**, *23*, 147. [CrossRef] [PubMed]
5. Chaban, Y.; Boekema, E.J.; Dudkina, N.V. Structures of mitochondrial oxidative phosphorylation supercomplexes and mechanisms for their stabilisation. *Biochim. Biophys. Acta* **2014**, *1837*, 418–426. [CrossRef]
6. Calvo, S.E.; Clauser, K.R.; Mootha, V.K. MitoCarta2.0: An updated inventory of mammalian mitochondrial proteins. *Nucleic Acids Res.* **2015**, *44*, D1251–D1257. [CrossRef]
7. Chinnery, P.F.; Hudson, G. Mitochondrial genetics. *Br. Med. Bull.* **2013**, *106*, 135–159. [CrossRef]
8. Falkenberg, M.; Larsson, N.-G.; Gustafsson, C.M. DNA Replication and Transcription in Mammalian Mitochondria. *Annu. Rev. Biochem.* **2007**, *76*, 679–699. [CrossRef]
9. Gustafsson, C.M.; Falkenberg, M.; Larsson, N.-G. Maintenance and Expression of Mammalian Mitochondrial DNA. *Annu. Rev. Biochem.* **2016**, *85*, 133–160. [CrossRef]
10. Falkenberg, M. Mitochondrial DNA replication in mammalian cells: Overview of the pathway. *Essays Biochem.* **2018**, *62*, 287–296.
11. Berk, A.J.; Clayton, D.A. Mechanism of mitochondrial DNA replication in mouse L-cells: Asynchronous replication of strands, segregation of circular daughter molecules, aspects of topology and turnover of an initiation sequence. *J. Mol. Biol.* **1974**, *86*, 801–824. [CrossRef]
12. Falkenberg, M.; Gaspari, M.; Rantanen, A.; Trifunovic, A.; Larsson, N.-G.; Gustafsson, C.M. Mitochondrial transcription factors B1 and B2 activate transcription of human mtDNA. *Nat. Genet.* **2002**, *31*, 289–294. [CrossRef]
13. Hillen, H.S.; Morozov, Y.I.; Sarfallah, A.; Temiakov, D.; Cramer, P. Structural Basis of Mitochondrial Transcription Initiation. *Cell* **2017**, *171*, 1072–1081. [CrossRef] [PubMed]
14. Minczuk, M.; He, J.; Duch, A.M.; Ettema, T.J.; Chlebowski, A.; Dzionek, K.; Nijtmans, L.G.; Huynen, M.A.; Holt, I.J. TEFM (c17orf42) is necessary for transcription of human mtDNA. *Nucleic Acids Res.* **2011**, *39*, 4284–4299. [CrossRef] [PubMed]
15. Posse, V.; Shahzad, S.; Falkenberg, M.; Hallberg, B.M.; Gustafsson, C.M. TEFM is a potent stimulator of mitochondrial transcription elongation in vitro. *Nucleic Acids Res.* **2015**, *43*, 2615–2624. [CrossRef] [PubMed]
16. Posse, V.; Al-Behadili, A.; Uhler, J.P.; Clausen, A.R.; Reyes, A.; Zeviani, M.; Falkenberg, M.; Gustafsson, C.M. RNase H1 directs origin-specific initiation of DNA replication in human mitochondria. *PLoS Genet.* **2019**, *15*, e1007781. [CrossRef]
17. Wanrooij, P.H.; Uhler, J.P.; Simonsson, T.; Falkenberg, M.; Gustafsson, C.M. G-quadruplex structures in RNA stimulate mitochondrial transcription termination and primer formation. *Proc. Natl. Acad. Sci. USA* **2010**, *107*, 16072–16077. [CrossRef]
18. Wanrooij, P.H.; Uhler, J.P.; Shi, Y.; Westerlund, F.; Falkenberg, M.; Gustafsson, C.M. A hybrid G-quadruplex structure formed between RNA and DNA explains the extraordinary stability of the mitochondrial R-loop. *Nucleic Acids Res.* **2012**, *40*, 10334–10344. [CrossRef]
19. Gray, H.; Wong, T.W. Purification and identification of subunit structure of the human mitochondrial DNA polymerase. *J. Biol. Chem.* **1992**, *267*, 5835–5841.
20. Yakubovskaya, E.; Chen, Z.; Carrodeguas, J.A.; Kisker, C.; Bogenhagen, D.F. Functional human mitochondrial dna polymerase forms a heterotrimer. *J. Biol. Chem.* **2005**, *281*, 374–382. [CrossRef]
21. Lee, Y.-S.; Kennedy, W.D.; Yin, Y.W. Structural insight into processive human mitochondrial DNA synthesis and disease-related polymerase mutations. *Cell* **2009**, *139*, 312–324. [CrossRef] [PubMed]
22. Carrodeguas, J.A.; Pinz, K.G.; Bogenhagen, D.F. DNA binding properties of human pol, γB. *J. Biol. Chem.* **2002**, *277*, 50008–50014. [CrossRef] [PubMed]

23. Korhonen, J.A.; Pham, X.H.; Pellegrini, M.; Falkenberg, M. Reconstitution of a minimal mtDNA replisome in vitro. *EMBO J.* **2004**, *23*, 2423–2429. [CrossRef] [PubMed]
24. Korhonen, J.A.; Gaspari, M.; Falkenberg, M. TWINKLE Has 5′ -> 3′ DNA Helicase Activity and Is Specifically Stimulated by Mitochondrial Single-stranded DNA-binding Protein. *J. Biol. Chem.* **2003**, *278*, 48627–48632. [CrossRef]
25. Martens, P.A.; Clayton, D.A. Mechanism of mitochondrial DNA replication in mouse L-cells: Localization and sequence of the light-strand origin of replication. *J. Mol. Biol.* **1979**, *135*, 327–351. [CrossRef]
26. Fusté, J.M.; Wanrooij, S.; Jemt, E.; Granycome, C.E.; Cluett, T.J.; Shi, Y.; Atanassova, N.; Holt, I.J.; Gustafsson, C.M.; Falkenberg, M. Mitochondrial RNA Polymerase Is Needed for Activation of the Origin of Light-Strand DNA Replication. *Mol. Cell* **2010**, *37*, 67–78. [CrossRef]
27. Wanrooij, S.; Fuste, J.M.; Farge, G.; Shi, Y.; Gustafsson, C.M.; Falkenberg, M. Human mitochondrial RNA polymerase primes lagging-strand DNA synthesis in vitro. *Proc. Natl. Acad. Sci. USA* **2008**, *105*, 11122–11127. [CrossRef]
28. Reyes, A.; Kazak, L.; Wood, S.R.; Yasukawa, T.; Jacobs, H.T.; Holt, I.J. Mitochondrial DNA replication proceeds via a 'bootlace' mechanism involving the incorporation of processed transcripts. *Nucleic Acids Res.* **2013**, *41*, 5837–5850. [CrossRef]
29. Yasukawa, T.; Kang, N. An overview of mammalian mitochondrial DNA replication mechanisms. *J. Biochem.* **2018**, *164*, 183–193. [CrossRef]
30. Holt, I.J.; Jacobs, H.T. Unique features of DNA replication in mitochondria: A functional and evolutionary perspective. *BioEssays* **2014**, *36*, 1024–1031. [CrossRef]
31. Holt, I.; Lorimer, H.E.; Jacobs, H.T. Coupled Leading- and Lagging-Strand Synthesis of Mammalian Mitochondrial DNA. *Cell* **2000**, *100*, 515–524. [CrossRef]
32. Holmes, J.B.; Akman, G.; Wood, S.R.; Sakhuja, K.; Cerritelli, S.M.; Moss, C.; Bowmaker, M.R.; Jacobs, H.T.; Crouch, R.J.; Holt, I.J.; et al. Primer retention owing to the absence of RNase H1 is catastrophic for mitochondrial DNA replication. *Proc. Natl. Acad. Sci. USA* **2015**, *112*, 9334–9339. [CrossRef] [PubMed]
33. Al-Behadili, A.; Uhler, J.P.; Berglund, A.K.; Peter, B.; Doimo, M.; Reyes, A.; Wanrooij, S.; Zeviani, M.; Falkenberg, M. A two-nuclease pathway involving RNase H1 is required for primer removal at human mitochondrial, O.r.i.L. *Nucleic Acids Res.* **2018**, *46*, 9471–9483. [CrossRef] [PubMed]
34. Cerritelli, S.M.; Frolova, E.G.; Feng, C.; Grinberg, A.; Love, P.E.; Crouch, R.J. Failure to Produce Mitochondrial DNA Results in Embryonic Lethality in Rnaseh1 Null Mice. *Mol. Cell* **2003**, *11*, 807–815. [CrossRef]
35. Lakshmipathy, U.; Campbell, C. The Human DNA Ligase III Gene Encodes Nuclear and Mitochondrial Proteins. *Mol. Cell. Biol.* **1999**, *19*, 3869–3876. [CrossRef]
36. Uhler, J.P.; Thörn, C.; Nicholls, T.J.; Matic, S.; Milenkovic, D.; Gustafsson, C.M.; Falkenberg, M. MGME1 processes flaps into ligatable nicks in concert with DNA polymerase gamma during mtDNA replication. *Nucleic Acids Res.* **2016**, *44*, 5861–5871. [CrossRef]
37. Nicholls, T.J.; Zsurka, G.; Peeva, V.; Schöler, S.; Szczesny, R.J.; Cysewski, D.; Reyes, A.; Kornblum, C.; Sciacco, M.; Moggio, M.; et al. Linear mtDNA fragments and unusual mtDNA rearrangements associated with pathological deficiency of MGME1 exonuclease. *Hum. Mol. Genet.* **2014**, *23*, 6147–6162. [CrossRef]
38. Nicholls, T.J.; Nadalutti, C.A.; Motori, E.; Sommerville, E.W.; Gorman, G.S.; Basu, S.; Hoberg, E.; Turnbull, D.M.; Chinnery, P.F.; Larsson, N.G.; et al. Topoisomerase 3alpha Is Required for Decatenation Segregation of Human mtDNA. *Mol. Cell* **2018**, *69*, 9–23.e6. [CrossRef]
39. Gorman, G.S.; Chinnery, P.F.; DiMauro, S.; Hirano, M.; Koga, Y.; McFarland, R.; Suomalainen, A.; Thorburn, D.R.; Zeviani, M.; Turnbull, D.M. Mitochondrial diseases. *Nat. Rev. Dis. Primers* **2016**, *2*, 16080. [CrossRef]
40. Schon, E.A.; Rizzuto, R.; Moraes, C.T.; Nakase, H.; Zeviani, M.; DiMauro, S. A direct repeat is a hotspot for large-scale deletion of human mitochondrial DNA. *Science* **1989**, *244*, 346–349. [CrossRef]
41. Viscomi, C.; Zeviani, M. MtDNA-maintenance defects: Syndromes and genes. *J. Inherit. Metab. Dis.* **2017**, *40*, 587–599. [CrossRef] [PubMed]
42. Bender, A.; Krishnan, K.J.; Morris, C.M.; Taylor, G.A.; Reeve, A.K.; Perry, R.H.; Jaros, E.; Hersheson, J.S.; Betts, J.; Klopstock, T.; et al. High levels of mitochondrial DNA deletions in substantia nigra neurons in aging and Parkinson disease. *Nat. Genet.* **2006**, *38*, 515–517. [CrossRef]

43. Kraytsberg, Y.; Kudryavtseva, E.; McKee, A.C.; Geula, C.; Kowall, N.W.; Khrapko, K. Mitochondrial DNA deletions are abundant and cause functional impairment in aged human substantia nigra neurons. *Nat. Genet.* **2006**, *38*, 518–520. [CrossRef] [PubMed]
44. Lawless, C.; Greaves, L.C.; Reeve, A.K.; Turnbull, D.M.; Vincent, A.E. The rise and rise of mitochondrial DNA mutations. *Open Biol.* **2020**, *10*, 200061. [CrossRef] [PubMed]
45. Nissanka, N.; Minczuk, M.; Moraes, C.T. Mechanisms of Mitochondrial DNA Deletion Formation. *Trends Genet.* **2019**, *35*, 235–244. [CrossRef] [PubMed]
46. Shoffner, J.M.; Lott, M.T.; Voljavec, A.S.; Soueidan, S.A.; Costigan, D.A.; Wallace, D.C. Spontaneous Kearns-Sayre/chronic external ophthalmoplegia plus syndrome associated with a mitochondrial DNA deletion: A slip-replication model and metabolic therapy. *Proc. Natl. Acad. Sci. USA* **1989**, *86*, 7952–7956. [CrossRef]
47. Persson, Ö.; Muthukumar, Y.; Basu, S.; Jenninger, L.; Uhler, J.P.; Berglund, A.-K.; McFarland, R.; Taylor, R.W.; Gustafsson, C.M.; Larsson, E. Copy-choice recombination during mitochondrial L-strand synthesis causes DNA deletions. *Nat. Commun.* **2019**, *10*, 1–10.
48. Nissanka, N.; Bacman, S.R.; Plastini, M.; Moraes, C.T. The mitochondrial DNA polymerase gamma degrades linear DNA fragments precluding the formation of deletions. *Nat. Commun.* **2018**, *9*, 2491. [CrossRef]
49. Srivastava, S.; Moraes, C.T. Double-strand breaks of mouse muscle mtDNA promote large deletions similar to multiple mtDNA deletions in humans. *Hum. Mol. Genet.* **2005**, *14*, 893–902. [CrossRef]
50. Bacman, S.R.; Williams, S.L.; Moraes, C.T. Intra-and inter-molecular recombination of mitochondrial DNA after in vivo induction of multiple double-strand breaks. *Nucleic Acids Res.* **2009**, *37*, 4218–4226. [CrossRef]
51. Krishnan, K.J.; Reeve, A.K.; Samuels, D.C.; Chinnery, P.F.; Blackwood, J.K.; Taylor, R.W.; Wanrooij, S.; Spelbrink, J.N.; Chrzanowska-Lightowlers, Z.M.; Turnbull, D.M. What causes mitochondrial DNA deletions in human cells? *Nat. Genet.* **2008**, *40*, 275–279. [CrossRef] [PubMed]
52. Farge, G.; Mehmedovic, M.; Baclayon, M.; Wildenberg, S.M.V.D.; Roos, W.H.; Gustafsson, C.M.; Wuite, G.J.; Falkenberg, M. In Vitro-Reconstituted Nucleoids Can Block Mitochondrial DNA Replication and Transcription. *Cell Rep.* **2014**, *8*, 66–74. [CrossRef] [PubMed]
53. Brown, T.A.; Tkachuk, A.N.; Shtengel, G.; Kopek, B.G.; Bogenhagen, D.F.; Hess, H.F.; Clayton, D.A. Superresolution fluorescence imaging of mitochondrial nucleoids reveals their spatial range, limits, and membrane interaction. *Mol. Cell. Biol.* **2011**, *31*, 4994–5010. [CrossRef] [PubMed]
54. Kukat, C.; Wurm, C.A.; Spahr, H.; Falkenberg, M.; Larsson, N.G.; Jakobs, S. Super-resolution microscopy reveals that mammalian mitochondrial nucleoids have a uniform size and frequently contain a single copy of mtDNA. *Proc. Natl. Acad. Sci. USA* **2011**, *108*, 13534–13539. [CrossRef]
55. Kaufman, B.A.; Durisic, N.; Mativetsky, J.M.; Costantino, S.; Hancock, M.A.; Grutter, P.; Shoubridge, E.A. The Mitochondrial Transcription Factor TFAM Coordinates the Assembly of Multiple DNA Molecules into Nucleoid-like Structures. *Mol. Biol. Cell* **2007**, *18*, 3225–3236. [CrossRef]
56. Ekstrand, M.I.; Falkenberg, M.; Rantanen, A.; Park, C.B.; Gaspari, M.; Hultenby, K.; Rustin, P.; Gustafsson, C.M.; Larsson, N.-G. Mitochondrial transcription factor A regulates mtDNA copy number in mammals. *Hum. Mol. Genet.* **2004**, *13*, 935–944. [CrossRef]
57. Bogenhagen, D.F. Mitochondrial DNA nucleoid structure. *Biochim. Biophys. Acta* **2012**, *1819*, 914–920. [CrossRef]
58. Takamatsu, C.; Umeda, S.; Ohsato, T.; Ohno, T.; Abe, Y.; Fukuoh, A.; Shinagawa, H.; Hamasaki, N.; Kang, D. Regulation of mitochondrial D-loops by transcription factor A and single-stranded DNA-binding protein. *EMBO Rep.* **2002**, *3*, 451–456. [CrossRef]
59. Wang, Y.; Bogenhagen, D.F. Human Mitochondrial DNA Nucleoids Are Linked to Protein Folding Machinery and Metabolic Enzymes at the Mitochondrial Inner Membrane. *J. Biol. Chem.* **2006**, *281*, 25791–25802. [CrossRef]
60. He, J.; Cooper, H.M.; Reyes, A.; Di Re, M.; Sembongi, H.; Litwin, T.R.; Gao, J.; Neuman, K.C.; Fearnley, I.M.; Spinazzola, A. Mitochondrial nucleoid interacting proteins support mitochondrial protein synthesis. *Nucleic Acids Res.* **2012**, *40*, 6109–6121. [CrossRef]
61. Han, S.; Udeshi, N.D.; Deerinck, T.J.; Svinkina, T.; Ellisman, M.H.; Carr, S.A.; Ting, A.Y. Proximity Biotinylation as a Method for Mapping Proteins Associated with mtDNA in Living Cells. *Cell Chem. Biol.* **2017**, *24*, 404–414. [CrossRef] [PubMed]

62. Bogenhagen, D.F.; Rousseau, D.; Burke, S. The Layered Structure of Human Mitochondrial DNA Nucleoids. *J. Biol. Chem.* **2007**, *283*, 3665–3675. [CrossRef] [PubMed]
63. Couvillion, M.T.; Soto, I.C.; Shipkovenska, G.; Churchman, L.S. Synchronized mitochondrial and cytosolic translation programs. *Nature* **2016**, *533*, 499–503. [CrossRef] [PubMed]
64. Richter-Dennerlein, R.; Oeljeklaus, S.; Lorenzi, I.; Ronsör, C.; Bareth, B.; Schendzielorz, A.B.; Wang, C.; Warscheid, B.; Rehling, P.; Dennerlein, S. Mitochondrial Protein Synthesis Adapts to Influx of Nuclear-Encoded Protein. *Cell* **2016**, *167*, 471–483.e10. [CrossRef]
65. Gilkerson, R.W.; Schon, E.A.; Hernandez, E.; Davidson, M.M. Mitochondrial nucleoids maintain genetic autonomy but allow for functional complementation. *J. Cell Biol.* **2008**, *181*, 1117–1128. [CrossRef]
66. Ishihara, T.; Ban-Ishihara, R.; Maeda, M.; Matsunaga, Y.; Ichimura, A.; Kyogoku, S.; Aoki, H.; Katada, S.; Nakada, K.; Nomura, M. Dynamics of Mitochondrial DNA Nucleoids Regulated by Mitochondrial Fission Is Essential for Maintenance of Homogeneously Active Mitochondria during Neonatal Heart Development. *Mol. Cell. Biol.* **2014**, *35*, 211–223. [CrossRef]
67. Nass, M.M. Mitochondrial DNA. I. Intramitochondrial distribution and structural relations of single- \and double-length circular DNA. *J. Mol. Biol.* **1969**, *42*, 521–528. [CrossRef]
68. Kopek, B.G.; Shtengel, G.; Xu, C.S.; Clayton, D.A.; Hess, H.F. Correlative 3D superresolution fluorescence and electron microscopy reveal the relationship of mitochondrial nucleoids to membranes. *Proc. Natl. Acad. Sci. USA* **2012**, *109*, 6136–6141. [CrossRef]
69. Iborra, F.J.; Kimura, H.; Cook, P.R. The functional organisation of mitochondrial genomes in human cells. *BMC Biol.* **2004**, *2*, 9. [CrossRef]
70. Albring, M.; Griffith, J.; Attardi, G. Association of a protein structure of probable membrane derivation with HeLa cell mitochondrial DNA near its origin of replication. *Proc. Natl. Acad. Sci. USA* **1977**, *74*, 1348–1352. [CrossRef]
71. Collins, T.J.; Berridge, M.J.; Lipp, P.; Bootman, M.D. Mitochondria are morphologically and functionally heterogeneous within cells. *EMBO J.* **2002**, *21*, 1616–1627. [CrossRef] [PubMed]
72. Gomes, L.C.G.; Scorrano, L. Mitochondrial morphology in mitophagy and macroautophagy. *Biochim. Biophys. Acta* **2013**, *1833*, 205–212. [CrossRef] [PubMed]
73. Wang, C.; Youle, R.J. The role of mitochondria in apoptosis. *Annu. Rev. Genet.* **2009**, *43*, 95–118. [CrossRef] [PubMed]
74. Chapman, J.; Fielder, E.; Passos, J.F. Mitochondrial dysfunction and cell senescence: Deciphering a complex relationship. *FEBS Lett.* **2019**, *593*, 1566–1579. [CrossRef]
75. Lewis, S.C.; Uchiyama, L.; Nunnari, J. ER-mitochondria contacts couple mtDNA synthesis with mitochondrial division in human cells. *Science* **2016**, *353*, aaf5549. [CrossRef]
76. Murley, A.; Lackner, L.L.; Osman, C.; West, M.; Voeltz, G.K.; Walter, P.; Nunnari, J.; Youle, R.J. ER-associated mitochondrial division links the distribution of mitochondria and mitochondrial DNA in yeast. *ELife* **2013**, *2*, e00422. [CrossRef] [PubMed]
77. Ban-Ishihara, R.; Ishihara, T.; Sasaki, N.; Mihara, K.; Ishihara, N. Dynamics of nucleoid structure regulated by mitochondrial fission contributes to cristae reformation and release of cytochrome c. *Proc. Natl. Acad. Sci. USA* **2013**, *110*, 11863–11868. [CrossRef]
78. Friedman, J.R.; Lackner, L.L.; West, M.; DiBenedetto, J.R.; Nunnari, J.; Voeltz, G.K. ER Tubules Mark Sites of Mitochondrial Division. *Science* **2011**, *334*, 358–362. [CrossRef]
79. Korobova, F.; Ramabhadran, V.; Higgs, H.N. An Actin-Dependent Step in Mitochondrial Fission Mediated by the ER-Associated Formin INF2. *Science* **2013**, *339*, 464–467. [CrossRef]
80. Manor, U.; Bartholomew, S.; Golani, G.; Christenson, E.; Kozlov, M.; Higgs, H.N.; Spudich, J.; Lippincott-Schwartz, J. A mitochondria-anchored isoform of the actin-nucleating spire protein regulates mitochondrial division. *ELife* **2015**, *4*. [CrossRef]
81. Smirnova, E.; Griparic, L.; Shurland, D.-L.; Van Der Bliek, A.M. Dynamin-related Protein Drp1 Is Required for Mitochondrial Division in Mammalian Cells. *Mol. Biol. Cell* **2001**, *12*, 2245–2256. [CrossRef] [PubMed]
82. Otera, H.; Wang, C.; Cleland, M.M.; Setoguchi, K.; Yokota, S.; Youle, R.J.; Mihara, K. Mff is an essential factor for mitochondrial recruitment of Drp1 during mitochondrial fission in mammalian cells. *J. Cell Biol.* **2010**, *191*, 1141–1158. [CrossRef] [PubMed]
83. James, D.I.; Parone, P.A.; Mattenberger, Y.; Martinou, J.-C. hFis1, a Novel Component of the Mammalian Mitochondrial Fission Machinery. *J. Biol. Chem.* **2003**, *278*, 36373–36379. [CrossRef] [PubMed]

84. Palmer, C.S.; Osellame, L.D.; Laine, D.; Koutsopoulos, O.S.; Frazier, A.E.; Ryan, M.T. MiD49 and MiD51, new components of the mitochondrial fission machinery. *EMBO Rep.* **2011**, *12*, 565–573. [CrossRef]
85. Losón, O.C.; Song, Z.; Chen, H.; Chan, D.C. Fis1, Mff, MiD49, and MiD51 mediate Drp1 recruitment in mitochondrial fission. *Mol. Biol. Cell* **2013**, *24*, 659–667. [CrossRef]
86. Kalia, R.; Wang, R.Y.-R.; Yusuf, A.; Thomas, P.V.; Agard, D.A.; Shaw, J.M.; Frost, A. Structural basis of mitochondrial receptor binding and constriction by DRP1. *Nature* **2018**, *558*, 401–405. [CrossRef]
87. Kamerkar, S.; Kraus, F.; Sharpe, A.; Pucadyil, T.J.; Ryan, M.T. Dynamin-related protein 1 has membrane constricting and severing abilities sufficient for mitochondrial and peroxisomal fission. *Nat. Commun.* **2018**, *9*, 5239. [CrossRef]
88. Fonseca, T.B.; Sánchez-Guerrero, Á.; Milosevic, I.; Raimundo, N. Mitochondrial fission requires DRP1 but not dynamins. *Nature* **2019**, *570*, E34–E42. [CrossRef]
89. Wong, Y.C.; Ysselstein, D.; Krainc, D. Mitochondria–lysosome contacts regulate mitochondrial fission via RAB7 GTP hydrolysis. *Nature* **2018**, *554*, 382–386. [CrossRef]
90. Nagashima, S.; Tábara, L.-C.; Tilokani, L.; Paupe, V.; Anand, H.; Pogson, J.H.; Zunino, R.; McBride, H.M.; Prudent, J. Golgi-derived PI(4)P-containing vesicles drive late steps of mitochondrial division. *Science* **2020**, *367*, 1366–1371. [CrossRef]
91. Garrido, N.; Griparic, L.; Jokitalo, E.; Wartiovaara, J.; Van Der Bliek, A.M.; Spelbrink, J.N. Composition and Dynamics of Human Mitochondrial Nucleoids. *Mol. Biol. Cell* **2003**, *14*, 1583–1596. [CrossRef]
92. Waterham, H.R.; Koster, J.; Van Roermund, C.W.; Mooyer, P.A.; Wanders, R.J.; Leonard, J.V. A Lethal Defect of Mitochondrial and Peroxisomal Fission. *N. Engl. J. Med.* **2007**, *356*, 1736–1741. [CrossRef] [PubMed]
93. Baxter, R.V.; Ben Othmane, K.; Rochelle, J.M.; Stajich, J.E.; Hulette, C.; Dew-Knight, S.; Hentati, F.; Ben Hamida, M.; Bel, S.; Stenger, J.E. Ganglioside-induced differentiation-associated protein-1 is mutant in Charcot-Marie-Tooth disease type 4A/8q21. *Nat. Genet.* **2001**, *30*, 21–22. [CrossRef] [PubMed]
94. Brown, E.J.; Schlöndorff, J.S.; Becker, D.J.; Tsukaguchi, H.; Tonna, S.J.; Uscinski, A.L.; Higgs, H.N.; Henderson, J.M.; Pollak, M.R. Mutations in the formin gene INF2 cause focal segmental glomerulosclerosis. *Nat. Genet.* **2010**, *42*, 72–76. [CrossRef] [PubMed]
95. Shamseldin, H.E.; Alshammari, M.; Al-Sheddi, T.; Salih, M.A.M.; Alkhalidi, H.; Kentab, A.; Repetto, G.M.; Hashem, M.O.; Alkuraya, F.S. Genomic analysis of mitochondrial diseases in a consanguineous population reveals novel candidate disease genes. *J. Med. Genet.* **2012**, *49*, 234–241. [CrossRef] [PubMed]
96. Abrams, A.J.; Hufnagel, R.B.; Rebelo, A.; Zanna, C.; Patel, N.; Gonzalez, M.A.; Campeanu, I.J.; Griffin, L.B.; Groenewald, S.; Strickland, A.V.; et al. Mutations in SLC25A46, encoding a UGO1-like protein, cause an optic atrophy spectrum disorder. *Nat. Genet.* **2015**, *47*, 926–932. [CrossRef]
97. Vanstone, J.R.; Smith, A.M.; McBride, S.; Naas, T.; Holčík, M.; Antoun, G.; Harper, M.-E.; Michaud, J.; Sell, E.; Chakraborty, P.; et al. DNM1L-related mitochondrial fission defect presenting as refractory epilepsy. *Eur. J. Hum. Genet.* **2015**, *24*, 1084–1088. [CrossRef]
98. Koch, J.; Feichtinger, R.G.; Freisinger, P.; Pies, M.; Schrödl, F.; Iuso, A.; Sperl, W.; Mayr, J.A.; Prokisch, H.; Haack, T.B. Disturbed mitochondrial and peroxisomal dynamics due to loss of MFF causes Leigh-like encephalopathy, optic atrophy and peripheral neuropathy. *J. Med. Genet.* **2016**, *53*, 270–278. [CrossRef]
99. Janer, A.; Prudent, J.; Paupe, V.; Fahiminiya, S.; Majewski, J.; Sgarioto, N.; Rosiers, C.D.; Forest, A.; Lin, Z.; Gingras, A.-C.; et al. SLC 25A46 is required for mitochondrial lipid homeostasis and cristae maintenance and is responsible for Leigh syndrome. *EMBO Mol. Med.* **2016**, *8*, 1019–1038. [CrossRef]
100. Bitoun, M.; Maugenre, S.; Jeannet, P.Y.; Lacène, E.; Ferrer, X.; Laforêt, P.; Martin, J.J.; Laporte, J.; Lochmüller, H.; Beggs, A.H.; et al. Mutations in dynamin 2 cause dominant centronuclear myopathy. *Nat. Genet.* **2005**, *37*, 1207–1209. [CrossRef]
101. Claeys, K.G.; Züchner, S.; Kennerson, M.; Berciano, J.; Garcia, A.; Verhoeven, K.; Storey, E.; Merory, J.R.; Bienfait, H.M.E.; Lammens, M.; et al. Phenotypic spectrum of dynamin 2 mutations in Charcot-Marie-Tooth neuropathy. *Brain* **2009**, *132 Pt 7*, 1741–1752. [CrossRef]
102. Böhm, J.; Biancalana, V.; Dechene, E.T.; Bitoun, M.; Pierson, C.R.; Schaefer, E.; Karasoy, H.; Dempsey, M.A.; Klein, F.; Dondaine, N.; et al. Mutation spectrum in the large GTPase dynamin 2, and genotype-phenotype correlation in autosomal dominant centronuclear myopathy. *Hum. Mutat.* **2012**, *33*, 949–959. [CrossRef] [PubMed]

103. Gal, A.; Inczedy-Farkas, G.; Pál, E.; Remenyi, V.; Bereznai, B.; Gellér, L.; Szelid, Z.; Merkely, B.; Molnar, M.J. The coexistence of dynamin 2 mutation and multiple mitochondrial DNA (mtDNA) deletions in the background of severe cardiomyopathy and centronuclear myopathy. *Clin. Neuropathol.* **2015**, *34*, 89–95. [CrossRef] [PubMed]
104. Boyer, O.; Nevo, F.; Plaisier, E.; Funalot, B.; Gribouval, O.; Benoit, G.; Huynh Cong, E.; Arrondel, C.; Tête, M.J.; Montjean, R.; et al. INF2Mutations in Charcot–Marie–Tooth Disease with Glomerulopathy. *N. Engl. J. Med.* **2011**, *365*, 2377–2388. [CrossRef] [PubMed]
105. Caridi, G.; Lugani, F.; Dagnino, M.; Gigante, M.; Iolascon, A.; Falco, M.; Graziano, C.; Benetti, E.; Dugo, M.; Del Prete, D.; et al. Novel INF2 mutations in an Italian cohort of patients with focal segmental glomerulosclerosis, renal failure and Charcot-Marie-Tooth neuropathy. *Nephrol. Dial. Transpl.* **2014**, *29* (Suppl. S4), iv80–iv86. [CrossRef]
106. Crimella, C.; Tonelli, A.; Airoldi, G.; Baschirotto, C.; D'Angelo, M.G.; Bonato, S.; Losito, L.; Trabacca, A.; Bresolin, N.; Bassi, M.T.; et al. The GST domain of GDAP1 is a frequent target of mutations in the dominant form of axonal Charcot Marie Tooth type, 2.K. *J. Med. Genet.* **2010**, *47*, 712–716. [CrossRef]
107. Nasca, A.; Nardecchia, F.; Commone, A.; Semeraro, M.; Legati, A.; Garavaglia, B.; Ghezzi, D.; Leuzzi, V. Clinical and Biochemical Features in a Patient With Mitochondrial Fission Factor Gene Alteration. *Front. Genet.* **2018**, *9*. [CrossRef]
108. Pedrola, L.; Espert, A.; Wu, X.; Claramunt, R.; Shy, M.E.; Palau, F. GDAP1, the protein causing Charcot-Marie-Tooth disease type 4A, is expressed in neurons and is associated with mitochondria. *Hum. Mol. Genet.* **2005**, *14*, 1087–1094. [CrossRef]
109. Pezzini, I.; Geroldi, A.; Capponi, S.; Gulli, R.; Schenone, A.; Grandis, M.; Doria-Lamba, L.; La Piana, C.; Cremonte, M.; Pisciotta, C.; et al. GDAP1 mutations in Italian axonal Charcot–Marie–Tooth patients: Phenotypic features and clinical course. *Neuromuscul. Disord.* **2016**, *26*, 26–32. [CrossRef]
110. Parone, P.A.; Da Cruz, S.; Tondera, D.; Mattenberger, Y.; James, D.I.; Maechler, P.; Barja, F.; Martinou, J.C. Preventing Mitochondrial Fission Impairs Mitochondrial Function and Leads to Loss of Mitochondrial DNA. *PLoS ONE* **2008**, *3*, e3257. [CrossRef]
111. Ishihara, N.; Nomura, M.; Jofuku, A.; Kato, H.; Suzuki, S.O.; Masuda, K.; Otera, H.; Nakanishi, Y.; Nonaka, I.; Goto, Y.; et al. Mitochondrial fission factor Drp1 is essential for embryonic development and synapse formation in mice. *Nature* **2009**, *11*, 958–966. [CrossRef] [PubMed]
112. Wakabayashi, J.; Zhang, Z.; Wakabayashi, N.; Tamura, Y.; Fukaya, M.; Kensler, T.W.; Iijima, M.; Sesaki, H. The dynamin-related GTPase Drp1 is required for embryonic and brain development in mice. *J. Cell Biol.* **2009**, *186*, 805–816. [CrossRef] [PubMed]
113. Chen, H.; Ren, S.; Clish, C.; Jain, M.; Mootha, V.; McCaffery, J.M.; Chan, D.C. Titration of mitochondrial fusion rescues Mff-deficient cardiomyopathy. *J. Cell Biol.* **2015**, *211*, 795–805. [CrossRef] [PubMed]
114. Kameoka, S.; Adachi, Y.; Okamoto, K.; Iijima, M.; Sesaki, H. Phosphatidic Acid and Cardiolipin Coordinate Mitochondrial Dynamics. *Trends Cell Biol.* **2017**, *28*, 67–76. [CrossRef]
115. Macdonald, P.J.; Stepanyants, N.; Mehrotra, N.; Mears, J.A.; Qi, X.; Sesaki, H.; Ramachandran, R. A dimeric equilibrium intermediate nucleates Drp1 reassembly on mitochondrial membranes for fission. *Mol. Biol. Cell* **2014**, *25*, 1905–1915. [CrossRef]
116. Francy, C.A.; Alvarez, F.J.D.; Zhou, L.; Ramachandran, R.; Mears, J.A. The Mechanoenzymatic Core of Dynamin-related Protein 1 Comprises the Minimal Machinery Required for Membrane Constriction. *J. Biol. Chem.* **2015**, *290*, 11692–11703. [CrossRef]
117. Francy, C.A.; Clinton, R.W.; Fröhlich, C.; Murphy, C.; Mears, J.A. Cryo-EM Studies of Drp1 Reveal Cardiolipin Interactions that Activate the Helical Oligomer. *Sci. Rep.* **2017**, *7*, 10744. [CrossRef]
118. Stepanyants, N.; Macdonald, P.J.; Francy, C.A.; Mears, J.A.; Qi, X.; Ramachandran, R. Cardiolipin's propensity for phase transition and its reorganisation by dynamin-related protein 1 form a basis for mitochondrial membrane fission. *Mol. Biol. Cell* **2015**, *26*, 3104–3116. [CrossRef]
119. dachi, Y.; Itoh, K.; Yamada, T.; Cerveny, K.L.; Suzuki, T.L.; Macdonald, P.; Frohman, M.A.; Ramachandran, R.; Iijima, M.; Sesaki, H. Coincident Phosphatidic Acid Interaction Restrains Drp1 in Mitochondrial Division. *Mol. Cell* **2016**, *63*, 1034–1043.
120. Luévano-Martínez, L.A.; Forni, M.F.; Dos Santos, V.T.; Souza-Pinto, N.C.; Kowaltowski, A.J. Cardiolipin is a key determinant for mtDNA stability and segregation during mitochondrial stress. *Biochim. Biophys. Acta* **2015**, *1847*, 587–598. [CrossRef]

121. Zhong, Q.; Gohil, V.M.; Ma, L.; Greenberg, M.L. Absence of Cardiolipin Results in Temperature Sensitivity, Respiratory Defects, and Mitochondrial DNA Instability Independent of pet56. *J. Biol. Chem.* **2004**, *279*, 32294–32300. [CrossRef] [PubMed]
122. Chen, S.; Liu, D.; Finley, R.L.; Greenberg, M.L., Jr. Loss of mitochondrial DNA in the yeast cardiolipin synthase crd1 mutant leads to up-regulation of the protein kinase Swe1p that regulates the G2/M transition. *J. Biol. Chem.* **2010**, *285*, 10397–10407. [CrossRef] [PubMed]
123. Acehan, D.; Malhotra, A.; Xu, Y.; Ren, M.; Stokes, D.L.; Schlame, M. Cardiolipin Affects the Supramolecular Organisation of ATP Synthase in Mitochondria. *Biophys. J.* **2011**, *100*, 2184–2192. [CrossRef] [PubMed]
124. Ono, T.; Isobe, K.; Nakada, K.; Hayashi, J.-I. Human cells are protected from mitochondrial dysfunction by complementation of DNA products in fused mitochondria. *Nat. Genet.* **2001**, *28*, 272–275. [CrossRef]
125. Cao, Y.L.; Meng, S.; Chen, Y.; Feng, J.X.; Gu, D.D.; Yu, B.; Li, Y.J.; Yang, J.Y.; Liao, S.; Chan, D.C.; et al. MFN1 structures reveal nucleotide-triggered dimerization critical for mitochondrial fusion. *Nature* **2017**, *542*, 372–376. [CrossRef]
126. Yan, L.; Qi, Y.; Huang, X.; Yu, C.; Lan, L.; Guo, X.; Rao, Z.; Hu, J.; Lou, Z. Structural basis for GTP hydrolysis and conformational change of MFN1 in mediating membrane fusion. *Nat. Struct. Mol. Biol.* **2018**, *25*, 233–243. [CrossRef]
127. Qi, Y.; Yan, L.; Yu, C.; Guo, X.; Zhou, X.; Hu, X.; Huang, X.; Rao, Z.; Lou, Z.; Hu, J. Structures of human mitofusin 1 provide insight into mitochondrial tethering. *J. Cell Biol.* **2016**, *215*, 621–629. [CrossRef]
128. Ishihara, N.; Eura, Y.; Mihara, K. Mitofusin 1 and 2 play distinct roles in mitochondrial fusion reactions via GTPase activity. *J. Cell Sci.* **2004**, *117*, 6535–6546. [CrossRef]
129. De Brito, O.M.; Scorrano, L. Mitofusin 2 tethers endoplasmic reticulum to mitochondria. *Nature* **2008**, *456*, 605–610. [CrossRef]
130. Filadi, R.; Greotti, E.; Turacchio, G.; Luini, A.; Pozzan, T.; Pizzo, P. Mitofusin 2 ablation increases endoplasmic reticulum-mitochondria coupling. *Proc. Natl. Acad. Sci. USA* **2015**, *112*, E2174–E2181. [CrossRef]
131. Wai, T.; Garcia-Prieto, J.; Baker, M.J.; Merkwirth, C.; Benit, P.; Rustin, P.; Ruperez, F.J.; Barbas, C.; Ibanez, B.; Langer, T. Imbalanced OPA1 processing and mitochondrial fragmentation cause heart failure in mice. *Science* **2015**, *350*, aad0116. [CrossRef] [PubMed]
132. Griparic, L.; Kanazawa, T.; Van Der Bliek, A.M. Regulation of the mitochondrial dynamin-like protein Opa1 by proteolytic cleavage. *J. Cell Biol.* **2007**, *178*, 757–764. [CrossRef] [PubMed]
133. Ehses, S.; Raschke, I.; Mancuso, G.; Bernacchia, A.; Geimer, S.; Tondera, D.; Martinou, J.C.; Westermann, B.; Rugarli, E.I.; Langer, T. Regulation of OPA1 processing and mitochondrial fusion by m-AAA protease isoenzymes and OMA1. *J. Cell Biol.* **2009**, *187*, 1023–1036. [CrossRef] [PubMed]
134. Ramos, E.S.; Motori, E.; Bruser, C.; Kuhl, I.; Yeroslaviz, A.; Ruzzenente, B.; Kauppila, J.H.K.; Busch, J.D.; Hultenby, K.; Habermann, B.H.; et al. Mitochondrial fusion is required for regulation of mitochondrial DNA replication. *PLoS Genet.* **2019**, *15*, e1008085. [CrossRef]
135. Elachouri, G.; Vidoni, S.; Zanna, C.; Pattyn, A.; Boukhaddaoui, H.; Gaget, K.; Yu-Wai-Man, P.; Gasparre, G.; Sarzi, E.; Delettre, C.; et al. OPA1 links human mitochondrial genome maintenance to mtDNA replication and distribution. *Genome Res.* **2010**, *21*, 12–20. [CrossRef]
136. Chen, H.; Vermulst, M.; Wang, Y.E.; Chomyn, A.; Prolla, T.A.; McCaffery, J.M.; Chan, D.C. Mitochondrial Fusion Is Required for mtDNA Stability in Skeletal Muscle and Tolerance of mtDNA Mutations. *Cell* **2010**, *141*, 280–289. [CrossRef]
137. Sugiura, A.; Nagashima, S.; Tokuyama, T.; Amo, T.; Matsuki, Y.; Ishido, S.; Kudo, Y.; McBride, H.M.; Fukuda, T.; Matsushita, N.; et al. MITOL Regulates Endoplasmic Reticulum-Mitochondria Contacts via Mitofusin2. *Mol. Cell* **2013**, *51*, 20–34. [CrossRef]
138. Chen, Y.; Csordás, G.; Jowdy, C.; Schneider, T.G.; Csordás, N.; Wang, W.; Liu, Y.; Kohlhaas, M.; Meiser, M.; Bergem, S.; et al. Mitofusin 2-containing mitochondrial-reticular microdomains direct rapid cardiomyocyte bioenergetic responses via interorganelle Ca($^{2+}$) crosstalk. *Circ. Res.* **2012**, *111*, 863–875. [CrossRef]
139. Yu-Wai-Man, P.; Griffiths, P.G.; Burke, A.; Sellar, P.W.; Clarke, M.P.; Gnanaraj, L.; Ah-Kine, D.; Hudson, G.; Czermin, B.; Taylor, R.W.; et al. The prevalence and natural history of dominant optic atrophy due to OPA1 mutations. *Ophthalmology* **2010**, *117*, 1538–1546. [CrossRef]
140. Yu-Wai-Man, P.; Chinnery, P.F. Dominant optic atrophy: Novel OPA1 mutations and revised prevalence estimates. *Ophthalmology* **2013**, *120*, 1712. [CrossRef]

141. Hudson, G.; Amati-Bonneau, P.; Blakely, E.L.; Stewart, J.D.; He, L.; Schaefer, A.M.; Griffiths, P.G.; Ahlqvist, K.; Suomalainen, A.; Reynier, P.; et al. Mutation of OPA1 causes dominant optic atrophy with external ophthalmoplegia, ataxia, deafness and multiple mitochondrial DNA deletions: A novel disorder of mtDNA maintenance. *Brain* **2008**, *131 Pt 2*, 329–337. [CrossRef]
142. Yu-Wai-Man, P.; Griffiths, P.G.; Gorman, G.S.; Lourenco, C.M.; Wright, A.F.; Auer-Grumbach, M.; Toscano, A.; Musumeci, O.; Valentino, M.L.; Caporali, L.; et al. Multi-system neurological disease is common in patients with OPA1 mutations. *Brain* **2010**, *133 Pt 3*, 771–786. [CrossRef]
143. Carelli, V.; Musumeci, O.; Caporali, L.; Zanna, C.; La Morgia, C.; Del Dotto, V.; Porcelli, A.M.; Rugolo, M.; Valentino, M.L.; Iommarini, L.; et al. Syndromic parkinsonism and dementia associated with OPA1 missense mutations. *Ann. Neurol.* **2015**, *78*, 21–38. [CrossRef] [PubMed]
144. Braathen, G.J.; Sand, J.C.; Lobato, A.; Høyer, H.; Russell, M.B. Genetic epidemiology of Charcot–Marie–Tooth in the general population. *Eur. J. Neurol.* **2010**, *18*, 39–48. [CrossRef]
145. Braathen, G.J.; Sand, J.C.; Lobato, A.; Høyer, H.; Russell, M.B. MFN2 point mutations occur in 3.4% of Charcot-Marie-Tooth families. An investigation of 232 Norwegian CMT families. *BMC Med. Genet.* **2010**, *11*, 48. [CrossRef]
146. Feely, S.M.; Laura, M.; Siskind, C.E.; Sottile, S.; Davis, M.; Gibbons, V.S.; Reilly, M.M.; Shy, M.E. MFN2 mutations cause severe phenotypes in most patients with, C.M.T.2.A. *Neurology* **2011**, *76*, 1690–1696. [CrossRef] [PubMed]
147. Casari, G.; De Fusco, M.; Ciarmatori, S.; Zeviani, M.; Mora, M.; Fernandez, P.; De Michele, G.; Filla, A.; Cocozza, S.; Marconi, R.; et al. Spastic paraplegia and OXPHOS impairment caused by mutations in paraplegin, a nuclear-encoded mitochondrial metalloprotease. *Cell* **1998**, *93*, 973–983. [CrossRef]
148. Pfeffer, G.; Pyle, A.; Griffin, H.; Miller, J.; Wilson, V.; Turnbull, L.; Fawcett, K.; Sims, D.; Eglon, G.; Hadjivassiliou, M.; et al. SPG7 mutations are a common cause of undiagnosed ataxia. *Neurology* **2015**, *84*, 1174–1176. [CrossRef]
149. Hewamadduma, C.A.; Hoggard, N.; O'Malley, R.; Robinson, M.K.; Beauchamp, N.J.; Segamogaite, R.; Martindale, J.; Rodgers, T.; Rao, G.; Sarrigiannis, P.; et al. Novel genotype-phenotype and MRI correlations in a large cohort of patients with SPG7 mutations. *Neurol. Genet.* **2018**, *4*, e279. [CrossRef]
150. De la Casa-Fages, B.; Fernández-Eulate, G.; Gamez, J.; Barahona-Hernando, R.; Morís, G.; García-Barcina, M.; Infante, J.; Zulaica, M.; Fernández-Pelayo, U.; Muñoz-Oreja, M.; et al. Parkinsonism and spastic paraplegia type 7: Expanding the spectrum of mitochondrial Parkinsonism. *Mov. Disord.* **2019**, *34*, 1547–1561. [CrossRef]
151. Di Bella, D.; Lazzaro, F.; Brusco, A.; Plumari, M.; Battaglia, G.; Pastore, A.; Finardi, A.; Cagnoli, C.; Tempia, F.; Frontali, M.; et al. Mutations in the mitochondrial protease gene AFG3L2 cause dominant hereditary ataxia SCA28. *Nat. Genet.* **2010**, *42*, 313–321. [CrossRef] [PubMed]
152. Gorman, G.S.; Pfeffer, G.; Griffin, H.; Blakely, E.L.; Kurzawa-Akanbi, M.; Gabriel, J.; Sitarz, K.; Roberts, M.; Schoser, B.; Pyle, A.; et al. Clonal expansion of secondary mitochondrial DNA deletions associated with spinocerebellar ataxia type 28. *JAMA Neurol.* **2015**, *72*, 106–111. [CrossRef] [PubMed]
153. Pfeffer, G.; Gorman, G.S.; Griffin, H.; Kurzawa-Akanbi, M.; Blakely, E.L.; Wilson, I.; Sitarz, K.; Moore, D.; Murphy, J.L.; Alston, C.L.; et al. Mutations in the SPG7 gene cause chronic progressive external ophthalmoplegia through disordered mitochondrial DNA maintenance. *Brain* **2014**, *137 Pt 5*, 1323–1336. [CrossRef]
154. Vielhaber, S.; Debska-Vielhaber, G.; Peeva, V.; Schoeler, S.; Kudin, A.P.; Minin, I.; Schreiber, S.; Dengler, R.; Kollewe, K.; Zuschratter, W.; et al. Mitofusin 2 mutations affect mitochondrial function by mitochondrial DNA depletion. *Acta Neuropathol.* **2013**, *125*, 245–256. [CrossRef] [PubMed]
155. Bonnen, P.E.; Yarham, J.W.; Besse, A.; Wu, P.; Faqeih, E.A.; Al-Asmari, A.M.; Saleh, M.A.; Eyaid, W.; Hadeel, A.; He, L.; et al. Mutations in FBXL4 cause mitochondrial encephalopathy and a disorder of mitochondrial DNA maintenance. *Am. J. Hum. Genet.* **2013**, *93*, 471–481. [CrossRef]
156. Di Mauro, S.; Hirano, M.M.E.R.R.F. *GeneReviews®*; Adam, M.P., Ardinger, H.H., Pagon, R.A., Wallace, S.E., Bean, L.J.H., Stephens, K., Amemiya, A., Eds.; University of Washington: Seattle, WA, USA, 1993. Available online: https://www.ncbi.nlm.nih.gov/books/NBK1520/ (accessed on 24 August 2020).
157. Sawyer, S.L.; Cheuk-Him Ng, A.; Innes, A.M.; Wagner, J.D.; Dyment, D.A.; Tetreault, M.; Majewski, J.; Boycott, K.M.; Screaton, R.A.; Nicholson, G. Homozygous mutations in MFN2 cause multiple symmetric lipomatosis associated with neuropathy. *Hum. Mol. Genet.* **2015**, *24*, 5109–5114. [CrossRef]

158. Capel, E.; Vatier, C.; Cervera, P.; Stojkovic, T.; Disse, E.; Cottereau, A.S.; Auclair, M.; Verpont, M.C.; Mosbah, H.; Gourdy, P.; et al. MFN2-associated lipomatosis: Clinical spectrum and impact on adipose tissue. *J. Clin. Lipidol.* **2018**, *12*, 1420–1435. [CrossRef]
159. Guan, K.; Farh, L.; Marshall, T.K.; Deschenes, R.J. Normal mitochondrial structure and genome maintenance in yeast requires the dynamin-like product of the MGM1 gene. *Curr. Genet.* **1993**, *24*, 141–148. [CrossRef]
160. Rapaport, D.; Brunner, M.; Neupert, W.; Westermann, B. Fzo1p is a mitochondrial outer membrane protein essential for the biogenesis of functional mitochondria in Saccharomyces cerevisiae. *J. Biol. Chem.* **1998**, *273*, 20150–20155. [CrossRef]
161. Chen, L.; Liu, T.; Tran, A.; Lu, X.; Tomilov, A.A.; Davies, V.; Cortopassi, G.; Chiamvimonvat, N.; Bers, D.M.; Votruba, M.; et al. OPA1 mutation and late-onset cardiomyopathy: Mitochondrial dysfunction and mtDNA instability. *J. Am. Heart Assoc.* **2012**, *1*, e003012. [CrossRef]
162. Chen, H.; Detmer, S.A.; Ewald, A.J.; Griffin, E.E.; Fraser, S.E.; Chan, D.C. Mitofusins Mfn1 and Mfn2 coordinately regulate mitochondrial fusion and are essential for embryonic development. *J. Cell Biol.* **2003**, *160*, 189–200. [CrossRef] [PubMed]
163. Ban, T.; Ishihara, T.; Kohno, H.; Saita, S.; Ichimura, A.; Maenaka, K.; Ishihara, N. Molecular basis of selective mitochondrial fusion by heterotypic action between OPA1 and cardiolipin. *Nat. Cell Biol.* **2017**, *19*, 856–863. [CrossRef] [PubMed]
164. DeVay, R.M.; Dominguez-Ramirez, L.; Lackner, L.L.; Hoppins, S.; Stahlberg, H.; Nunnari, J. Coassembly of Mgm1 isoforms requires cardiolipin and mediates mitochondrial inner membrane fusion. *J. Cell Biol.* **2009**, *186*, 793–803. [CrossRef] [PubMed]
165. Ban, T.; Heymann, J.A.; Song, Z.; Hinshaw, J.E.; Chan, D.C. OPA1 disease alleles causing dominant optic atrophy have defects in cardiolipin-stimulated GTP hydrolysis and membrane tubulation. *Hum. Mol. Genet.* **2010**, *19*, 2113–2122. [CrossRef]
166. Choi, S.Y.; Huang, P.; Jenkins, G.M.; Chan, D.C.; Schiller, J.; Frohman, M.A. A common lipid links Mfn-mediated mitochondrial fusion and SNARE-regulated exocytosis. *Nat. Cell Biol.* **2006**, *8*, 1255–1262. [CrossRef]
167. Ban-Ishihara, R.; Tomohiro-Takamiya, S.; Tani, M.; Baudier, J.; Ishihara, N.; Kuge, O. COX assembly factor ccdc56 regulates mitochondrial morphology by affecting mitochondrial recruitment of Drp1. *FEBS Lett.* **2015**, *589 Pt B*, 3126–3132. [CrossRef]
168. Cooper, H.M.; Yang, Y.; Ylikallio, E.; Khairullin, R.; Woldegebriel, R.; Lin, K.-L.; Euro, L.; Palin, E.; Wolf, A.; Trokovic, R.; et al. ATPase-deficient mitochondrial inner membrane protein ATAD3A disturbs mitochondrial dynamics in dominant hereditary spastic paraplegia. *Hum. Mol. Genet.* **2017**, *26*, 1432–1443. [CrossRef]
169. Gilquin, B.; Taillebourg, E.; Cherradi, N.; Hubstenberger, A.; Gay, O.; Merle, N.; Assard, N.; Fauvarque, M.O.; Tomohiro, S.; Kuge, O.; et al. The AAA+ ATPase ATAD3A controls mitochondrial dynamics at the interface of the inner and outer membranes. *Mol. Cell Biol.* **2010**, *30*, 1984–1996. [CrossRef]
170. Zhao, Y.; Sun, X.; Hu, D.; Prosdocimo, D.A.; Hoppel, C.; Jain, M.K.; Ramachandran, R.; Qi, X. ATAD3A oligomerization causes neurodegeneration by coupling mitochondrial fragmentation and bioenergetics defects. *Nat. Commun.* **2019**, *10*, 1317. [CrossRef]
171. Desai, R.; Frazier, A.E.; Durigon, R.; Patel, H.; Jones, A.W.; Dalla Rosa, I.; Lake, N.J.; Compton, A.G.; Mountford, H.S.; Tucker, E.J.; et al. ATAD3 gene cluster deletions cause cerebellar dysfunction associated with altered mitochondrial DNA and cholesterol metabolism. *Brain* **2017**, *140*, 1595–1610. [CrossRef]
172. Gunning, A.C.; Strucinska, K.; Muñoz Oreja, M.; Parrish, A.; Caswell, R.; Stals, K.L.; Durigon, R.; Durlacher-Betzer, K.; Cunningham, M.H.; Grochowski, C.M.; et al. Recurrent De Novo NAHR Reciprocal Duplications in the ATAD3 Gene Cluster Cause a Neurogenetic Trait with Perturbed Cholesterol and Mitochondrial Metabolism. *Am. J. Hum. Genet.* **2020**, *106*, 272–279. [CrossRef] [PubMed]
173. Peralta, S.; Goffart, S.; Williams, S.L.; Diaz, F.; Garcia, S.; Nissanka, N.; Area-Gomez, E.; Pohjoismaki, J.; Moraes, C.T. ATAD3 controls mitochondrial cristae structure in mouse muscle, influencing mtDNA replication and cholesterol levels. *J. Cell Sci.* **2018**, *131*. [CrossRef] [PubMed]
174. Amati-Bonneau, P.; Valentino, M.L.; Reynier, P.; Gallardo, M.E.; Bornstein, B.; Boissiere, A.; Campos, Y.; Rivera, H.; de la Aleja, J.G.; Carroccia, R.; et al. OPA1 mutations induce mitochondrial DNA instability and optic atrophy 'plus' phenotypes. *Brain* **2008**, *131 Pt 2*, 338–351. [CrossRef]

175. Lodi, R.; Tonon, C.; Valentino, M.L.; Manners, D.; Testa, C.; Malucelli, E.; La Morgia, C.; Barboni, P.; Carbonelli, M.; Schimpf, S.; et al. Defective mitochondrial adenosine triphosphate production in skeletal muscle from patients with dominant optic atrophy due to OPA1 mutations. *Arch. Neurol.* **2011**, *68*, 67–73. [CrossRef] [PubMed]
176. Rouzier, C.; Bannwarth, S.; Chaussenot, A.; Chevrollier, A.; Verschueren, A.; Bonello-Palot, N.; Bonello-Palot, N.; Fragaki, K.; Cano, A.; Pouget, J.; et al. The MFN2 gene is responsible for mitochondrial DNA instability and optic atrophy 'plus' phenotype. *Brain* **2012**, *135 Pt 1*, 23–34. [CrossRef]
177. Yu, T.; Robotham, J.L.; Yoon, Y. Increased production of reactive oxygen species in hyperglycemic conditions requires dynamic change of mitochondrial morphology. *Proc. Natl. Acad. Sci. USA* **2006**, *103*, 2653–2658. [CrossRef]
178. Trifunovic, A.; Wredenberg, A.; Falkenberg, M.; Spelbrink, J.N.; Rovio, A.T.; Bruder, C.E.; Bohlooly, Y.M.; Gidlof, S.; Oldfors, A.; Wibom, R.; et al. Premature ageing in mice expressing defective mitochondrial DNA polymerase. *Nature* **2004**, *429*, 417–423. [CrossRef]
179. Legros, F.; Malka, F.; Frachon, P.; Lombes, A.; Rojo, M. Organisation and dynamics of human mitochondrial DNA. *J. Cell Sci.* **2004**, *117 Pt 13*, 2653–2662. [CrossRef]
180. Kandul, N.P.; Zhang, T.; Hay, B.A.; Guo, M. Selective removal of deletion-bearing mitochondrial DNA in heteroplasmic Drosophila. *Nat. Commun.* **2016**, *7*, 13100. [CrossRef]
181. Malena, A.; Loro, E.; Di Re, M.; Holt, I.J.; Vergani, L. Inhibition of mitochondrial fission favours mutant over wild-type mitochondrial DNA. *Hum. Mol. Genet.* **2009**, *18*, 3407–3416. [CrossRef]
182. Pfanner, N.; Meijer, M. The Tom and Tim machine. *Curr. Biol.* **1997**, *7*, R100–R103. [CrossRef]
183. Herrmann, J.M.; Neupert, W. Protein insertion into the inner membrane of mitochondria. *IUBMB Life* **2003**, *55*, 219–225. [CrossRef]
184. Wilkens, V.; Kohl, W.; Busch, K. Restricted diffusion of OXPHOS complexes in dynamic mitochondria delays their exchange between cristae and engenders a transitory mosaic distribution. *J. Cell Sci.* **2013**, *126 Pt 1*, 103–116. [CrossRef]
185. Perkins, G.; Renken, C.; Martone, M.E.; Young, S.J.; Ellisman, M.; Frey, T. Electron tomography of neuronal mitochondria: Three-dimensional structure and organisation of cristae and membrane contacts. *J. Struct. Biol.* **1997**, *119*, 260–272. [CrossRef] [PubMed]
186. Perkins, G.; Song, J.Y.; Tarsa, L.; Deerinck, T.J.; Ellisman, M.H.; Frey, T.G. Electron tomography of mitochondria from brown adipocytes reveals crista junctions. *J. Bioenergy Biomembr.* **1998**, *30*, 431–442. [CrossRef]
187. Gilkerson, R.W.; Selker, J.M.; Capaldi, R.A. The cristal membrane of mitochondria is the principal site of oxidative phosphorylation. *FEBS Lett.* **2003**, *546*, 355–358. [CrossRef]
188. Stephan, T.; Roesch, A.; Riedel, D.; Jakobs, S. Live-cell STED nanoscopy of mitochondrial cristae. *Sci. Rep.* **2019**, *9*, 12419. [CrossRef] [PubMed]
189. Jajoo, R.; Jung, Y.; Huh, D.; Viana, M.P.; Rafelski, S.M.; Springer, M.; Paulsson, J. Accurate concentration control of mitochondria and nucleoids. *Science* **2016**, *351*, 169–172. [CrossRef]
190. Prachar, J. Ultrastructure of mitochondrial nucleoid and its surroundings. *Gen. Physiol. Biophys.* **2016**, *35*, 273–286. [CrossRef]
191. Rajala, N.; Gerhold, J.; Martinsson, P.; Klymov, A.; Spelbrink, J.N. Replication factors transiently associate with mtDNA at the mitochondrial inner membrane to facilitate replication. *Nucleic Acids Res.* **2013**, *42*, 952–967. [CrossRef]
192. Khalifat, N.; Puff, N.; Bonneau, S.; Fournier, J.-B.; Angelova, M.I. Membrane Deformation under Local pH Gradient: Mimicking Mitochondrial Cristae Dynamics. *Biophys. J.* **2008**, *95*, 4924–4933. [CrossRef] [PubMed]
193. Thomas, A.; Barriere, S.; Broseus, L.; Brooke, J.; Lorenzi, C.; Villemin, J.P.; Beurier, G.; Sabatier, R.; Reynes, C.; Mancheron, A.; et al. GECKO is a genetic algorithm to classify and explore high throughput sequencing data. *Commun. Biol.* **2019**, *2*, 222. [CrossRef]
194. Pfeiffer, K.; Gohil, V.; Stuart, R.A.; Hunte, C.; Brandt, U.; Greenberg, M.L.; Schägger, H. Cardiolipin stabilizes respiratory chain supercomplexes. *J. Biol. Chem.* **2003**, *278*, 52873–52880. [CrossRef]
195. Tasseva, G.; Bai, H.D.; Davidescu, M.; Haromy, A.; Michelakis, E.; Vance, J.E. Phosphatidylethanolamine Deficiency in Mammalian Mitochondria Impairs Oxidative Phosphorylation and Alters Mitochondrial Morphology. *J. Biol. Chem.* **2012**, *288*, 4158–4173. [CrossRef] [PubMed]
196. Fleischer, S.; Rouser, G.; Fleischer, B.; Casu, A.; Kritchevsky, G. Lipid composition of mitochondria from bovine heart, liver, and kidney. *J. Lipid Res.* **1967**, *8*, 170–180. [PubMed]

197. Gerhold, J.M.; Cansiz-Arda, S.; Lohmus, M.; Engberg, O.; Reyes, A.; van Rennes, H.; Sanz, A.; Holt, I.J.; Cooper, H.M.; Spelbrink, J.N. Human Mitochondrial DNA-Protein Complexes Attach to a Cholesterol-Rich Membrane Structure. *Sci. Rep.* **2015**, *5*, 15292. [CrossRef] [PubMed]
198. Hung, V.; Lam, S.S.; Udeshi, N.D.; Svinkina, T.; Guzman, G.; Mootha, V.K.; Carr, S.A.; Ting, A.Y. Proteomic mapping of cytosol-facing outer mitochondrial and ER membranes in living human cells by proximity biotinylation. *Elife* **2017**, *6*, 1206. [CrossRef]
199. Hubstenberger, A.; Merle, N.; Charton, R.; Brandolin, G.; Rousseau, D. Topological analysis of ATAD3A insertion in purified human mitochondria. *J. Bioenergy Biomembr.* **2010**, *42*, 143–150. [CrossRef]
200. Issop, L.; Fan, J.; Lee, S.; Rone, M.B.; Basu, K.; Mui, J.; Papadopoulos, V.; Fan, J. Mitochondria-Associated Membrane Formation in Hormone-Stimulated Leydig Cell Steroidogenesis: Role of ATAD3. *Endocrinology* **2015**, *156*, 334–345. [CrossRef]
201. Olichon, A.; Baricault, L.; Gas, N.; Guillou, E.; Valette, A.; Belenguer, P.; Lenaers, G. Loss of OPA1 perturbates the mitochondrial inner membrane structure and integrity, leading to cytochrome c release and apoptosis. *J. Biol. Chem.* **2003**, *278*, 7743–7746. [CrossRef]
202. Kasashima, K.; Sumitani, M.; Satoh, M.; Endo, H. Human prohibitin 1 maintains the organisation and stability of the mitochondrial nucleoids. *Exp. Cell Res.* **2008**, *314*, 988–996. [CrossRef] [PubMed]
203. Osman, C.; Haag, M.; Potting, C.; Rodenfels, J.; Dip, P.V.; Wieland, F.T.; Brügger, B.; Westermann, B.; Langer, T. The genetic interactome of prohibitins: Coordinated control of cardiolipin and phosphatidylethanolamine by conserved regulators in mitochondria. *J. Cell Biol.* **2009**, *184*, 583–596. [CrossRef]
204. Merkwirth, C.; Dargazanli, S.; Tatsuta, T.; Geimer, S.; Löwer, B.; Wunderlich, F.T.; von Kleist-Retzow, J.C.; Waisman, A.; Westermann, B.; Langer, T. Prohibitins control cell proliferation and apoptosis by regulating OPA1-dependent cristae morphogenesis in mitochondria. *Genes Dev.* **2008**, *22*, 476–488. [CrossRef] [PubMed]
205. Richter-Dennerlein, R.; Korwitz, A.; Haag, M.; Tatsuta, T.; Dargazanli, S.; Baker, M.; Decker, T.; Lamkemeyer, T.; Rugarli, E.I.; Langer, T. DNAJC19, a mitochondrial cochaperone associated with cardiomyopathy, forms a complex with prohibitins to regulate cardiolipin remodeling. *Cell Metab.* **2014**, *20*, 158–171. [CrossRef]
206. Wolf, D.M.; Segawa, M.; Kondadi, A.K.; Anand, R.; Bailey, S.T.; Reichert, A.S.; van der Bliek, A.M.; Shackelford, D.B.; Liesa, M.; Shirihai, O.S. Individual cristae within the same mitochondrion display different membrane potentials and are functionally independent. *EMBO J.* **2019**, *38*, e101056. [CrossRef]
207. Friedman, J.R.; Mourier, A.; Yamada, J.; McCaffery, J.M.; Nunnari, J. MICOS coordinates with respiratory complexes and lipids to establish mitochondrial inner membrane architecture. *Elife* **2015**, *4*, e07739. [CrossRef] [PubMed]
208. Korner, C.; Barrera, M.; Dukanovic, J.; Eydt, K.; Harner, M.; Rabl, R.; Vogel, F.; Rapaport, D.; Neupert, W.; Reichert, A.S. The C-terminal domain of Fcj1 is required for formation of crista junctions and interacts with the TOB/SAM complex in mitochondria. *Mol. Biol. Cell* **2012**, *23*, 2143–2155. [CrossRef] [PubMed]
209. Glytsou, C.; Calvo, E.; Cogliati, S.; Mehrotra, A.; Anastasia, I.; Rigoni, G.; Raimondi, A.; Shintani, N.; Loureiro, M.; Vazquez, J.; et al. Optic Atrophy 1 Is Epistatic to the Core MICOS Component MIC60 in Mitochondrial Cristae Shape Control. *Cell Rep.* **2016**, *17*, 3024–3034. [CrossRef]
210. Li, H.; Ruan, Y.; Zhang, K.; Jian, F.; Hu, C.; Miao, L.; Gong, L.; Sun, L.; Zhang, X.; Chen, S.; et al. Mic60/Mitofilin determines MICOS assembly essential for mitochondrial dynamics and mtDNA nucleoid organisation. *Cell Death Differ.* **2016**, *23*, 380–392. [CrossRef]
211. Barbot, M.; Jans, D.C.; Schulz, C.; Denkert, N.; Kroppen, B.; Hoppert, M.; Jakobs, S.; Meinecke, M. Mic10 oligomerizes to bend mitochondrial inner membranes at cristae junctions. *Cell Metab.* **2015**, *21*, 756–763. [CrossRef]
212. Bohnert, M.; Zerbes, R.M.; Davies, K.M.; Muhleip, A.W.; Rampelt, H.; Horvath, S.E.; Boenke, T.; Kram, A.; Perschil, I.; Veenhuis, M.; et al. Central role of Mic10 in the mitochondrial contact site and cristae organizing system. *Cell Metab.* **2015**, *21*, 747–755. [CrossRef] [PubMed]
213. Koob, S.; Barrera, M.; Anand, R.; Reichert, A.S. The non-glycosylated isoform of MIC26 is a constituent of the mammalian MICOS complex and promotes formation of crista junctions. *Biochim. Biophys. Acta* **2015**, *1853*, 1551–1563. [CrossRef] [PubMed]
214. Weber, T.A.; Koob, S.; Heide, H.; Wittig, I.; Head, B.; van der Bliek, A.; Brandt, U.; Mittelbronn, M.; Reichert, A.S. APOOL is a cardiolipin-binding constituent of the Mitofilin/MINOS protein complex determining cristae morphology in mammalian mitochondria. *PLoS ONE* **2013**, *8*, e63683. [CrossRef] [PubMed]

215. Davies, K.M.; Anselmi, C.; Wittig, I.; Faraldo-Gómez, J.D.; Kühlbrandt, W. Structure of the yeast F1Fo-ATP synthase dimer and its role in shaping the mitochondrial cristae. *Proc. Natl. Acad. Sci. USA* **2012**, *109*, 13602–13607. [CrossRef]
216. Paumard, P.; Vaillier, J.; Coulary, B.; Schaeffer, J.; Soubannier, V.; Mueller, D.M.; Brethes, D.; di Rago, J.P.; Velours, J. The ATP synthase is involved in generating mitochondrial cristae morphology. *EMBO J.* **2002**, *21*, 221–230. [CrossRef]
217. Eydt, K.; Davies, K.J.; Behrendt, C.; Wittig, I.; Reichert, A.S. Cristae architecture is determined by an interplay of the MICOS complex and the F1Fo ATP synthase via Mic27 and Mic10. *Microb. Cell* **2017**, *4*, 259–272. [CrossRef]
218. Quintana-Cabrera, R.; Quirin, C.; Glytsou, C.; Corrado, M.; Urbani, A.; Pellattiero, A.; Calvo, E.; Vazquez, J.; Enriquez, J.A.; Gerle, C.; et al. The cristae modulator Optic atrophy 1 requires mitochondrial ATP synthase oligomers to safeguard mitochondrial function. *Nat. Commun.* **2018**, *9*, 3399. [CrossRef]
219. Meeusen, S.; DeVay, R.; Block, J.; Cassidy-Stone, A.; Wayson, S.; McCaffery, J.M.; Nunnari, J. Mitochondrial inner-membrane fusion and crista maintenance requires the dynamin-related GTPase Mgm1. *Cell* **2006**, *127*, 383–395. [CrossRef]
220. Cipolat, S.; Rudka, T.; Hartmann, D.; Costa, V.; Serneels, L.; Craessaerts, K.; Metzger, K.; Frezza, C.; Annaert, W.; D'Adamio, L.; et al. Mitochondrial rhomboid PARL regulates cytochrome c release during apoptosis via OPA1-dependent cristae remodeling. *Cell* **2006**, *126*, 163–175. [CrossRef]
221. Khacho, M.; Tarabay, M.; Patten, D.; Khacho, P.; MacLaurin, J.G.; Guadagno, J.; Bergeron, R.; Cregan, S.P.; Harper, M.E.; Park, D.S.; et al. Acidosis overrides oxygen deprivation to maintain mitochondrial function and cell survival. *Nat. Commun.* **2014**, *5*, 3550. [CrossRef]
222. Patten, D.A.; Wong, J.; Khacho, M.; Soubannier, V.; Mailloux, R.J.; Pilon-Larose, K.; MacLaurin, J.G.; Park, D.S.; McBride, H.M.; Trinkle-Mulcahy, L.; et al. OPA1-dependent cristae modulation is essential for cellular adaptation to metabolic demand. *EMBO J.* **2014**, *33*, 2676–2691. [CrossRef] [PubMed]
223. Darshi, M.; Mendiola, V.L.; Mackey, M.R.; Murphy, A.N.; Koller, A.; Perkins, G.A.; Ellisman, M.H.; Taylor, S.S. ChChd3, an inner mitochondrial membrane protein, is essential for maintaining crista integrity and mitochondrial function. *J. Biol. Chem.* **2011**, *286*, 2918–2932. [CrossRef] [PubMed]
224. Perciavalle, R.M.; Stewart, D.P.; Koss, B.; Lynch, J.; Milasta, S.; Bathina, M.; Temirov, J.; Cleland, M.M.; Pelletier, S.; Schuetz, J.D.; et al. Anti-apoptotic MCL-1 localizes to the mitochondrial matrix and couples mitochondrial fusion to respiration. *Nat. Cell Biol.* **2012**, *14*, 575–583. [CrossRef]
225. Böttinger, L.; Horvath, S.E.; Kleinschroth, T.; Hunte, C.; Daum, G.; Pfanner, N.; Becker, T. Phosphatidylethanolamine and cardiolipin differentially affect the stability of mitochondrial respiratory chain supercomplexes. *J. Mol. Biol.* **2012**, *423*, 677–686. [CrossRef]
226. Desmurs, M.; Foti, M.; Raemy, E.; Vaz, F.M.; Martinou, J.C.; Bairoch, A.; Lane, L. C11orf83, a mitochondrial cardiolipin-binding protein involved in bc1 complex assembly and supercomplex stabilization. *Mol. Cell Biol.* **2015**, *35*, 1139–1156. [CrossRef] [PubMed]
227. Marusich, M.F.; Robinson, B.H.; Taanman, J.W.; Kim, S.J.; Schillace, R.; Smith, J.L.; Capaldi, R.A. Expression of mtDNA and nDNA encoded respiratory chain proteins in chemically and genetically-derived Rho0 human fibroblasts: A comparison of subunit proteins in normal fibroblasts treated with ethidium bromide and fibroblasts from a patient with mtDNA depletion syndrome. *Biochim. Biophys. Acta* **1997**, *1362*, 145–159.
228. Fukuyama, R.; Nakayama, A.; Nakase, T.; Toba, H.; Mukainaka, T.; Sakaguchi, H.; Saiwaki, T.; Sakurai, H.; Wada, M.; Fushiki, S. A newly established neuronal rho-0 cell line highly susceptible to oxidative stress accumulates iron and other metals. Relevance to the origin of metal ion deposits in brains with neurodegenerative disorders. *J. Biol. Chem.* **2002**, *277*, 41455–41462. [CrossRef]
229. Srinivasan, S.; Guha, M.; Kashina, A.; Avadhani, N.G. Mitochondrial dysfunction and mitochondrial dynamics-The cancer connection. *Biochim. Biophys. Acta Bioenergy* **2017**, *1858*, 602–614. [CrossRef]
230. Larsson, N.G.; Wang, J.; Wilhelmsson, H.; Oldfors, A.; Rustin, P.; Lewandoski, M.; Barsh, G.S.; Clayton, D.A. Mitochondrial transcription factor A is necessary for mtDNA maintenance and embryogenesis in mice. *Nat. Genet.* **1998**, *18*, 231–236. [CrossRef]
231. Wang, J.; Wilhelmsson, H.; Graff, C.; Li, H.; Oldfors, A.; Rustin, P.; Bruning, J.C.; Kahn, C.R.; Clayton, D.A.; Barsh, G.S.; et al. Dilated cardiomyopathy and atrioventricular conduction blocks induced by heart-specific inactivation of mitochondrial DNA gene expression. *Nat. Genet.* **1999**, *21*, 133–137. [CrossRef]

232. Wredenberg, A.; Wibom, R.; Wilhelmsson, H.; Graff, C.; Wiener, H.H.; Burden, S.J.; Oldfors, A.; Westerblad, H.; Larsson, N.G. Increased mitochondrial mass in mitochondrial myopathy mice. *Proc. Natl. Acad. Sci. USA* **2002**, *99*, 15066–15071. [CrossRef] [PubMed]
233. Kim, J.Y.; Hwang, J.M.; Ko, H.S.; Seong, M.W.; Park, B.J.; Park, S.S. Mitochondrial DNA content is decreased in autosomal dominant optic atrophy. *Neurology* **2005**, *64*, 966–972. [CrossRef]
234. Frezza, C.; Cipolat, S.; Martins de Brito, O.; Micaroni, M.; Beznoussenko, G.V.; Rudka, T.; Bartoli, D.; Polishuck, R.S.; Danial, N.N.; De Strooper, B.; et al. OPA1 controls apoptotic cristae remodeling independently from mitochondrial fusion. *Cell* **2006**, *126*, 177–189. [CrossRef] [PubMed]
235. Tezze, C.; Romanello, V.; Desbats, M.A.; Fadini, G.P.; Albiero, M.; Favaro, G.; Ciciliot, S.; Soriano, M.E.; Morbidoni, V.; Cerqua, C.; et al. Age-Associated Loss of OPA1 in Muscle Impacts Muscle Mass, Metabolic Homeostasis, Systemic Inflammation, and Epithelial Senescence. *Cell Metab.* **2017**, *25*, 1374–1389.e6. [CrossRef] [PubMed]
236. Alavi, M.V.; Bette, S.; Schimpf, S.; Schuettauf, F.; Schraermeyer, U.; Wehrl, H.F.; Ruttiger, L.; Beck, S.C.; Tonagel, F.; Pichler, B.J.; et al. A splice site mutation in the murine Opa1 gene features pathology of autosomal dominant optic atrophy. *Brain* **2007**, *130 Pt 4*, 1029–1042. [CrossRef]
237. Rabl, R.; Soubannier, V.; Scholz, R.; Vogel, F.; Mendl, N.; Vasiljev-Neumeyer, A.; Korner, C.; Jagasia, R.; Keil, T.; Baumeister, W.; et al. Formation of cristae and crista junctions in mitochondria depends on antagonism between Fcj1 and Su e/g. *J. Cell Biol.* **2009**, *185*, 1047–1063. [CrossRef]
238. Itoh, K.; Tamura, Y.; Iijima, M.; Sesaki, H. Effects of Fcj1-Mos1 and mitochondrial division on aggregation of mitochondrial DNA nucleoids and organelle morphology. *Mol. Biol. Cell* **2013**, *24*, 1842–1851. [CrossRef]
239. Genin, E.C.; Plutino, M.; Bannwarth, S.; Villa, E.; Cisneros-Barroso, E.; Roy, M.; Ortega-Vila, B.; Fragaki, K.; Lespinasse, F.; Pinero-Martos, E.; et al. CHCHD10 mutations promote loss of mitochondrial cristae junctions with impaired mitochondrial genome maintenance and inhibition of apoptosis. *EMBO Mol. Med.* **2016**, *8*, 58–72. [CrossRef]
240. Wang, C.; Taki, M.; Sato, Y.; Tamura, Y.; Yaginuma, H.; Okada, Y.; Yamaguchi, S. A photostable fluorescent marker for the superresolution live imaging of the dynamic structure of the mitochondrial cristae. *Proc. Natl. Acad. Sci. USA* **2019**, *116*, 15817–15822. [CrossRef]
241. Kondadi, A.K.; Anand, R.; Hänsch, S.; Urbach, J.; Zobel, T.; Wolf, D.M.; Segawa, M.; Liesa, M.; Shirihai, O.S.; Weidtkamp-Peters, S.; et al. Cristae undergo continuous cycles of membrane remodelling in a MICOS-dependent manner. *EMBO Rep.* **2020**, *21*, e49776. [CrossRef]
242. Guarani, V.; Jardel, C.; Chrétien, D.; Lombès, A.; Bénit, P.; Labasse, C.; Lacène, E.; Bourillon, A.; Imbard, A.; Benoist, J.F.; et al. QIL1 mutation causes MICOS disassembly and early onset fatal mitochondrial encephalopathy with liver disease. *Elife* **2016**, *5*, e17163. [CrossRef] [PubMed]
243. Zeharia, A.; Friedman, J.R.; Tobar, A.; Saada, A.; Konen, O.; Fellig, Y.; Shaag, A.; Nunnari, J.; Elpeleg, O. Mitochondrial hepato-encephalopathy due to deficiency of QIL1/MIC13 (C19orf70), a MICOS complex subunit. *Eur. J. Hum. Genet.* **2016**, *24*, 1778–1782. [CrossRef] [PubMed]
244. Gödiker, J.; Grüneberg, M.; DuChesne, I.; Reunert, J.; Rust, S.; Westermann, C.; Wada, Y.; Classen, G.; Langhans, C.D.; Schlingmann, K.P.; et al. QIL1-dependent assembly of MICOS complex-lethal mutation in C19ORF70 resulting in liver disease and severe neurological retardation. *J. Hum. Genet.* **2018**, *63*, 707–716. [CrossRef] [PubMed]
245. Benincá, C.; Zanette, V.; Brischigliaro, M.; Johnson, M.; Reyes, A.; Valle, D.A.D.; Robinson, A.R.; Degiorgi, A.; Yeates, A.; Telles, B.A.; et al. Mutation in the MICOS subunit gene APOO (MIC26) associated with an X-linked recessive mitochondrial myopathy, lactic acidosis, cognitive impairment and autistic features. *J. Med. Genet.* **2020**. [CrossRef]
246. Valente, E.M.; Abou-Sleiman, P.M.; Caputo, V.; Muqit, M.M.; Harvey, K.; Gispert, S.; Ali, Z.; Del Turco, D.; Bentivoglio, A.R.; Healy, D.G.; et al. Hereditary early-onset Parkinson's disease caused by mutations in PINK1. *Science* **2004**, *304*, 1158–1160. [CrossRef]
247. Pankratz, N.; Foroud, T. Genetics of Parkinson disease. *Genet. Med.* **2007**, *9*, 801–811. [CrossRef]
248. Pickrell, A.M.; Youle, R.J. The roles of PINK1, parkin, and mitochondrial fidelity in Parkinson's disease. *Neuron* **2015**, *85*, 257–273. [CrossRef]
249. Tsai, P.I.; Lin, C.H.; Hsieh, C.H.; Papakyrikos, A.M.; Kim, M.J.; Napolioni, V.; Schoor, C.; Couthouis, J.; Wu, R.M.; Wszolek, Z.K.; et al. PINK1 Phosphorylates MIC60/Mitofilin to Control Structural Plasticity of Mitochondrial Crista Junctions. *Mol. Cell* **2018**, *69*, 744–756.e6. [CrossRef]

250. Ogaki, K.; Koga, S.; Heckman, M.G.; Fiesel, F.C.; Ando, M.; Labbé, C.; Lorenzo-Betancor, O.; Moussaud-Lamodière, E.L.; Soto-Ortolaza, A.I.; Walton, R.L.; et al. Mitochondrial targeting sequence variants of the CHCHD2 gene are a risk for Lewy body disorders. *Neurology* **2015**, *85*, 2016–2025. [CrossRef]
251. Bannwarth, S.; Ait-El-Mkadem, S.; Chaussenot, A.; Genin, E.C.; Lacas-Gervais, S.; Fragaki, K.; Berg-Alonso, L.; Kageyama, Y.; Serre, V.; Moore, D.G.; et al. A mitochondrial origin for frontotemporal dementia and amyotrophic lateral sclerosis through CHCHD10 involvement. *Brain* **2014**, *137 Pt 8*, 2329–2345. [CrossRef]
252. Project MinE ALS Sequencing Consortium. CHCHD10 variants in amyotrophic lateral sclerosis: Where is the evidence? *Ann. Neurol.* **2018**, *84*, 110–116. [CrossRef] [PubMed]
253. Harel, T.; Yoon, W.H.; Garone, C.; Gu, S.; Coban-Akdemir, Z.; Eldomery, M.K.; Posey, J.E.; Jhangiani, S.N.; Rosenfeld, J.A.; Cho, M.T.; et al. Recurrent De Novo and Biallelic Variation of ATAD3A, Encoding a Mitochondrial Membrane Protein, Results in Distinct Neurological Syndromes. *Am. J. Hum. Genet.* **2016**, *99*, 831–845. [CrossRef] [PubMed]
254. Bione, S.; D'Adamo, P.; Maestrini, E.; Gedeon, A.K.; Bolhuis, P.A.; Toniolo, D. A novel X-linked gene, G4.5. is responsible for Barth syndrome. *Nat. Genet.* **1996**, *12*, 385–389. [CrossRef] [PubMed]
255. Clarke, S.L.; Bowron, A.; Gonzalez, I.L.; Groves, S.J.; Newbury-Ecob, R.; Clayton, N.; Martin, R.P.; Tsai-Goodman, B.; Garratt, V.; Ashworth, M.; et al. Barth syndrome. *Orphanet J. Rare Dis.* **2013**, *8*, 23. [CrossRef]
256. Mayr, J.A.; Havlícková, V.; Zimmermann, F.; Magler, I.; Kaplanová, V.; Jesina, P.; Pecinová, A.; Nusková, H.; Koch, J.; Sperl, W.; et al. Mitochondrial ATP synthase deficiency due to a mutation in the ATP5E gene for the F1 epsilon subunit. *Hum. Mol. Genet.* **2010**, *19*, 3430–3439. [CrossRef]
257. Sheffer, R.; Douiev, L.; Edvardson, S.; Shaag, A.; Tamimi, K.; Soiferman, D.; Meiner, V.; Saada, A. Postnatal microcephaly and pain insensitivity due to a de novo heterozygous DNM1L mutation causing impaired mitochondrial fission and function. *Am. J. Med. Genet. A* **2016**, *170*, 1603–1607. [CrossRef]
258. Gerber, S.; Charif, M.; Chevrollier, A.; Chaumette, T.; Angebault, C.; Kane, M.S.; Paris, A.; Alban, J.; Quiles, M.; Delettre, C.; et al. Mutations in DNM1L, as in OPA1, result in dominant optic atrophy despite opposite effects on mitochondrial fusion and fission. *Brain* **2017**, *140*, 2586–2596. [CrossRef]
259. Jungbluth, H.; Cullup, T.; Lillis, S.; Zhou, H.; Abbs, S.; Sewry, C.; Muntoni, F. Centronuclear myopathy with cataracts due to a novel dynamin 2 (DNM2) mutation. *Neuromuscul. Disord.* **2010**, *20*, 49–52. [CrossRef]
260. Susman, R.D.; Quijano-Roy, S.; Yang, N.; Webster, R.; Clarke, N.F.; Dowling, J.; Kennerson, M.; Nicholson, G.; Biancalana, V.; Ilkovski, B.; et al. Expanding the clinical, pathological and MRI phenotype of DNM2-related centronuclear myopathy. *Neuromuscul. Disord.* **2010**, *20*, 229–237. [CrossRef]
261. Mademan, I.; Deconinck, T.; Dinopoulos, A.; Voit, T.; Schara, U.; Devriendt, K.; Meijers, B.; Lerut, E.; De Jonghe, P.; Baets, J. De novo INF2 mutations expand the genetic spectrum of hereditary neuropathy with glomerulopathy. *Neurology* **2013**, *81*, 1953–1958. [CrossRef]
262. Wan, J.; Steffen, J.; Yourshaw, M.; Mamsa, H.; Andersen, E.; Rudnik-Schöneborn, S.; Pope, K.; Howell, K.B.; McLean, C.A.; Kornberg, A.J.; et al. Loss of function of SLC25A46 causes lethal congenital pontocerebellar hypoplasia. *Brain* **2016**, *139*, 2877–2890. [CrossRef] [PubMed]
263. Hammer, M.B.; Ding, J.; Mochel, F.; Eleuch-Fayache, G.; Charles, P.; Coutelier, M.; Gibbs, J.R.; Arepalli, S.K.; Chong, S.B.; Hernandez, D.G.; et al. SLC25A46 Mutations Associated with Autosomal Recessive Cerebellar Ataxia in North African Families. *Neurodegener Dis.* **2017**, *17*, 208–212. [CrossRef] [PubMed]
264. White, K.E.; Davies, V.J.; Hogan, V.E.; Piechota, M.J.; Nichols, P.P.; Turnbull, D.M.; Votruba, M. OPA1 deficiency associated with increased autophagy in retinal ganglion cells in a murine model of dominant optic atrophy. *Investig. Ophthalmol. Vis. Sci.* **2009**, *50*, 2567–2571. [CrossRef] [PubMed]
265. Spiegel, R.; Saada, A.; Flannery, P.J.; Burté, F.; Soiferman, D.; Khayat, M.; Eisner, V.; Vladovski, E.; Taylor, R.W.; Bindoff, L.A.; et al. Fatal infantile mitochondrial encephalomyopathy, hypertrophic cardiomyopathy and optic atrophy associated with a homozygous OPA1 mutation. *J. Med. Genet.* **2016**, *53*, 127–131. [CrossRef] [PubMed]
266. Delettre, C.; Lenaers, G.; Griffoin, J.M.; Gigarel, N.; Lorenzo, C.; Belenguer, P.; Pelloquin, L.; Grosgeorge, J.; Turc-Carel, C.; Perret, E.; et al. Nuclear gene OPA1, encoding a mitochondrial dynamin-related protein, is mutated in dominant optic atrophy. *Nat. Genet.* **2000**, *26*, 207–210. [CrossRef]
267. Alexander, C.; Votruba, M.; Pesch, U.E.; Thiselton, D.L.; Mayer, S.; Moore, A.; Rodriguez, M.; Kellner, U.; Leo-Kottler, B.; Auburger, G.; et al. OPA1, encoding a dynamin-related GTPase, is mutated in autosomal dominant optic atrophy linked to chromosome 3q28. *Nat. Genet.* **2000**, *26*, 211–215. [CrossRef]

268. Bonneau, D.; Colin, E.; Oca, F.; Ferré, M.; Chevrollier, A.; Guéguen, N.; Desquiret-Dumas, V.; N'Guyen, S.; Barth, M.; Zanlonghi, X.; et al. Early-onset Behr syndrome due to compound heterozygous mutations in OPA1. *Brain* **2014**, *137*, e301. [CrossRef]
269. Schaaf, C.P.; Blazo, M.; Lewis, R.A.; Tonini, R.E.; Takei, H.; Wang, J.; Wong, L.J.; Scaglia, F. Early-onset severe neuromuscular phenotype associated with compound heterozygosity for OPA1 mutations. *Mol. Genet. Metab.* **2011**, *103*, 383–387. [CrossRef]
270. Rubegni, A.; Pisano, T.; Bacci, G.; Tessa, A.; Battini, R.; Procopio, E.; Giglio, S.; Pasquariello, R.; Santorelli, F.M.; Guerrini, R.; et al. Leigh-like neuroimaging features associated with new biallelic mutations in OPA1. *Eur. J. Paediatr. Neurol.* **2017**, *21*, 671–677. [CrossRef]
271. Züchner, S.; Mersiyanova, I.V.; Muglia, M.; Bissar-Tadmouri, N.; Rochelle, J.; Dadali, E.L.; Zappia, M.; Nelis, E.; Patitucci, A.; Senderek, J.; et al. Mutations in the mitochondrial GTPase mitofusin 2 cause Charcot-Marie-Tooth neuropathy type, 2.A. *Nat. Genet.* **2004**, *36*, 449–451. [CrossRef]
272. Kijima, K.; Numakura, C.; Izumino, H.; Umetsu, K.; Nezu, A.; Shiiki, T.; Ogawa, M.; Ishizaki, Y.; Kitamura, T.; Shozawa, Y.; et al. Mitochondrial GTPase mitofusin 2 mutation in Charcot-Marie-Tooth neuropathy type, 2.A. *Hum. Genet.* **2005**, *116*, 23–27. [CrossRef]
273. Nicholson, G.A.; Magdelaine, C.; Zhu, D.; Grew, S.; Ryan, M.M.; Sturtz, F.; Vallat, J.M.; Ouvrier, R.A. Severe early-onset axonal neuropathy with homozygous and compound heterozygous MFN2 mutations. *Neurology* **2008**, *70*, 1678–1681. [CrossRef] [PubMed]
274. Polke, J.M.; Laurá, M.; Pareyson, D.; Taroni, F.; Milani, M.; Bergamin, G.; Gibbons, V.S.; Houlden, H.; Chamley, S.C.; Blake, J.; et al. Recessive axonal Charcot-Marie-Tooth disease due to compound heterozygous mitofusin 2 mutations. *Neurology* **2011**, *77*, 168–173. [CrossRef]
275. Renaldo, F.; Amati-Bonneau, P.; Slama, A.; Romana, C.; Forin, V.; Doummar, D.; Barnerias, C.; Bursztyn, J.; Mayer, M.; Khouri, N.; et al. MFN2, a new gene responsible for mitochondrial DNA depletion. *Brain* **2012**, *135 Pt 8*, e223, 1–4; Author reply e4, 1–3. [CrossRef]
276. Barøy, T.; Pedurupillay, C.R.; Bliksrud, Y.T.; Rasmussen, M.; Holmgren, A.; Vigeland, M.D.; Hughes, T.; Brink, M.; Rodenburg, R.; Nedregaard, B.; et al. A novel mutation in FBXL4 in a Norwegian child with encephalomyopathic mitochondrial DNA depletion syndrome 13. *Eur. J. Med. Genet.* **2016**, *59*, 342–346. [CrossRef] [PubMed]
277. Antoun, G.; McBride, S.; Vanstone, J.R.; Naas, T.; Michaud, J.; Redpath, S.; McMillan, H.J.; Brophy, J.; Daoud, H.; Chakraborty, P.; et al. Detailed Biochemical and Bioenergetic Characterization of FBXL4-Related Encephalomyopathic Mitochondrial DNA Depletion. *JIMD Rep.* **2016**, *27*, 1–9. [PubMed]
278. Gai, X.; Ghezzi, D.; Johnson, M.A.; Biagosch, C.A.; Shamseldin, H.E.; Haack, T.B.; Reyes, A.; Tsukikawa, M.; Sheldon, C.A.; Srinivasan, S.; et al. Mutations in FBXL4, encoding a mitochondrial protein, cause early-onset mitochondrial encephalomyopathy. *Am. J. Hum. Genet.* **2013**, *93*, 482–495. [CrossRef]
279. Huemer, M.; Karall, D.; Schossig, A.; Abdenur, J.E.; Al Jasmi, F.; Biagosch, C.; Distelmaier, F.; Freisinger, P.; Graham, B.H.; Haack, T.B.; et al. Clinical, morphological, biochemical, imaging and outcome parameters in 21 individuals with mitochondrial maintenance defect related to FBXL4 mutations. *J. Inherit. Metab. Dis.* **2015**, *38*, 905–914. [CrossRef]
280. Nasca, A.; Scotton, C.; Zaharieva, I.; Neri, M.; Selvatici, R.; Magnusson, O.T.; Gal, A.; Weaver, D.; Rossi, R.; Armaroli, A.; et al. Recessive mutations in MSTO1 cause mitochondrial dynamics impairment, leading to myopathy and ataxia. *Hum. Mutat.* **2017**, *38*, 970–977. [CrossRef]
281. Gal, A.; Balicza, P.; Weaver, D.; Naghdi, S.; Joseph, S.K.; Várnai, P.; Gyuris, T.; Horváth, A.; Nagy, L.; Seifert, E.L.; et al. MSTO1 is a cytoplasmic pro-mitochondrial fusion protein, whose mutation induces myopathy and ataxia in humans. *EMBO Mol. Med.* **2017**, *9*, 967–984. [CrossRef]
282. Iwama, K.; Takaori, T.; Fukushima, A.; Tohyama, J.; Ishiyama, A.; Ohba, C.; Mitsuhashi, S.; Miyatake, S.; Takata, A.; Miyake, N.; et al. Novel recessive mutations in MSTO1 cause cerebellar atrophy with pigmentary retinopathy. *J. Hum. Genet.* **2018**, *63*, 263–270. [CrossRef] [PubMed]
283. Hartmann, B.; Wai, T.; Hu, H.; MacVicar, T.; Musante, L.; Fischer-Zirnsak, B.; Stenzel, W.; Gräf, R.; van den Heuvel, L.; Ropers, H.H.; et al. Homozygous YME1L1 mutation causes mitochondriopathy with optic atrophy and mitochondrial network fragmentation. *Elife* **2016**, *5*, 5. [CrossRef]
284. Wilkinson, P.A.; Crosby, A.H.; Turner, C.; Bradley, L.J.; Ginsberg, L.; Wood, N.W.; Schapira, A.H.; Warner, T.T. A clinical, genetic and biochemical study of SPG7 mutations in hereditary spastic paraplegia. *Brain* **2004**, *127 Pt 5*, 973–980. [CrossRef]

285. Elleuch, N.; Depienne, C.; Benomar, A.; Hernandez, A.M.; Ferrer, X.; Fontaine, B.; Grid, D.; Tallaksen, C.M.; Zemmouri, R.; Stevanin, G.; et al. Mutation analysis of the paraplegin gene (SPG7) in patients with hereditary spastic paraplegia. *Neurology* **2006**, *66*, 654–659. [CrossRef] [PubMed]
286. Banfi, S.; Bassi, M.T.; Andolfi, G.; Marchitiello, A.; Zanotta, S.; Ballabio, A.; Casari, G.; Franco, B. Identification and characterization of AFG3L2, a novel paraplegin-related gene. *Genomics* **1999**, *59*, 51–58. [CrossRef]
287. Edener, U.; Wöllner, J.; Hehr, U.; Kohl, Z.; Schilling, S.; Kreuz, F.; Bauer, P.; Bernard, V.; Gillessen-Kaesbach, G.; Zühlke, C. Early onset and slow progression of SCA28, a rare dominant ataxia in a large four-generation family with a novel AFG3L2 mutation. *Eur. J. Hum. Genet.* **2010**, *18*, 965–968. [CrossRef] [PubMed]
288. Mariotti, C.; Brusco, A.; Di Bella, D.; Cagnoli, C.; Seri, M.; Gellera, C.; Di Donato, S.; Taroni, F. Spinocerebellar ataxia type 28: A novel autosomal dominant cerebellar ataxia characterized by slow progression and ophthalmoparesis. *Cerebellum* **2008**, *7*, 184–188. [CrossRef]
289. Penttilä, S.; Jokela, M.; Bouquin, H.; Saukkonen, A.M.; Toivanen, J.; Udd, B. Late onset spinal motor neuronopathy is caused by mutation in CHCHD10. *Ann. Neurol.* **2015**, *77*, 163–172. [CrossRef]
290. Auranen, M.; Ylikallio, E.; Shcherbii, M.; Paetau, A.; Kiuru-Enari, S.; Toppila, J.P.; Tyynismaa, H. CHCHD10 variant p.(Gly66Val) causes axonal Charcot-Marie-Tooth disease. *Neurol. Genet.* **2015**, *1*, e1. [CrossRef]
291. Ajroud-Driss, S.; Fecto, F.; Ajroud, K.; Lalani, I.; Calvo, S.E.; Mootha, V.K.; Deng, H.X.; Siddique, N.; Tahmoush, A.J.; Heiman-Patterson, T.D.; et al. Mutation in the novel nuclear-encoded mitochondrial protein CHCHD10 in a family with autosomal dominant mitochondrial myopathy. *Neurogenetics* **2015**, *16*, 1–9. [CrossRef]
292. Chaussenot, A.; Le Ber, I.; Ait-El-Mkadem, S.; Camuzat, A.; de Septenville, A.; Bannwarth, S.; Genin, E.C.; Serre, V.; Augé, G.; Brice, A.; et al. Screening of CHCHD10 in a French cohort confirms the involvement of this gene in frontotemporal dementia with amyotrophic lateral sclerosis patients. *Neurobiol. Aging* **2014**, *35*, 2884.e1–e4. [CrossRef]
293. Morais, V.A.; Verstreken, P.; Roethig, A.; Smet, J.; Snellinx, A.; Vanbrabant, M.; Haddad, D.; Frezza, C.; Mandemakers, W.; Vogt-Weisenhorn, D. Parkinson's disease mutations in PINK1 result in decreased Complex I activity and deficient synaptic function. *EMBO Mol. Med.* **2009**, *1*, 99–111. [CrossRef] [PubMed]
294. Chishti, M.A.; Bohlega, S.; Ahmed, M.; Loualich, A.; Carroll, P.; Sato, C.; St George-Hyslop, P.; Westaway, D.; Rogaeva, E. T313M PINK1 mutation in an extended highly consanguineous Saudi family with early-onset Parkinson disease. *Arch Neurol.* **2006**, *63*, 1483–1485. [CrossRef]
295. Ibáñez, P.; Lesage, S.; Lohmann, E.; Thobois, S.; De Michele, G.; Borg, M.; Agid, Y.; Dürr, A.; Brice, A.; The French Parkinson's Disease Genetics Study Group. Mutational analysis of the PINK1 gene in early-onset parkinsonism in Europe and North Africa. *Brain* **2006**, *129 Pt 3*, 686–694.
296. Silvestri, L.; Caputo, V.; Bellacchio, E.; Atorino, L.; Dallapiccola, B.; Valente, E.M.; Casari, G. Mitochondrial import and enzymatic activity of PINK1 mutants associated to recessive parkinsonism. *Hum. Mol. Genet.* **2005**, *14*, 3477–3492. [CrossRef] [PubMed]
297. Rogaeva, E.; Johnson, J.; Lang, A.E.; Gulick, C.; Gwinn-Hardy, K.; Kawarai, T.; Sato, C.; Morgan, A.; Werner, J.; Nussbaum, R.; et al. Analysis of the PINK1 gene in a large cohort of cases with Parkinson disease. *Arch. Neurol.* **2004**, *61*, 1898–1904. [CrossRef] [PubMed]
298. Lieber, D.S.; Calvo, S.E.; Shanahan, K.; Slate, N.G.; Liu, S.; Hershman, S.G.; Gold, N.B.; Chapman, B.A.; Thorburn, D.R.; Berry, G.T.; et al. Targeted exome sequencing of suspected mitochondrial disorders. *Neurology* **2013**, *80*, 1762–1770. [CrossRef]
299. Jonckheere, A.I.; Renkema, G.H.; Bras, M.; van den Heuvel, L.P.; Hoischen, A.; Gilissen, C.; Nabuurs, S.B.; Huynen, M.A.; de Vries, M.C.; Smeitink, J.A.; et al. A complex V ATP5A1 defect causes fatal neonatal mitochondrial encephalopathy. *Brain* **2013**, *136 Pt 5*, 1544–1554. [CrossRef]
300. Saric, A.; Andreau, K.; Armand, A.S.; Møller, I.M.; Petit, P.X. Barth Syndrome: From Mitochondrial Dysfunctions Associated with Aberrant Production of Reactive Oxygen Species to Pluripotent Stem Cell Studies. *Front. Genet.* **2015**, *6*, 359. [CrossRef]
301. Barth, P.G.; Scholte, H.R.; Berden, J.A.; Van der Klei-Van Moorsel, J.M.; Luyt-Houwen, I.E.; Van't Veer-Korthof, E.T.; Van der Harten, J.J.; Sobotka-Plojhar, M.A. An X-linked mitochondrial disease affecting cardiac muscle, skeletal muscle and neutrophil leucocytes. *J. Neurol. Sci.* **1983**, *62*, 327–355. [CrossRef]

302. D'Adamo, P.; Fassone, L.; Gedeon, A.; Janssen, E.A.; Bione, S.; Bolhuis, P.A.; Barth, P.G.; Wilson, M.; Haan, E.; Orstavik, K.H.; et al. The X-linked gene G4.5 is responsible for different infantile dilated cardiomyopathies. *Am. J. Hum. Genet.* **1997**, *61*, 862–867. [CrossRef] [PubMed]
303. Ichida, F.; Tsubata, S.; Bowles, K.R.; Haneda, N.; Uese, K.; Miyawaki, T.; Dreyer, W.J.; Messina, J.; Li, H.; Bowles, N.E.; et al. Novel gene mutations in patients with left ventricular noncompaction or Barth syndrome. *Circulation* **2001**, *103*, 1256–1263. [CrossRef] [PubMed]
304. Spencer, C.T.; Bryant, R.M.; Day, J.; Gonzalez, I.L.; Colan, S.D.; Thompson, W.R.; Berthy, J.; Redfearn, S.P.; Byrne, B.J. Cardiac and clinical phenotype in Barth syndrome. *Pediatrics* **2006**, *118*, e337–e346. [CrossRef] [PubMed]

 life

MDPI

Article

Dependence of Leydig Cell's Mitochondrial Physiology on Luteinizing Hormone Signaling

Marija L. J. Medar [1,†], Dijana Z. Marinkovic [1,†], Zvezdana Kojic [2], Alisa P. Becin [1], Isidora M. Starovlah [1], Tamara Kravic-Stevovic [3], Silvana A. Andric [1] and Tatjana S. Kostic [1,*]

1 Department of Biology and Ecology, Faculty of Sciences, University of Novi Sad, 21102 Novi Sad, Serbia; marija.medar@dbe.uns.ac.rs (M.L.J.M.); dijana.marinkovic@dbe.uns.ac.rs (D.Z.M.); alisa.becin@dbe.uns.ac.rs (A.P.B.); isidora.starovlah@dbe.uns.ac.rs (I.M.S.); silvana.andric@dbe.uns.ac.rs (S.A.A.)

2 Institute of Physiology, Faculty of Medicine, University of Belgrade, 11000 Belgrade, Serbia; zvezdana.kojic@med.bg.ac.rs

3 Institute of Histology and Embriology, Faculty of Medicine, University of Belgrade, 11000 Belgrade, Serbia; tamara.kravic-stevovic@med.bg.ac.rs

* Correspondence: tatjana.kostic@dbe.uns.ac.rs; Tel.: +381-63-514-716

† These authors equally contributed to study development.

Abstract: Knowledge about the relationship between steroidogenesis and the regulation of the mitochondrial bioenergetics and dynamics, in steroidogenic cells, is not completely elucidated. Here we employed in vivo and ex vivo experimental models to analyze mitochondrial physiology in Leydig cells depending on the different LH-cAMP environments. Activation of LH-receptor in rat Leydig cells ex and in vivo triggered cAMP, increased oxygen consumption, mitoenergetic and steroidogenic activities. Increased mitoenergetic activity i.e., ATP production is achieved through augmented glycolytic ATP production and a small part of oxidative phosphorylation (OXPHOS). Transcription of major genes responsible for mitochondrial dynamics was upregulated for *Ppargc1a* (regulator of mitogenesis and function) and downregulated for *Drp1* (main fission marker), *Prkn*, *Pink1* and *Tfeb* (mitophagy markers). Leydig cells from gonadotropin-treated rats show increased mitogenesis confirmed by increased mitochondrial mass, increased mtDNA, more frequent mitochondria observed by a transmission electron microscope and increased expression of subunits of respiratory proteins *Cytc*/CYTC and COX4. Opposite, Leydig cells from hypogonadotropic-hypogonadal rats characterized by low LH-cAMP, testosterone, and ATP production, reduced markers of mitogenesis and mitofusion (*Mfn1/2*, *Opa1*) associated with reduced mtDNA content. Altogether results underline LH-cAMP signaling as an important regulator of mitochondrial physiology arranging mitochondrial dynamics, bioenergetic and steroidogenic function in Leydig cells.

Keywords: mitochondrial dynamics; mitoenergetics; mitosteroidogenesis; LH; cAMP; Leydig cell

Citation: Medar, M.L.J.; Marinkovic, D.Z.; Kojic, Z.; Becin, A.P.; Starovlah, I.M.; Kravic-Stevovic, T.; Andric, S.A.; Kostic, T.S. Dependence of Leydig Cell's Mitochondrial Physiology on Luteinizing Hormone Signaling. *Life* **2021**, *11*, 19. https://doi.org/10.3390/life11010019

Received: 24 November 2020
Accepted: 29 December 2020
Published: 31 December 2020

1. Introduction

Mitochondria are multifunctional cellular organelles with essential roles in vitally important processes including cellular energy production, reactive oxygen species (ROS) synthesis, apoptosis signaling, maintaining calcium homeostasis, and metabolic integration. In steroidogenic cells, the initial step of steroid hormones synthesis occurs in mitochondria. Despite the importance of the mitochondria in cellular and metabolic health, the details about their organization and synchronization are not well characterized.

In Leydig cells, mitosteroidogenesis is enabled by cholesterol availability and mitochondrial targeting including steroidogenic enzyme localization in mitochondria. It is controlled by the interaction of luteinizing hormone (LH) with its specific receptor which triggers the cyclic adenosine monophosphate–protein kinase A (cAMP-PRKA) signaling [1,2] that activates mitochondria-targeted StAR protein [3] and other proteins of transduceosome responsible for cholesterol transport into inner mitochondrial membrane. Once cholesterol comes to the inner mitochondrial membrane it is converted to pregnenolone by CYP11A1 to start steroidogenesis [4]. The CYP11A1 cleaves the 20, 22 bound in the insoluble cholesterol to become soluble pregnenolone which activates a downstream HSD3B and other steroidogenic enzymes to produce testosterone. Activation of mitochondrial CYP11A1 is a hormonally regulated and rate-limiting step and is considered a determinant of the steroidogenic capacity of the cells [5].

The mitoenergetic role arises from the generation of ATP by oxidizing hydrogens derived from carbohydrates and fats. Electrons from NADH are sent to multimeric mitochondrial protein Complex I (NADH dehydrogenase) which forwards to ubiquinone (coenzyme Q/CoQ) and then to Complex III (ubiquinol/cytochrome c oxidoreductase) which further shuttles the electrons to cytochrome c. Cytochrome c transfers electrons to cytochrome-containing Complex IV (cytochrome c oxidase, COX), the terminal protein complex of the mitochondrial respiratory chain, which uses the electrons to reduce O_2 to yield H_2O. The free energy of electron movement through the OXPHOS pathway is used to pump protons (H^+) out of the mitochondrial matrix into the mitochondrial intermembrane space, creating a capacitance across the inner membrane i.e., mitochondrial membrane potential (Δψm). This potential energy is utilized to drive ATP synthesis by the Complex V (ATP synthase) OXPHOS [6].

Subsequently, ATP is also synthesized by cytosolic glycolysis. Although glycolysis produces much less ATP per cycle than OXPHOS, it nonetheless plays an important role in stimulated and proliferated cells with high metabolic demand [7,8].

For efficient steroidogenesis in Leydig cells, functional mitochondria are required for both Δψm and ATP synthesis [9,10]. Cellular bioenergetic demands are closely linked to mitochondrial dynamics whose maintenance is particularly important for efficient steroidogenesis [11]. Mitochondria are dynamic organelles able to rearrange themselves, through coordinated cycles of fusion and fission, to properly adapt to cellular metabolic needs. Mitochondrial fusion is responsible for the formation of the mitochondrial network, allowing more efficient operation in both bioenergetic and metabolic senses. By contrast, fission is responsible for fragmenting the mitochondrial network, which is normally associated with resting cells, or disposing of damaged organelles through mitophagy/autophagy processes [12,13]. Additionally, cells activate mitochondrial biogenesis to form new mitochondria but also rearrange mitochondrial architecture to adapt to changes in metabolic demands [14]. The balance between these states is crucial for maintaining a physiological mitochondrial network. The mitochondrial dynamics are enabled by the coordinated action of many genes and following proteins that can be considered as a marker of mitochondrial dynamics: the *Ppargc1a*/PGC1a plays a role in mitochondrial biogenesis through transcriptional regulation of its downstream genes such as *Nrf1*, *Nrf2*, and *Tfam* leading to synthesis of mitochondrial DNA, proteins, and generation of new mitochondria [15,16]; two key multifunctional gene products, *Pink1*/PINK1 and *Prkn*/PARKIN, jointly promote the degradation of defective mitochondria providing quality control by mitophagy [17]; mitochondrial fusion/fission is regulated by several essential genes such as *Mfn1*, *Mfn2*, *Opa1*, and *Drp1* [14].

It is known that the cAMP/PRKA signaling at the outer membrane of mitochondria regulates several processes, such as mitochondrial protein import [18,19], apoptosis [20], autophagy [21], mitophagy [22] and mitochondrial fission and fusion [11,23–25]. In addition, mitochondrial cAMP signaling has been involved in modulation of OXPHOS [26,27] and regulation of ATP synthesis [27–30].

However, the details of the regulation of mitochondrial fission–fusion dynamics together with biogenesis and mitophagy/autophagy remain to be completely elucidated, and is particularly poorly characterized in the hormonal regulation of steroidogenic function. Despite the key role of mitochondria in steroid synthesis, the reports exploring the relationship between steroidogenesis, bioenergetics, and mitochondrial dynamics i.e., mitochondrial fitness in respect to tropic stimulation of Leydig cells are insufficient. Therefore, in this study, we analyzed the Leydig cell's mitochondrial response/fitness on in vivo and ex vivo LH stimulation. Results were analyzed from the point of view of the cAMP and testosterone changes.

2. Materials and Methods

2.1. Animals

All experiments were conducted using male Wistar rats which were bred and raised in the animal facility of the Faculty of Sciences, University of Novi Sad (Novi Sad, Serbia). The animals were sustained in carefully regulated environmental conditions (22 ± 2 °C; 14 h light–10 h dark cycle) with unlimited access to water and food (ad libitum). The experiments were approved by local Ethical Committee on Animal Care and Use of the University of Novi Sad (statement no. 01-201/3) operating under the rules of the National Council for animal welfare and the National Low for Animal Welfare (March 2009) and in accordance with the National Research Council publication Guide for the Care and Use of Laboratory Animals (copyright 1996, National Academy of Science, Washington, DC, USA) and European Convention for the Protection of Vertebrate Animals used for Experimental and other Scientific Purposes (Council of Europe No 123, Strasbourg 1985).

2.2. In Vivo Experiments

The animals underwent two different in vivo treatments mimicking acute and chronic manipulation of intracellular cAMP level: (1) treatment of intact male rats with agonist of LH receptors in order to increase cAMP signaling in Leydig cells, (2) experimental model of hypogonadotropic hypogonadism in order to downregulate reproductive axis and lower cAMP signaling in Leydig cells. In the first approach, a group of adult male rats were treated with a subcutaneous injection of Pregnyl (Organon, Holland, active component is human chorionic gonadotropin (hCG)), 40 IU/50 μL per 100 g of animal weight (a dose that effectively increases cAMP in rat Leydig cells [6]. The control group received the same amount of 0.9% NaCl. Animals were decapitated 2 h and 6 h after treatment. In the second approach, adult male rats were injected intramuscularly with the long-lasting GnRH analog diphereline (PharmaSwiss, Belgrade, Serbia; 0.29 mg/50 μL/100 g). Control rats were injected with the same amount of vehicle. After one month, animals were sacrificed in the morning. In both experimental approaches, trunk blood and testes were collected and used for further analysis.

2.3. Collection of Testicular Interstitial Fluid (TIF) and Preparation of Purified Leydig Cells

Leydig cells were isolated following the same protocol previously described by our research group [9,31–33]. Briefly, after isolation, testes were decapsulated and main blood vessel removed. With the intention to collect testicular interstitial fluid (TIF), testes were placed in the Falcon Mash 100 μm (Sigma, St. Louise, MO, USA) in 50 mL tubes and centrifuged at 100× g/7 min. TIF samples were stored at −80 °C until utilization. Testicular tissue was further used and placed in 50 mL plastic tubes containing 0.25 mg/mL collagenase; 1.5%-bovine serum albumin (BSA); 20 mM 4-(2-hydroxyethyl)-1-piperazineethanesulfonic acid (HEPES)-M199, Sigma, St. Louise, Missouri (2 testes per tube). Cell isolation was continued by placing plastic tubes into shaking-water bath (15 min/34 °C/120 cycles per min). In order to stop enzymatic reaction 40 mL of cold medium was added and seminiferous tubules were removed during filtration through Mesh cell strainer 100 μm (Sigma Inc.). Remaining interstitial cell suspension was centrifuged (160× g for 5 min) and resuspended in 8 mL per tube Dulbecco's Modified Eagle Medium/Nutrient Mixture F-12 (DMEM-F12) medium (Sigma, St. Louise, MO, USA). With the purpose of isolating Leydig cells from interstitial cells different densities (1.080, 1.065 and 1.045 g/mL) of Percoll gradient were used. Interstitial cells were centrifuged 1100× g for 28 min (brake free). Gradient fragments that contained Leydig cells (1.080/1.065 g/mL and 1.065/1.045 g/mL) were collected, washed in M199-0.1% BSA and centrifuged at 200× g/5 min. Cell precipitate was resuspended in 5 mL DMEM/F12 and the proportion of Leydig cells present in culture was determined by staining for HSD3B activity using nitro blue tetrazolium (NBT) chloride test [34] with some modifications). Briefly, Leydig cells were incubated in presence of 10 μM pregnenolone, 1 mg/mL NBT and 3 mg/mL β-NAD + for 1.5 h/34 °C/500 rpm/O_2, in final volume of 200 μL. At the end of the incubation period, the suspension was centrifuged for 7 min/500× *g*. Supernatant was discarded while cells pellet resuspended in a drop of medium and put on the microscopic slide (two slides per tube). Positive (blue) cells were counted using Image Tool Ink, Ver 3.00 software. According to HSD3B staining, presence of Leydig cells in the culture was more than 90%. As for the Trypan blue exclusion test, cell viability was greater than 95%.

2.4. *Ex Vivo Experiments*

Leydig cell primary culture was obtained by plating 3 × 10^6 cells in Petri dish (55 mm) and placed in a CO_2 incubator at 34 °C for 3 h to recover and attach. Cells were respectively treated with hCG (100 ng/mL, Sigma, St. Louise, MO, USA) and oligomycin (20 μM, Sigma, St. Louise, MO, USA) or CMI (a PRKA inhibitor, 0.5 μM; 4-cyano-3-methylisoquinoline from Calbiochem, (San Diego, CA, USA)) [35]. After period of stimulation, cell medium and cells were collected and stored until analysis.

2.5. *Testosterone and cAMP Measurements*

Testosterone level was measured in serum using radioimmunoassay (RIA). Androgens detected in serum by RIA were referred to as testosterone + dihydrotestosterone (T + DHT) because the antitestosterone serum number 250 used in this study showed 100% cross-reactivity with DHT. All samples were measured in duplicate in one assay (sensitivity: 6 pg/tube; intra-assay coefficient of variation: 5–8%; inter-assay coefficient of variation: 7.5%). Cyclic nucleotide levels were measured by the cAMP enzyme-linked immunosorbent assay (ELISA) kit (Cayman Chemicals, Ann Arbor, MI, USA), with a quantification limit of 0.1 pmol/mL for acetylated cAMP samples.

2.6. *ATP Level, Mitochondrial Membrane Potential and O_2 Consumption Measurements*

Determination of ATP level was performed using the ATP Bioluminescence CLS II kit following manual instruction (Roche, lifescience.roche.com). Leydig cells (2 × 10^6/0.5 mL) were incubated in shaking-water bath (1 h/34 °C/80 cycles per min) and centrifuged (1200× *g*/5 min). Precipitated cells were resuspended in boiling water and Tris-Ethylenediaminetetraacetic acid (EDTA) (1:9), boiled for 2 more minutes, centrifuged (900× *g*/1 min) and final supernatant was used for ATP measurement. Sample/standard and Luciferase reagent were mixed 1:1 and luminescence was measured by the Biosystems/luminometar (Fluoroscan, Ascent, FL, USA). Mitochondrial abundance and mitochondrial membrane potential (ΔΨm) were determined by mitotrack green and tetramethylrhodamine (TMRE), respectively, staining for 20 min/34 °C/5% CO_2 were applied with subsequent fluorescence reading (Fluoroscan, Ascent, FL, USA). The excitation wavelength used for the each test was 485 and 550 nm while emission wavelengths were 510 and 590 nm respectively; after staining, cells were washed with 0.1% BSA-PBS [31,33,36]. Oxygen consumption by Leydig cells suspension was measured by Clark electrode at 34 °C and oxygen uptake and oxygraphic curves were obtained by digital multimeter VC820 (Conard Electronic, Hirsau, Bavaria, Germany) and software Digiscope for Windows (version 2.06) [31].

2.7. *Genomic DNA Purification, RNA Isolation, cDNA Synthesis and Real-Time Polymerase Chain Reaction (PCR) Relative Quantitative Analysis*

Genomic DNA from Leydig cells was purified by Wizard® Genomic DNA Purification Kit (www.promega.com, #TM050) and total RNA isolation was performed using the Rneasy Mini Kit (www.qiagen.com) according to manuals instructions. Concentration and purity of total RNA and DNA were measured using BioSpec-nano (Shimadzu Biotech, Kyoto, Japan) followed by DNAse I treatment of the isolated RNA (www.invitrogen.com). First strand of cDNA was synthesized using a High-Capacity cDNA Reverse Transcription Kit (Thermo

Fisher Scientific, Waltham, MA, USA) following manufacturer's protocol. Relative gene expression analysis was accomplished by quantitative real-time polymerase chain reaction (qRT-PCR) using SYBR-Green based technology (Applied Biosystems, Thermo Fisher Scientific, Waltham, MA, USA). The qRT-PCR reaction was performed in the presence of 25 ng/5 μL cDNA from reverse transcription reaction and 500 nM specific primers. The expression of several reference genes (*Actb, Rsp16, Rsp18, Gapdh*) were tested. The *Actb, Gapdh* and *Rsp16* were identified as the stable genes. In this study the relative quantification of all samples was analyzed in duplicate and *Gapdh* was used as endogenous control. The *MtNd1-B2m* ratio estimated in genomic DNA was used for mtDNA copies determination [37]. Sequences of all primers were arranged using www.ncbi.nih.gov/sites/entrez and showed in Table S1.

2.8. Protein Extraction and Western Blot Analysis

Platted cells (3×10^6) were lysed by buffer (20 mM HEPES, 10 mM EDTA, 2.5 mM $MgCl_2$, 1 mM dithiothreitol (DTT), 40 mM β-glicerophosphate, 1% NP-40, 2 μM leupeptin, 1 μM aprotinin, 0.5 mM 4-bensenesulfonyl fluoride hydrochloride (AEBSF), phosphatase inhibitor cocktail [PhosSTOP, www.roche.com]) and the lysates were uniformed in protein concentration by Bradford method. Samples were mixed with SDS protein gel loading solution (Quality Biological, Inc., Gaithersburg, MD, USA), boiled for 3 min, and separated by one-dimensional SDS–PAGE electrophoresis (www.bio-rad.com). Transfer of the proteins to the PVDF membrane was performed by electroblotting overnight at 4 °C/40 A and efficacy of transfer was checked by staining and distaining of the gels. The membranes were blocked by 3% BSA-1× TBS for 2 h/RT, incubation with primary antibody was done overnight at 4 °C, 0.1% Tween-1× TBS was used for washing membranes, and incubation with secondary antibody was done for 1 h/RT [31]. Immunoreactive bands were detected using luminol reaction and MyECL imager (www.fischer.sci)/films and densitometric measurements were performed by Image J program version 1.32 (https://rsbweb.nich.gov//ij/download.html). Antiserums for COX4 (Cat. No. sc-58348), and actin (ACTN) (sc8432) were purchased from Santa Cruz Biotechnology (Heidelberg, Germany), antiserum for cytochrome C (CYTC) (ABIN3024606) was obtained from Oncogene Research Products (San Diego, CA, USA).

2.9. Transmission Electron Microscope (TEM) Analysis of Leydig Cell Mitochondria

Testes were collected from control and hCG treated animals, and fragments of testicular tissue were fixed in 3% glutaraldehyde, post-fixed in 1% osmium tetroxide, dehydrated in graded alcohols, and then embedded in Epon 812. The ultrathin sections were stained in uranyl acetate and lead citrate and were examined using a Morgagni 268D electron microscope (FEI, Hillsboro, OR, USA).

2.10. Statistical Analysis

For in vivo studies the results represent group mean ± standard error of the mean (SEM) values of individual variations from 4 rats per group. For ex vivo measurements data represents mean ± SEM from three to five independent replicates. The results were analyzed by Mann–Whitney's unpaired nonparametric two-tailed test (for two point data experiments), or for group comparison one-way analysis of variance (ANOVA), followed by the Student–Newman–Keuls multiple range test. Correlation analysis was performed in the R studio using 95% confidence intervals.

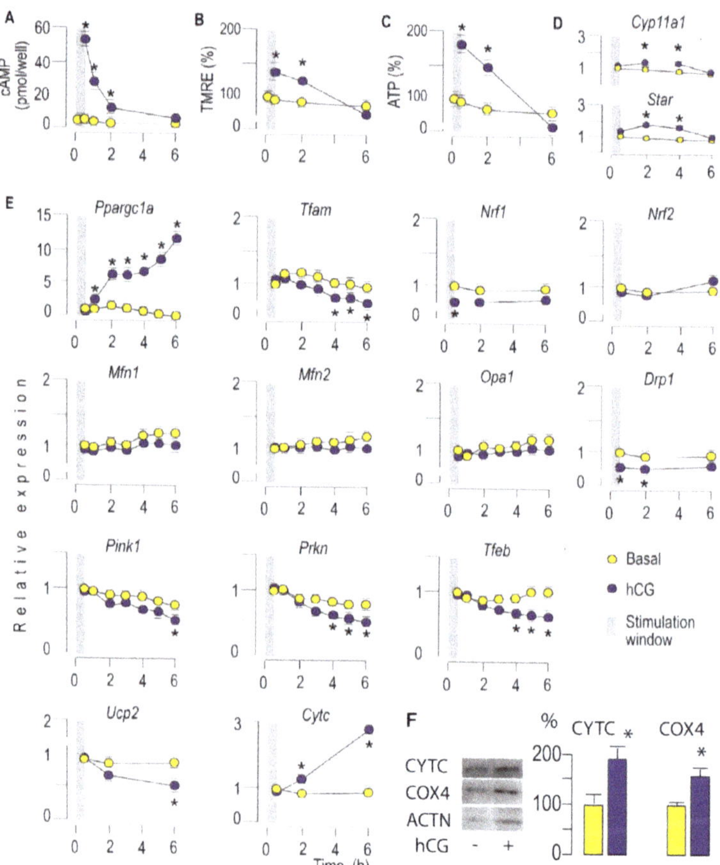

Figure 2. Activation of LHR changed expression of genes important for mitochondrial function in rat Leydig cells ex vivo. Leydig cells from adult rats were isolated and primary culture was formed. The cells were stimulated with hCG (100 ng/mL) for 30 min, media changed and cell incubation was continued; cells were collected every 30 min for 6 h; moment when stimulation started was referred to as 0h. Cells were collected in indicated time points and used for measurement of intracellular concentration of cAMP (**A**), $\Delta\Psi m$ (**B**) and ATP (**C**). RNA was isolated and quantitative real-time polymerase chain reaction (qRT-PCR) was performed to detect expression level of steroidogenic (**D**) or genes important for mitochondrial function/dynamics (**E**). Abundances of CYTC and COX4 in Leydig cells stimulated w/wo hCG for 2 h were evaluated by Western blot (**F**). Representative blots are shown. The bars near the respective bands show values obtained by scanning densitometry and normalized on the ACTN. Data points are group mean ± SEM values of three independent experiments, $n = 3$. * Statistical significance between the groups for the same time point ($p < 0.05$).

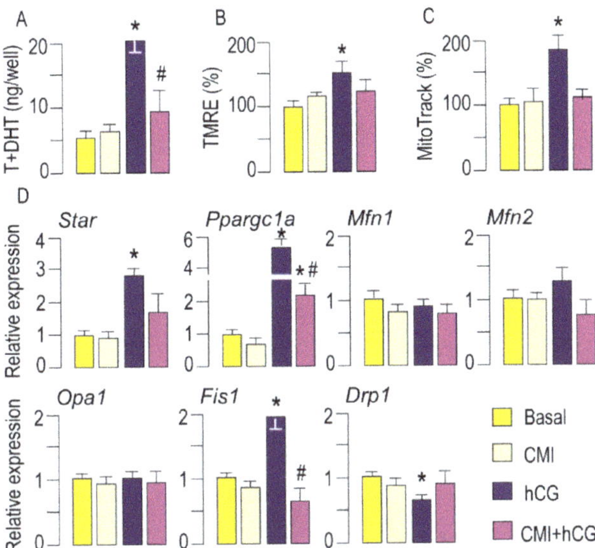

Figure 3. Activation of cAMP-PRKA signaling increased mitochondrial mass and changed expression of genes important for regulation of mitobiogenesis and mitofission. Primary culture of rat Leydig cell was stimulated with hCG (100 ng/mL) in presence or absence of PRKA inhibitor (CMI, 0.5 μM) for 30 min. After 30 min culture media were changed with fresh media without hCG but with CMI. Cells were collected after 2 h from beginning of stimulation. T + DHT was measured in culture media (**A**) while cells were used for ΔΨm (**B**), and mitochondrial mass (**C**) determination. RNA was isolated and qRT-PCR was performed to detect expression level of *Star* and genes important for mitochondrial function/dynamics (**D**). Data bars are group mean ± SEM values of three independent experiments, $n = 3$. * Statistical significance between the control and treated group; # difference in respect to hCG-treated group, ($p < 0.05$).

3.3. Activation of LHR Changed the Expression of Genes Important for Mitochondrial Function in Rat Leydig Cells In Vivo

To investigate in vivo effects of LHR activation on the mitochondrial physiology in Leydig cells, adult male rats were treated with hCG, and effects were estimated 2 and 6 h following the treatment. Blood testosterone level (Figure 4A) and cAMP concentration in TIF (Figure 4B), were significantly higher in treated rats in both examined time points indicating treatment efficiency. Elevated testosterone, Δψm in both time points (Figure 4C) and rise ATP production in the 2nd hour after hCG injection (Figure 4D) suggested increased mitochondrial engagement in both processes, energy production, and steroidogenesis.

As expected, treatment increased genes responsible for mitochondrial steroidogenesis (*Star* and *Cyp11a1*), in both time points (Figure 4E). Changes in the expression of the main regulators of mitochondrial dynamics after in vivo administration of hCG were very similar to those occurring after ex vivo treatment (Figure 4F–I). Results suggested altered mitochondrial biogenesis following hCG injection: *Ppargc1a* increased following 2 h treatment but was below control values in the 6th hour; transcription of *Nrf1* decreased 2 h after injection and then approached the control values; *Nrf2*, remained unchanged while the reduction of *Tfam* occurred in the 6th hour (Figure 4F). Expression of genes involved in the regulation of mitofusion, *Mfn1*, *Mfn2*, and *Opa1*, remained unchanged, and the mitofission regulator, *Drp1* was reduced in the 6th hour after hCG injection (Figure 4G). However, results pointed to altered mitophagy in Leydig cells following LHR activation: *Pink1* and *Prkn* decreased in the 6th hour as well as a marker of autophagy, *Tfeb* (Figure 4H). The trend of decreased transcription after hCG treatment was also noticed in *Ucp2 and Cox4/2* expression while *Cytc* increased in both time points (Figure 4I).

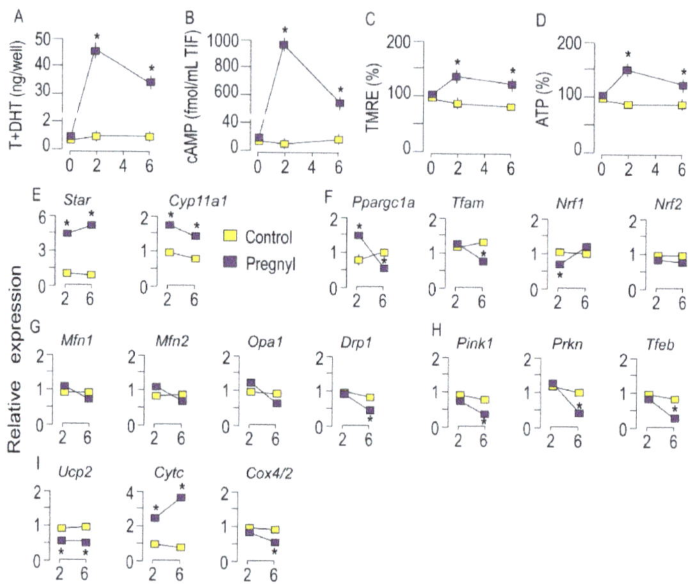

Figure 4. Activation of LHR changed expression of genes important for mitochondrial function in rat Leydig cells in vivo. Rats were sacrificed 2 h and 6 h after hCG/vehicle application. Serum and TIF were collected for measurement of testosterone (**A**) and cAMP (**B**), respectively. Leydig cells were isolated and used for $\Delta\Psi m$ (**C**) and ATP (**D**) determination. RNA was isolated from Leydig cells and used in qRT-PCR analysis in order to examine expression level of genes that regulate mitochondrial function (**E–I**). Data points/bars are group mean ± SEM values of three independent experiments, $n = 3$. * Statistical significance between the groups for the same time point ($p < 0.05$).

Increased mitochondrial biogenesis in Leydig cells following gonadotropin injection was analyzed (Figure 5A). The Mitotracker-green dye has high affinity and specificity for lipid mitochondrial membranes and produces green fluorescence which was employed to examine mitochondrial abundance. The gonadotropin injection caused a 32% increase in mitochondrial abundance compared with control values (Figure 5A). Additionally, increased mitochondrial abundance due to LHR stimulation was confirmed by the estimation of mtDNA content (Figure 5B).

The ultrastructural appearances of mitochondria in Leydig cells from hCG treated and control rats were assessed by transmission electron microscopy (TEM) analysis. Mitochondria were more frequently detected in hCG treated Leydig cells (Figure 5E,F), compared to control Leydig cells (Figure 5C,D). Treated and control Leydig cells mitochondria had both tubular and lamellar cristae, and occasionally mitochondria in gonadotropin stimulated cells had more electron dense matrix (Figure 5E,F). Leydig cells treated with hCG had more electron lucent chromatin (euchromatin) (Figure 5E,F) while more regions of condensed chromatin (heterochromatin) were seen in nuclei of control cells (Figure 5C,D).

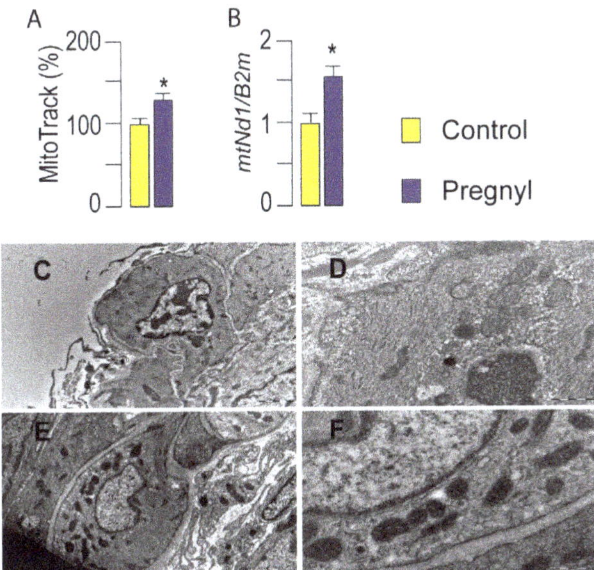

Figure 5. Activation of LHR increased mitochondrial mass and changed Leydig cell's mitochondrial ultrastructure. Two hours after hCG injection testes were isolated, fragments of testicular tissue immediately fixed for transmission electron microscopy (TEM) analysis while the rest of the tissue was used for Leydig cells isolation. Cells, individually isolated, were plated for mitochondrial abundance determination (10^5/well) using Mitotracker green (**A**) or for genomic DNA isolation (10^6/well) (**B**). The mtNd1/B2m ratio was used for mtDNA content determination. Ultramicrographs of Leydig cell from control rats at low (89,000×; **C**) and high magnification (28,000×; **D**) with well preserved mitochondria, and Leydig cell from hCG-treated rats at low (56,000×; **E**) and high magnification (28,000×; **F**) with numerous mitochondria with electron dense matrix. Data bar represents group means ± SEM values from three rats (n = 3). * Statistical significance at level $p < 0.05$ compared to the untreated group.

3.4. Mitochondrial Function Was Dampened in Leydig Cells from Hypogonadotropic Hypogonadal Rats

Given the high performance of mitochondria in Leydig cells after LHR stimulation, it was of interest to investigate the effects of a low level of LH and intracellular cAMP on mitochondrial physiology. To do this, the rats were treated with the long-lasting GnRH analog [32]. Such treatment in male rats causes hypogonadotropic hypogonadism, which is characterized by a low level of testosterone (Figure 6A) in circulation as well as a low level of cAMP in Leydig cells (Figure 6B). Additionally, treatment reduced $\Delta\psi m$ (Figure 6C) and ATP production (Figure 6D). Expression of the main steroidogenic genes involved in steroid production in mitochondria *Star* and *Cyp11a1* were decreased in Leydig cells from hypogonadal rats (Figure 6E). Furthermore, genes involved in the regulation of mitochondrial biogenesis *Ppargc1a*, *Nrf1* and *Nrf2* were reduced but *Tfam* remained unchanged (Figure 6F). Hypogonadism has affected mitofission and mitofusion in Leydig cells by reducing transcription of *Drp1* as well as *Mfn1* and *Mfn2* without changing *Opa1* (Figure 6G). Hypogonadism elevated *Tfeb* transcription in Leydig cells but not *Pink1* and *Prkn* (Figure 6H). In Leydig cells from hypogonadal rats, transcription of *Cox4/2* and *Cytc* lessened but no changes were observed in *Ucp2* (Figure 6I). Finally, the expression of mitochondrial *Nd1* decreased in Leydig cells from hypogonadal rats (Figure 6J).

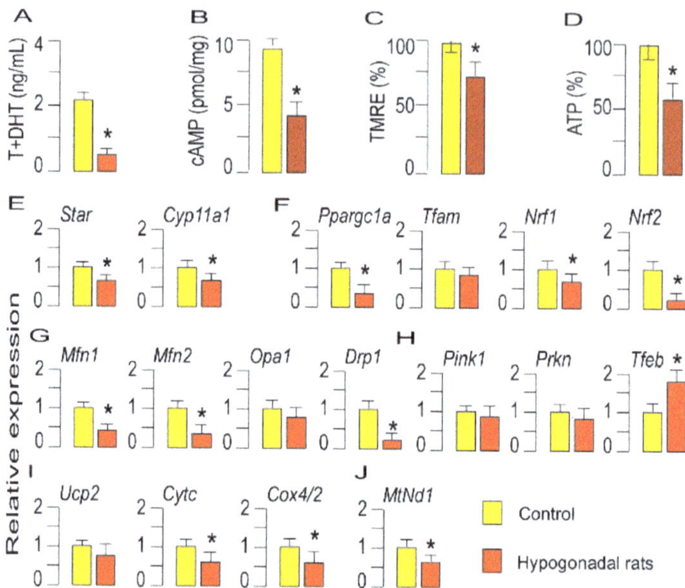

Figure 6. Mitochondrial function is altered in Leydig cells from hypogonadotropic hypogonadal rats. Serum from control and hypogonadotropic hypogonadal rats was prepared for testosterone measurement (**A**). Leydig cells were isolated and used for cAMP (**B**), $\Delta\Psi m$ (**C**) and ATP (**D**) determination. The RNA was isolated and used in qRT-PCR analysis in order to examine expression level of genes that regulate mitochondrial function (**E**–**J**). Data bars are group mean ± SEM values of three independent experiments, $n = 3$. * Statistical significance ($p < 0.05$) between control and hypogonadal rats.

4. Discussion

Since mitochondria are at the core of cellular needs, communication with the host cells is vitally important. The Leydig cell's endocrine function is governed by LH-cAMP-PRKA signaling which is the most essential signaling pathway involved in the regulation of mitosteroidogenesis and consequently testosterone production. The question addressed in this study is at what extend mitochondrial functions including energetic, steroidogenic, and overall mitochondrial fitness are regulated by LH signaling.

Results obtained indicated the regulatory impact of LH-cAMP on Leydig cell's mitochondrial energetic and steroidogenic function including regulation of mitochondrial dynamics and biogenesis which generally alleviate these processes. The main conclusions were derived based on in vivo and ex vivo models with a different LH environment and Leydig cells with different steroidogenic capacity.

Our results pointed out that the stimulation of LHR signaling in vivo and ex vivo can increase $\Delta\psi m$ and stimulate ATP and steroid production. This is in line with earlier studies on Leydig cell primary culture that emphasized the importance of maintaining the $\Delta\psi m$ for energetic and steroidogenic mitochondrial activities [10,38]. Furthermore, mitochondrial electron transport chain, which drives formation of the $\Delta\psi m$ utilized for ATP synthesis, is critical for LH-mediated testosterone production [39], as a number of steroidogenic steps were suppressed upon OXPHOS inhibition. The results presented here show that blocking of ATP synthase, besides ATP, blocks also the increase in testosterone caused by LHR-activation, indicating that mitochondrial respiration is important for testosterone synthesis. These events are supported by increased O_2 consumption which is necessary for both processes. Namely, LHR-stimulation of Leydig cells increases basal O_2 usage and rate of ATP production. However, LHR-stimulation of basal O_2 consumption could indicate a leak of the respiration rate [40] which can be interpreted as increased O_2 usage in expense for steroid and ROS formation.

In LH stimulated cells, ATP is predominantly produced in OXPHOS but showed an increased dependence on glycolysis and less dependence on oxidative metabolism. The ATP produced in glycolysis increased by 58% while ATP produced in mitochondrial respiration increased by 18% in gonadotropin-stimulated cells. The important metabolic coupling could be present. It is well documented that the cAMP-PRKA pathway plays a regulatory role in the glycolytic metabolism, exerting a stimulatory effect on glucose transport and utilization via phosphorylation of different downstream target proteins [41,42]. Increasing glycolytic activity in the cytosol could also lead to an adequate increase in NADH which are readily transferred to the matrix of mitochondria where they can be involved in the OXPHOS process [43].

However, in LH stimulated cells a mitosteroidogenesis and energetics do not appear to be parallel processes. Our results showed the stimulatory effect of LH on mitochondrial steroidogenesis lasts longer than the effect on ATP production. This may be the result of LH-cAMP-PRKA activation of genes important for mitosteroidogenesis (*Star*, *Cyp11a1*). Results from the hypogonadal model illustrate the described relationship among energetic and steroidogenic mitochondrial parameters but on a reduced scale i.e., low cAMP, followed by decreased $\Delta\psi m$, ATP, testosterone, and *Star* and *Cyp11a1* transcription.

Effective energetic, steroidogenic, and other mitochondrial functions are essentially dependent on a proper mitochondrial dynamic network regulated by mitochondrial fusion and fission [29,44]. Mitofusion represents a limiting step in the onset of processes required for the transport of intermediate products in/out mitochondria and is essential for cholesterol import into mitochondria and steroid formation [11,24]. Nonetheless, in Leydig cells with LH-activated steroidogenesis the expression of *Mfn1*, *Mfn2*, and *Opa1*, were unchanged suggesting regulatory action on the posttranslational level of these gene products. Under the same conditions, the decreased transcription of profission *Drp1* was observed. Since mitofusion is a prerequisite for steroid hormone synthesis in Leydig cells [11], reduced *Drp1* expression could facilitate mitochondrial fusion that might be important for the formation of the multiprotein complex that delivers cholesterol to the CYP11 system in addition to stimulation of oxidative phosphorylation. Indeed, blocking of DRP1 activity results in increased mitofusion in Leydig cells [25]. Moreover, PRKA phosphorylation of DRP1 inhibits mitofission enabling an optimal environment for steroids biosynthesis [25]. Results obtained by ex vivo PRKA inhibition prevented hCG-triggered decrease of *Drp1* transcription supporting PRKA-regulation of DRP1 expression.

It has been shown that inhibiting fission or promoting fusion decreases mitophagy, whereas enhanced fission precedes and facilitates it [45]. In LH-stimulated Leydig cells the expression of genes involved in quality control, mitophagy *Pink1*, *Prkn* and autophagy *Tfeb* were decreased. Accordingly, the activation of the cAMP-PRKA pathway is associated with an inhibitory effect on mitophagy [29]. By contrast, the increased transcription of mitobiogenesis marker, *Ppargc1a* in addition to *Cytc* as well as COX4 indicates LH-potential for new mitochondria formation. This is in agreement with observed increased PGC1a in Leydig cells cultivated with hCG for long [33]. Indeed, results obtained by Mitotrack clearly indicate LHR-regulated grow in mitochondrial mass. Moreover, TEM analysis revealed more abundant mitochondria in Leydig cells obtained from hCG-treated rats. Gonadotropin treatment of Leydig cells stimulates expression of CYTC and COX4 subunits of respiratory proteins which again points to increased biogenesis. However, the transcriptional pattern of *Cox4/2* over time, after in vivo hCG-treatment, differs from the protein pattern, indicating possible paracrine regulation as well as the existence of a post-transcriptional/posttranslational modulatory mechanism. It is known that genes that contribute to the same respiratory protein complex are from both mtDNA and nuclear DNA [46]. Thus, mitochondria and the nucleus finely coordinate the transcription, translation, and import of mitochondrial proteins to ensure the proper relationship between the various OXPHOS components. This communication occurs bidirectionally in time dependent manner indicating very sophisticated regulation of many signaling pathways. The nucleus can influence mitochondrial structure, replication, biogenesis, fission, and fusion, the number of mitochondria, and the mitophagy turnover rate [47]. Nonetheless, the regulation of respiratory proteins expression and different behavior of genes encoding the subunits together with proteins involved in mitochondrial dynamics including anterograde and retrograde signaling in steroidogenic cells need to be further investigated. Results from the study on the hypogonadal model show that decreased steroidogenesis in Leydig cells is enabled by mitochondrial dysfunction. This is connected to decreased mitochondrial engagement in steroidogenesis due to decreased expression of *Star* and *Cyp11a1*, thereby decreased cholesterol import and convert into biologically active steroid precursors; decreased aerobic energy production by mitochondria resulting in lowered ATP levels; decreased mitochondrial biogenesis illustrated by decreased expression of the main marker of mitochondrial biogenesis (*Ppargc1*); amended mitochondrial dynamics i.e., mitofusion due to decreased *Mfn1* and *Mfn2* as well as mitofission due to decreased *Drp1* expression with potential to increased autophagy as result of increased *Tfeb*.

Taken together, the results from all experimental models in this study revealed the dependence of mitochondrial physiology on LH-cAMP signaling in Leydig cells (Figure 7). Specifically, the expression level of genes that indicate the intensity of mitochondrial biogenesis (*Ppargc1a* and *Cytc*; Figure 7C), as well as genes responsible for mitosteroidogenesis (*Star*, *Cyp11a1*; Figure 7D), are positively correlated with the level of cAMP in the cell. On the contrary expression of genes involved in the regulation of mitophagy (*Prkn*) and autophagy (*Tfeb*) are in negative correlation with cAMP level (Figure 7A).

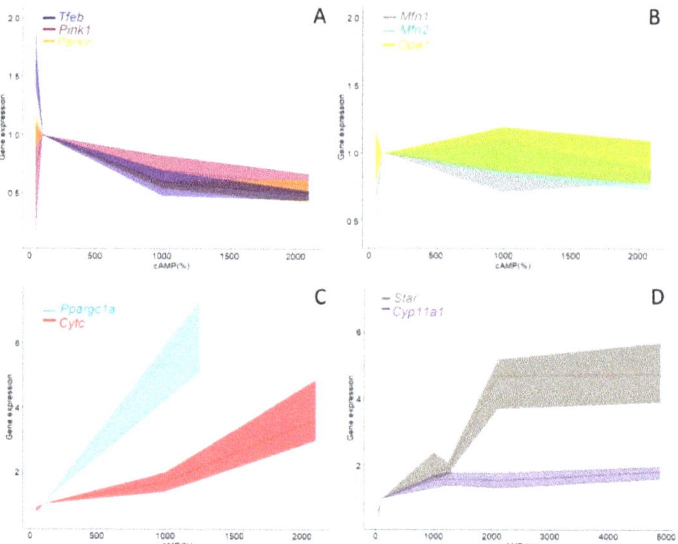

Figure 7. Relationships between the concentration of cAMP and expression of genes responsible for mitochondrial physiology in Leydig cells. Data shown are pool from 3 different experimental models (ex vivo and in vivo stimulation of LHR as well as experimental model for hypogonadotropic hypogonadism). Data for several groups of genes involved in regulation of mitophagy (**A**), mitofusion (**B**), mitochondrial biogenesis (**C**) and mitochondrial steroidogenesis (**D**), covaried with cAMP concentration. Ribbons represent 95% confidence intervals.

Results support the conclusion that the LH-cAMP pathway is involved in the control of mitochondrial biogenesis and dynamics coupled with mitosteroidogenesis and energetic function; therefore, it significantly influences mitochondrial fitness in Leydig cells.

Supplementary Materials: The following are available online at https://www.mdpi.com/2075-1729/11/1/19/s1, Table S1.

Author Contributions: M.L.J.M.—acquisition of the data; analysis and interpretation of the data; final approval of the version to be submitted; D.Z.M.—acquisition of the data; analysis and drafting the manuscript; revising manuscript critically for important intellectual content; final approval of the version to be submitted; A.P.B.—acquisition of the data; analysis and interpretation of the data; revising manuscript critically for important intellectual content; final approval of the version to be submitted; Z.K.—analyses of O_2 consumption; analysis and interpretation of the data; revising manuscript critically for important intellectual content; final approval of the version to be submitted; I.M.S.—acquisition of the data; final approval of the version to be submitted; T.K.-S.—TEM analysis; final approval of the version to be submitted; S.A.A.—acquisition of the data; analysis and interpretation of the data; revising manuscript critically for important intellectual content; final approval of the version to be submitted; T.S.K.—the conception and design of the research; acquisition of the data; analysis and interpretation of the data; drafting the manuscript; revising manuscript critically for important intellectual content; final approval of the version to be submitted. All authors—approved the final version of the manuscript; agree to be accountable for all aspects of the work in ensuring that questions related to the accuracy or integrity of any part of the work are appropriately investigated and resolved; qualify for authorship, and all those who qualify for authorship are listed.

Funding: This research was supported by the Serbian Ministry of Education and Technological Development grant no. 173057 and CeRes grant, and the Autonomic Province of Vojvodina grants no. 3822 and no. 2130.

Institutional Review Board Statement: The study was conducted according to the guidelines of the Declaration of Helsinki, and approved by the Ethical Committee on Animal Care and Use of the University of Novi Sad (statement no. 01-201/3).

Informed Consent Statement: Not applicable.

Data Availability Statement: All relevant data are available from the corresponding author on request. Further information and requests for data, resources and reagents should be directed to and will be fulfilled by the corresponding author, Tatjana Kostic (tatjana.kostic@dbe.uns.ac.rs).

Acknowledgments: We are very grateful to Gordon Niswender (Colorado State University) for supplying antibodies for radioimmunoassay analysis. Also we thank Aleksandar Baburski and Srdjan Sokanovic for technical assistance in conducting experiments.

Conflicts of Interest: The authors declare no conflict of interest.

References

1. Cooper, D.M.F. Regulation and organization of adenylyl cyclases and cAMP. *Biochem. J.* **2003**, *375*, 517–529. [CrossRef] [PubMed]
2. Dufau, M.L. The luteinizing hormone receptor. *Annu. Rev. Physiol.* **1998**, *60*, 461–496. [CrossRef]
3. Stocco, D.M.; Wang, X.; Jo, Y.; Manna, P.R. Multiple Signaling Pathways Regulating Steroidogenesis and Steroidogenic Acute Regulatory Protein Expression: More Complicated than We Thought. *Mol. Endocrinol.* **2005**, *19*, 2647–2659. [CrossRef] [PubMed]
4. Payne, A.H.; Hales, D.B. Overview of Steroidogenic Enzymes in the Pathway from Cholesterol to Active Steroid Hormones. *Endocr. Rev.* **2004**, *25*, 947–970. [CrossRef] [PubMed]
5. Miller, W.L. Steroid hormone synthesis in mitochondria. *Mol. Cell. Endocrinol.* **2013**, *379*, 62–73. [CrossRef]
6. Midzak, A.; Papadopoulos, V. Adrenal Mitochondria and Steroidogenesis: From Individual Proteins to Functional Protein Assemblies. *Front. Endocrinol.* **2016**, *7*, 106. [CrossRef]
7. Heiden, M.G.V.; Cantley, L.C.; Thompson, C.B. Understanding the Warburg Effect: The Metabolic Requirements of Cell Proliferation. *Science* **2009**, *324*, 1029–1033. [CrossRef]
8. Zheng, J. Energy metabolism of cancer: Glycolysis versus oxidative phosphorylation (Review). *Oncol. Lett.* **2012**, *4*, 1151–1157. [CrossRef]
9. Andric, S.A.; Janjic, M.M.; Stojkov, N.J.; Kostic, T.S. Protein kinase G-mediated stimulation of basal Leydig cell steroidogenesis. *Am. J. Physiol. Metab.* **2007**, *293*, E1399–E1408. [CrossRef]
10. Midzak, A.S.; Chen, H.; Aon, M.A.; Papadopoulos, V.; Zirkin, B.R. ATP Synthesis, Mitochondrial Function, and Steroid Biosynthesis in Rodent Primary and Tumor Leydig Cells. *Biol. Reprod.* **2011**, *84*, 976–985. [CrossRef]
11. Duarte, A.; Poderoso, C.; Cooke, M.; Soria, G.; Maciel, F.C.; Gottifredi, V.; Podesta, E.J. Mitochondrial Fusion Is Essential for Steroid Biosynthesis. *PLoS ONE* **2012**, *7*, e45829. [CrossRef] [PubMed]
12. Meyer, J.N.; Leuthner, T.C.; Luz, A.L. Mitochondrial fusion, fission, and mitochondrial toxicity. *Toxicology* **2017**, *391*, 42–53. [CrossRef] [PubMed]

13. Westermann, B. Bioenergetic role of mitochondrial fusion and fission. *Biochim. Biophys. Acta Bioenerg.* **2012**, *1817*, 1833–1838. [CrossRef] [PubMed]
14. Friedman, J.R.; Nunnari, J. Mitochondrial form and function. *Nature* **2014**, *505*, 335–343. [CrossRef]
15. Dominy, J.E.; Puigserver, P. Mitochondrial Biogenesis through Activation of Nuclear Signaling Proteins. *Cold Spring Harb. Perspect. Biol.* **2013**, *5*, a015008. [CrossRef]
16. Li, P.A.; Hou, X.; Hao, S. Mitochondrial biogenesis in neurodegeneration. *J. Neurosci. Res.* **2017**, *95*, 2025–2029. [CrossRef]
17. Palikaras, K.; Lionaki, E.; Tavernarakis, N. Mechanisms of mitophagy in cellular homeostasis, physiology and pathology. *Nat. Cell Biol.* **2018**, *20*, 1013–1022. [CrossRef]
18. Schmidt, O.; Harbauer, A.B.; Rao, S.; Eyrich, B.; Zahedi, R.P.; Stojanovski, D.; Schönfisch, B.; Guiard, B.; Sickmann, A.; Pfanner, N.; et al. Regulation of Mitochondrial Protein Import by Cytosolic Kinases. *Cell* **2011**, *144*, 227–239. [CrossRef]
19. Opalińska, M.; Meisinger, C. Metabolic control via the mitochondrial protein import machinery. *Curr. Opin. Cell Biol.* **2015**, *33*, 42–48. [CrossRef]
20. Moujalled, D.; Weston, R.; Anderton, H.; Ninnis, R.; Goel, P.; Coley, A.; Huang, D.C.S.; Wu, L.; Strasser, A.; Puthalakath, H. Cyclic-AMP-dependent protein kinase A regulates apoptosis by stabilizing the BH3-only protein Bim. *EMBO Rep.* **2011**, *12*, 77–83. [CrossRef]
21. Gomes, L.C.; Di Benedetto, G.; Scorrano, L. During autophagy mitochondria elongate, are spared from degradation and sustain cell viability. *Nat. Cell Biol.* **2011**, *13*, 589–598. [CrossRef] [PubMed]
22. Akabane, S.; Uno, M.; Tani, N.; Shimazaki, S.; Ebara, N.; Kato, H.; Kosako, H.; Oka, T. PKA Regulates PINK1 Stability and Parkin Recruitment to Damaged Mitochondria through Phosphorylation of MIC60. *Mol. Cell* **2016**, *62*, 371–384. [CrossRef] [PubMed]
23. Lefkimmiatis, K. cAMP signalling meets mitochondrial compartments. *Biochem. Soc. Trans.* **2014**, *42*, 265–269. [CrossRef] [PubMed]
24. Duarte, A.; Castillo, F.; Podesta, E.J.; Poderoso, C. Mitochondrial Fusion and ERK Activity Regulate Steroidogenic Acute Regulatory Protein Localization in Mitochondria. *PLoS ONE* **2014**, *9*, e100387. [CrossRef] [PubMed]
25. Park, J.-E.; Kim, Y.-J.; Lee, S.G.; Kim, J.Y.; Chung, J.-Y.; Jeong, S.-Y.; Koh, H.; Yun, J.; Park, H.T.; Yoo, Y.H.; et al. Drp1 Phosphorylation Is Indispensable for Steroidogenesis in Leydig Cells. *Endocrinology* **2019**, *160*, 729–743. [CrossRef] [PubMed]
26. De Rasmo, D.; Signorile, A.; Santeramo, A.; Larizza, M.; Lattanzio, P.; Capitanio, G.; Papa, S. Intramitochondrial adenylyl cyclase controls the turnover of nuclear-encoded subunits and activity of mammalian complex I of the respiratory chain. *Biochim. Biophys. Acta Bioenerg.* **2015**, *1853*, 183–191. [CrossRef]
27. De Rasmo, D.; Micelli, L.; Santeramo, A.; Signorile, A.; Lattanzio, P.; Papa, S. cAMP regulates the functional activity, coupling efficiency and structural organization of mammalian F O F 1 ATP synthase. *Biochim. Biophys. Acta Gen. Subj.* **2016**, *1857*, 350–358. [CrossRef]
28. Acin-Perez, R.; Salazar, E.; Kamenetsky, M.; Buck, J.; Levin, L.R.; Manfredi, G. Cyclic AMP Produced inside Mitochondria Regulates Oxidative Phosphorylation. *Cell Metab.* **2009**, *9*, 265–276. [CrossRef]
29. Di Benedetto, G.; Gerbino, A.; Lefkimmiatis, K. Shaping mitochondrial dynamics: The role of cAMP signalling. *Biochem. Biophys. Res. Commun.* **2018**, *500*, 65–74. [CrossRef]
30. Di Benedetto, G.; Scalzotto, E.; Mongillo, M.; Pozzan, T. Mitochondrial Ca^{2+} Uptake Induces Cyclic AMP Generation in the Matrix and Modulates Organelle ATP Levels. *Cell Metab.* **2013**, *17*, 965–975. [CrossRef]
31. Andric, S.A.; Kojic, Z.; Bjelic, M.M.; Mihajlović, A.I.; Baburski, A.Z.; Sokanovic, S.J.; Janjic, M.M.; Stojkov, N.J.; Stojilkovic, S.S.; Kostic, T.S. The opposite roles of glucocorticoid and α1-adrenergic receptors in stress triggered apoptosis of rat Leydig cells. *Am. J. Physiol. Metab.* **2013**, *304*, E51–E59. [CrossRef] [PubMed]
32. Baburski, A.Z.; Andric, S.A.; Kostic, T.S. Luteinizing hormone signaling is involved in synchronization of Leydig cell's clock and is crucial for rhythm robustness of testosterone production. *Biol. Reprod.* **2019**, *100*, 1406–1415. [CrossRef] [PubMed]
33. Gak, I.A.; Radovic, S.M.; Dukic, A.R.; Janjic, M.M.; Stojkov-Mimic, N.J.; Kostic, T.S.; Andric, S.A. Stress triggers mitochondrial biogenesis to preserve steroidogenesis in Leydig cells. *Biochim. Biophys. Acta Bioenerg.* **2015**, *1853*, 2217–2227. [CrossRef] [PubMed]
34. Payne, A.H.; Downing, J.R.; Wong, K.-L. Luteinizing Hormone Receptors and Testosterone Synthesis in Two Distinct Populations of Ley dig Cells. *Endocrinology* **1980**, *106*, 1424–1429. [CrossRef]
35. Kostic, T.S.; Andric, S.A.; Stojilkovic, S.S. Receptor-Controlled Phosphorylation of α_1 Soluble Guanylyl Cyclase Enhances Nitric Oxide-Dependent Cyclic Guanosine 5′-Monophosphate Production in Pituitary Cells. *Mol. Endocrinol.* **2004**, *18*, 458–470. [CrossRef]
36. Sokanovic, S.J.; Baburski, A.Z.; Kojic, Z.; Medar, M.L.J.; Andric, S.A.; Kostic, T.S. Aging-Related Increase of cGMP Disrupts Mitochondrial Homeostasis in Leydig Cells. *J. Gerontol. Ser. A Biol. Sci. Med. Sci.* **2020**. [CrossRef]
37. Rato, L.; Duarte, A.I.; Tomás, G.D.; Santos, M.S.; Moreira, P.I.; Socorro, S.; Cavaco, J.E.; Alves, M.G.; Oliveira, P.F. Pre-diabetes alters testicular PGC1-α/SIRT3 axis modulating mitochondrial bioenergetics and oxidative stress. *Biochim. Biophys. Acta Bioenerg.* **2014**, *1837*, 335–344. [CrossRef]
38. Allen, J.A.; Shankara, T.; Janus, P.; Buck, S.; Diemer, T.; Hales, K.H.; Hales, D.B. Energized, Polarized, and Actively Respiring Mitochondria Are Required for Acute Leydig Cell Steroidogenesis. *Endocrinology* **2006**, *147*, 3924–3935. [CrossRef]
39. Midzak, A.S.; Chen, H.; Papadopoulos, V.; Zirkin, B.R. Leydig cell aging and the mechanisms of reduced testosterone synthesis. *Mol. Cell. Endocrinol.* **2009**, *299*, 23–31. [CrossRef]

40. Salin, K.; Villasevil, E.M.; Anderson, G.J.; Selman, C.; Chinopoulos, C.; Metcalfe, N.B. The RCR and ATP/O Indices Can Give Contradictory Messages about Mitochondrial Efficiency. *Integr. Comp. Biol.* **2018**, *58*, 486–494. [CrossRef]
41. Palorini, R.; De Rasmo, D.; Gaviraghi, M.; Danna, L.S.; Signorile, A.; Cirulli, C.; Chiaradonna, F.; Alberghina, L.; Papa, S. Oncogenic K-ras expression is associated with derangement of the cAMP/PKA pathway and forskolin-reversible alterations of mitochondrial dynamics and respiration. *Oncogene* **2012**, *32*, 352–362. [CrossRef] [PubMed]
42. Yu, T.; Yang, G.; Hou, Y.; Tang, X.; Wu, C.; Wu, X.-A.; Guo, L.; Zhu, Q.; Luo, H.; Du, Y.-E.; et al. Cytoplasmic GPER translocation in cancer-associated fibroblasts mediates cAMP/PKA/CREB/glycolytic axis to confer tumor cells with multidrug resistance. *Oncogene* **2017**, *36*, 2131–2145. [CrossRef] [PubMed]
43. Glancy, B.; Kane, D.A.; Kavazis, A.N.; Goodwin, M.L.; Willis, W.T.; Gladden, L.B. Mitochondrial lactate metabolism: History and implications for exercise and disease. *J. Physiol.* **2020**. [CrossRef] [PubMed]
44. Chen, H.; Chan, D.C. Emerging functions of mammalian mitochondrial fusion and fission. *Hum. Mol. Genet.* **2005**, *14*, R283–R289. [CrossRef]
45. Twig, G.; Elorza, A.; Molina, A.A.J.; Mohamed, H.; Wikstrom, J.D.; Walzer, G.; Stiles, L.; Haigh, S.E.; Katz, S.; Las, G.; et al. Fission and selective fusion govern mitochondrial segregation and elimination by autophagy. *EMBO J.* **2008**, *27*, 433–446. [CrossRef] [PubMed]
46. Ben-Shachar, D. The interplay between mitochondrial complex I, dopamine and Sp1 in schizophrenia. *J. Neural Transm.* **2009**, *116*, 1383–1396. [CrossRef]
47. Wallace, D.C. The epigenome and the mitochondrion: Bioenergetics and the environment. *Genes Dev.* **2010**, *24*, 1571–1573. [CrossRef]

life

MDPI

Review

Natural and Artificial Mechanisms of Mitochondrial Genome Elimination

Elvira G. Zakirova [1], Vladimir V. Muzyka [1,2], Ilya O. Mazunin [3] and Konstantin E. Orishchenko [1,2,*]

1 Federal Research Center Institute of Cytology and Genetics, Siberian Branch of the Russian Academy of Sciences, 630090 Novosibirsk, Russia; zakirova@bionet.nsc.ru (E.G.Z.); muzyka@bionet.nsc.ru (V.V.M.)
2 Department of Genetic Technologies, Novosibirsk State University, 630090 Novosibirsk, Russia
3 Skolkovo Institute of Science and Technology, 143026 Skolkovo, Russia; I.Mazunin@skoltech.ru
* Correspondence: OrishchenkoKE@icg.sbras.ru

Abstract: The generally accepted theory of the genetic drift of mitochondrial alleles during mammalian ontogenesis is based on the presence of a selective bottleneck in the female germline. However, there is a variety of different theories on the pathways of genetic regulation of mitochondrial DNA (mtDNA) dynamics in oogenesis and adult somatic cells. The current review summarizes present knowledge on the natural mechanisms of mitochondrial genome elimination during mammalian development. We also discuss the variety of existing and developing methodologies for artificial manipulation of the mtDNA heteroplasmy level. Understanding of the basics of mtDNA dynamics will shed the light on the pathogenesis and potential therapies of human diseases associated with mitochondrial dysfunction.

Keywords: mitochondrial DNA segregation; heteroplasmy; selective elimination; mitophagy; mitochondrial engineered nucleases

check for updates

Citation: Zakirova, E.G.; Muzyka, V.V.; Mazunin, I.O.; Orishchenko, K.E. Natural and Artificial Mechanisms of Mitochondrial Genome Elimination. *Life* **2021**, *11*, 76. https://doi.org/10.3390/life11020076

Academic Editor: Francesco Bruni
Received: 18 December 2020
Accepted: 19 January 2021
Published: 20 January 2021

1. Introduction

Ten years after the discovery of the DNA double helix, Magrit and Sylvan Nass found that mitochondria harbor their own double-stranded DNA (mitochondrial DNA, mtDNA), whose structure differs from the nuclear DNA [1]. Currently, it is established that human circular mtDNA of 16,659 bp in length encodes 37 genes, which are essential for oxidative phosphorylation (OXPHOS) and stable energy production by the cell [2,3].

Highly compact mtDNA possesses many features distinct from the nuclear genome [4], some of which lead to the unique genetic drift of mtDNA. The transmission of the mitochondrial genome is strictly maternal. Uniparental inheritance was first shown in rats [5] and mice [6] and then confirmed in humans [7]. Although several studies debate the biparental mtDNA inheritance [8–11], such cases seem to be very rare, and their mechanisms remain largely unknown. Secondly, the mitochondrial genome exists in multiple copies. One somatic cell contains around 10^3–10^4 molecules of mtDNA [12,13]. Most frequently, a cell has only one type of mitochondrial genome. This condition—homoplasmy—usually occurs when all alleles of mtDNA are clonally expanded. However, considering the increased mutation rate of mtDNA comparing to the nuclear DNA, defective copies of mtDNA could coexist with wild-type alleles in a cell [14–17]. This phenomenon referred as heteroplasmy is specific to mitochondrial genome.

Different cells in a multicellular organism might have diverse mtDNA species. Moreover, the level of heteroplasmy may vary between the cells of the same tissue or organ, between the organs in one individual, and between individuals in a single family [18]. Such heterogeneity may occur at different developmental stages and proceed in an unpredictable direction. Defining the causes, dynamics, and driving forces of heteroplasmy will provide an insight on the inheritance and the progression of many mitochondrial diseases, the prevalence of which is approximately 1:4300 in the human population [19].

Here, we review several mechanisms of heteroplasmy level regulation and mtDNA segregation in somatic and germ cells, including the natural processes of mtDNA elimination. Additionally, we discuss how cells control the diversity of mtDNA species. Additionally, we review the state-of-the-art approaches (including those under active development) for the artificial modulation of mtDNA heteroplasmy. Finally, we provide specific examples of how these mechanisms can be used for uncovering the pathogenesis and rising novel therapies of mitochondrial dysfunction-associated diseases.

2. Mitochondrial Genome Heteroplasmy

It is believed that due to the proximity of the respiratory chain complexes, the absence of histones on mtDNA, and the lack of effective mechanisms of mtDNA repair, the mitochondrial genome is extremely vulnerable to the effects of reactive oxygen species (ROS). However, it is known that mitochondria possess several DNA repair mechanisms present in the nucleus. Most notably, there are base excision repair (BER) and microhomology-mediated end joining (MMEJ). The possibility of the presence of nucleotide excision repair (NER), mismatch repair (MMR), homologous recombination (HR,) and classical non-homologous end joining (NHEJ) is still under extensive debate [20]. In addition to that, even a rescue mechanism for the replication fork has been demonstrated in mitochondria [21]. However, the overall number of effectively proceeding DNA repair mechanisms in mitochondria is somewhat smaller than in the nucleus [20]. Moreover, although it is now known that mtDNA is not naked, the nucleoid structures formed to protect mtDNA are not exactly analogous to histones in the nuclear genome [22], and therefore, mitochondrial DNA is more available for mutagenic agents than the nuclear one. In turn, all this leads to a 10–100-fold increase of mutation rate compared to the nuclear DNA [14–17]. However, in a recent study, Kauppila et al. [23] demonstrated that the increase of ROS level in mitochondria does not lead to the growth in the number of de novo mutations in the mitochondrial genome. Moreover, point mutations in mtDNA most frequently form as transitions, suggesting the high error rate of mitochondrial DNA polymerase (POLG) during the replication. Based on that, one could infer that the main mutagenesis-inducing factor of mtDNA is an incorrect functioning of mitochondrial DNA replicative machinery. However, we cannot completely exclude the effect of ROS, intermolecular recombination, and specific mechanisms of mtDNA repair in mtDNA mutation formation.

MtDNA mutations could lead to the manifestation of a wide spectrum of neuromuscular and neurodegenerative diseases [24]. Regardless of the cause, the mutated copies of mtDNA can create the heteroplasmy within the cell. Depending on the mutation, cell, and tissue type, the critical heteroplasmy level for the manifestation of mitochondrial disease varies between 60% and 90% of defective copies [25]. The differential sensitivity of organs to physiological changes in mitochondria is a crucial parameter to determine the heteroplasmy threshold effect. The threshold in highly aerobic tissues such as muscle, heart, and central nervous system is usually lower than in less aerobically active tissues [26]. In addition to the tissue specificity of the heteroplasmy threshold level, it also depends on whether it is a point mutation (average critical level >80%) or a deletion (>60%) [27,28].

Until recently, it was considered that mtDNA heteroplasmy is an extremely rare event and that its threshold level is achieved due to a clonal expansion of mutant mtDNA molecules during the development [29]. However, ultra-deep resequencing revealed that the majority of human population, regardless of age, has low-level heteroplasmy in many tissues [30,31]. The universality and the prevalence of heteroplasmy cannot be explained only by spontaneous mutagenesis during early development. A portion of heterogenous mtDNA is inherited, which is proved by the presence of some of the low-frequency mutant alleles in both offspring and mothers [32]. Once emerged, somatic and inherited mutations might either reach a threshold level during life or be transmitted to the next generation preserving the low abundance, or even be completely eliminated after a certain time [33]. Therefore, the heteroplasmy is dynamic, and the behavior of mtDNA haplotypes segregation is rather unpredictable.

For the first time, a heteroplasmic shift within one generation was reported for the Holstein cows [34]. It is well established that one of the critical events affecting the genetic drift of mtDNA is the female germline bottleneck during the oogenesis. Resulting from it, the number of mtDNA copies decreases from 100,000 in the fertilized oocyte to 200 in a primordial germline cell [35–38]. Therefore, only a selected subpopulation of mtDNA would become prevalent in the formation of the mitochondrial bioenergetics function in the next generation.

Currently, a bottleneck theory is confirmed in multiple model organisms and systems, including early embryogenesis in humans [16,37]. Mathematical and statistical models demonstrated that the decrease in the amount of mtDNA in the germline cells is sufficient to shift the heteroplasmy level considerably between mother and offspring [39–42]. Importantly, a bottleneck theory predicts only a stochastic mtDNA segregation during embryogenesis, without considering a potentially significant change in heteroplasmy level during the entire span of individual development. The rate of the genetic drift of mtDNA might be affected by the vegetative segregation of mitotic cells, the preferential replication of certain mtDNA subpopulations, the subcompartmentalization of mtDNA into separate homoplasmic clusters, and a number of other less studied factors [18,36,43]. Notably, ROS have been shown to direct mtDNA segregation toward homoplasmy in human primary fibroblasts via a linear concatemer formation during mtDNA replication [44]. In addition, the intracellular mitochondrial quality control system might affect the consequent clonal expansion of certain mtDNA variants. Although it is currently impossible to determine which of the above mechanisms is the most relevant, it is obvious that together, they can lead to changes in the intracellular level of heteroplasmy during development, and, possibly, to the manifestation of mitochondrial disease.

3. Natural Mechanisms of mtDNA Elimination

3.1. A Selective Elimination of mtDNA in the Germline

Previously, it was considered that the animal mitochondrial genome undergoes neither positive nor negative selection during the evolution. However, after analyzing the inheritance of mtDNA in different model organisms, several groups established that mtDNA segregation proceeds under an intense negative selection during individual development [39,45,46]. After assessing the dependence of heteroplasmy level in the offspring on the mother's heteroplasmy level for several mutations, Stewart et al. demonstrated that pathological mutations are selectively eliminated during female oocyte maturation in mice [47]. Following this work, there have been multiple papers focused on the effects of certain mutations on mtDNA segregation in fruit fly [46,48], mouse [49], and human [16]. In case of the presence of two different mitochondrial mutations within mouse mitochondrial genome, a directional selection proceeded against a severe mutation causing an open reading frameshift in the *ND6* gene. At the same time, a milder missense mutation in *cytochrome c oxidase I* (*CoI*) gene was retained despite causing myopathy and cardiomyopathy in mice [49]. These results suggest the presence of effective complementation in the mitochondrial genome, leading to the elimination of mutations with a strong pathogenic effect while maintaining less pathogenic ones [46]. The mutation type and the possibility of its inheritance also correlate with the mutation localization in the mitochondrial genome. For instance, mtDNA variants with mutations in the D-loop are transmitted more frequently than the variants with mutations in ribosomal RNA genes, while mtDNA mutations of structural and transport RNA genes are transmitted to the next generation with similar frequencies [16]. As for protein-coding genes, mtDNA molecules with synonymous mutations are more frequently transferred to the next generation than nonsynonymous [47,50]. Overall, the above studies suggest that the negative selection in female germline cells helps to maintain and expand mtDNA variants that do not decrease the activity of the mitochondrial respiratory chain.

Potential Mechanisms of mtDNA Negative Selection

There is scarce knowledge of the negative mtDNA selection during female germline development. One study suggests that it proceeds before the germline bottleneck [51]. By eliminating oocytes with pathological mtDNA mutations, negative selection decreases the probability of a severe mitochondrial disease manifestation and promotes non-pathological mtDNA haplotypes to the next generation [52]. The mitochondrial quality control (MQC) process is considered to predominantly regulate mtDNA variant selection [16]. PTEN-induced kinase 1 (PINK1)/Parkin-mediated mitophagy is the most well-established way of MQC [53]. It is initiated in response to the increase of the proportion of defective components of the OXPHOS system, leading to selective elimination of the affected organelles with pathogenic mtDNA molecules [54]. In response to the mitochondrial dysfunction, PINK1 kinase stabilizes at the outer mitochondrial membrane (OMM) and consequently recruits ubiquitin E3 ligase Parkin [55,56] and several other autophagy receptors to the OMM [57]. Recruited enzymes ubiquitinate mitofusins (MFNs) and other proteins located at the OMM and are responsible for its fusion [58]. In turn, this inhibits the fusion of defective mitochondria to a single reticular network [59]. The suppression of mitochondrial fusion in heteroplasmic cells could lead to the physical isolation of mitochondria containing mutant mtDNA for preventing the functional complementation with wild-type molecules of mtDNA. In human cybrid cells derived from multiple different tissues and containing the *CoI* gene mutation, it has been shown that *Parkin* overexpression could induce mitophagy [53]. The overexpression of *Pink1* and *Parkin* also leads to a selective elimination of mutant mtDNA in muscles of adult *D. Melanogaster* [60]. In contrast, *Parkin* knockout in *C. elegans* cells containing 60% of mtDNA with large-scale deletion [61] leads to a heteroplasmy shift toward mutant mtDNA [62].

At first glance, negative selection via mitophagy matches well with a bottleneck hypothesis explaining the mtDNA amount reduction during the oogenesis. However, several researchers suggested that a bottleneck might form not only due to a reduction in the number of mtDNA molecules but also due to a focal replication of mtDNA subpopulation during oogenesis [51,63] and via the packaging of several mtDNA molecules into segregation units [36,64]. Both pathways visibly limit the effective size of the mtDNA population, leading to intensified segregation of the selected mtDNA pool and consequent heteroplasmic shift in developing oocytes [65].

To test the hypothesis of the elimination of defective mtDNA copies via selective wild-type mtDNA replication, Hill et al. analyzed mtDNA selection in early embryonic oogenesis in multiple generations of *D. melanogaster* [48]. Fruit flies with homoplasmic temperature-sensitive mutation *mt:CoIT300I* in *the cytochrome c oxidase* (*CoI*) gene have an impaired eclosion phase and early lethality (within 5 days) during the incubation at increased temperature (29 °C). Heteroplasmic flies are not killed at high temperature. Nevertheless, among the offspring, flies with wild-type mtDNA were predominant. The identified dependence of the fly development on the incubation temperature enabled determining how the degradation of the high temperature-sensitive mtDNA variants proceed at the molecular level. It was shown that mtDNA selection starts as soon as the replication in the late germarium. *mt:CoIT300I* mutation causes a reduction of cytochrome C oxidase activity at high temperature, consequently blocking mutant mtDNA replication in the germ cells. However, how is oxidative chain impairment associated with mtDNA replicative activity? This phenomenon has been explained by Stewart et al. [18]. It is established that all mitochondrial replicative proteins are encoded by nuclear genes and are imported from the cytoplasm. The transfer of these proteins depends on the mitochondrial membrane potential reflecting the functional state of mitochondria [66]. As a result, intermembrane protein transport is suppressed in response to the decrease of the membrane potential and of the electron transport chain (ETC) activity. In turn, this leads to the blockage of synthetic processes in defective organelles. Therefore, the formation of the germline cells of the next generation is under the nuclear genome control, and thus positive or negative mtDNA selection happens depending on the synergy between the nuclear and the mitochondrial genome. The driving forces of the negative selection are reviewed below.

According to the segregation units model, the bottleneck in the female germ cells proceeds without a reduction in the mtDNA content. It potentially takes place due to the packaging of mtDNA molecules in a small number of homoplasmic clusters, which segregate into daughter cells during mitosis [36,64]. Jenuth et al. have identified around 200 segregation units in mice [37]. This number resembles an absolute quantity of mtDNA molecules in forming mouse oocytes [40]. Considering the fact that on average, one nucleoid carries approximately 1.4 mtDNA molecules [67], one could assume that the nucleoid is the unit of a homoplasmic clustering and determines the bottleneck width. This parameter defines which mtDNA molecules would shape the functional characteristics of future oocytes.

The negative selection mechanisms presented above contribute to a decrease in the number of mtDNA molecules before or during the bottleneck. Moreover, these processes take place at the mitochondrial level. Meanwhile, mtDNA degeneration can occur at the cellular level. For example, primary oocytes with a high load of mtDNA mutations can also be selectively degraded (follicular atresia) [68].

Currently, there is no direct empirical evidence as to which selection mechanism is key in the germline bottleneck. Moreover, the functioning of these processes in somatic cells (for example, mitophagy) is not excluded. In any case, they all modulate the natural drift of mtDNA.

3.2. Dynamics of Heteroplasmy in Somatic Cells

Unlike in the germline, mtDNA mutations in somatic cells frequently demonstrate neutral patterns of segregation [69]. Recently, investigators have described several cases of both positive and negative somatic selection of mutant mtDNA subpopulations [70–72].

For instance, a gradual reduction in the number of mtDNA molecules with mt3243A>G mutation in a gene encoding *tRNALeu(UUR)* in the patient's hematopoietic stem cells has been reported [73]. For example, the reason for such an exponential decrease of mutant mtDNA content could be spontaneous mtDNA vegetative segregation between daughter cells or a directed selection against the pathogenic mutation [70]. One of the pathways of directed selection might potentially be through the fragmentation of the mitochondrial network and the elimination of impaired mitochondria by mitophagy. It becomes possible if the mutation severely affects cell function and viability. A similar reduction of the copy number of mtDNA with mt3243A>G mutation has been detected in epithelial [71] and buccal [74] cells during aging. The strong correlation between the age and the heteroplasmy level matches well with the presence of a negative selection in mitotic cells. The mt3243A>G mutation mentioned above for the dividing cells segregates differently in neuronal model in vitro. Surprisingly, compared to hematopoietic stem and epithelial cells, iPSC populations differentiated into neurons in vitro demonstrated stable heteroplasmy levels upon both serial passaging and differentiation [75]. Moreover, iPSC-derived neurons retain not only the heteroplasmy level but also the parental cell pathological phenotype [76]. Overall, mtDNA segregation process largely depends on the examined cell type revealing the oxidative chain performance as the major factor for such heterogeneity [77].

MtDNA segregation is a continuing process happening not only in developing gonads and in dividing cells but also in many post-mitotic tissues throughout the lifespan. Inherited at low frequency or acquired during the life, pathogenic mtDNA variants could reach a biochemical threshold level at any age due to the relaxed replication of mtDNA [78]. Works on *C. briggsae* [79] or human cybrid cells [72], containing mitochondria with partially deleted mtDNA, demonstrated the preferential accumulation of defective mtDNA molecules. The segregation toward mutant variants could be associated with reduced mtDNA size following a deletion because the repopulation of shorter molecules is much faster than of full-size wild-type mtDNA. Despite potential negative effects of mutations on mitochondrial function, a similar mechanism may be also inherent for the germline cells.

An unexpected positive effect of mtDNA negative selection failure was found in the oncogenesis of several cancers. Heteroplasmic mtDNA mutations affecting *ND1-6* and *ND4L* genes stimulate tumor growth by activating anti-apoptotic pathways [80–83]. At the same time, reaching a certain threshold heteroplasmy level or mutant mtDNA homoplasmy promotes the inhibition of cancer cell mitochondrial function and the tumor suppression [84–86]. In turn, this leads to a better recovery rate for patients with breast cancer or acute myeloid leukemia [87,88]. The negative correlation between mtDNA mutation burden and the tumor progression suggests the existence of a compensatory mechanism for cancer cell elimination and organism protection against metastasis and disease relapse. However, the practical application of directed mtDNA mutagenesis is limited due to the lack of the information on the formation and development of different tumor types.

All of the mechanisms that affect mtDNA segregation during development are summarized in Figure 1.

However, currently, we lack unambiguous answers to the following questions: how does directed mtDNA selection go; how many bottlenecks do cells undergo during the oogenesis; what is the dynamics of segregation for different mtDNA mutations? It is obvious that identified factors, which affect mtDNA dynamics, are not mutually exclusive and represent a part of a unified process, which requires further extensive investigation.

3.3. *Paternal mtDNA Degradation*

As mentioned earlier, mtDNA is inherited maternally in the majority of animal species. The meaning of this inheritance remain poorly explored. Previously, it was hypothesized that the difference between male and female gamete sizes could block the compatibility of paternal and maternal mitochondrial genomes in a zygote [89]. However, a unicellular algae *C. reinhardtii* has similar in size gametes of two sexes, but still, only the maternal mitochondrial genome is inherited [90]. Considering the "unified model of organelle inheritance", the growing evidence is accumulating for uniparental transmission of mtDNA to prevent the spread of selfish (fast replicating) alleles in the population [91,92]. Such elements can be mutant sperm genomes that replicate faster than the wild-type genome but do not adapt to the egg nucleus [93]. In addition to that, a possible reason for the maternal inheritance of mtDNA is a tendency of cells to maintain more metabolically effective variants in the homoplasmic state [77,94]. However, uniparental inheritance is evolutionarily unstable, because mitochondria are subject to Muller's ratchet [93].

To support a strictly maternal inheritance and to decrease the competition between parental mtDNA species, several mechanisms of paternal mtDNA degradation in both spermatogenesis and fertilization evolved during the evolution.

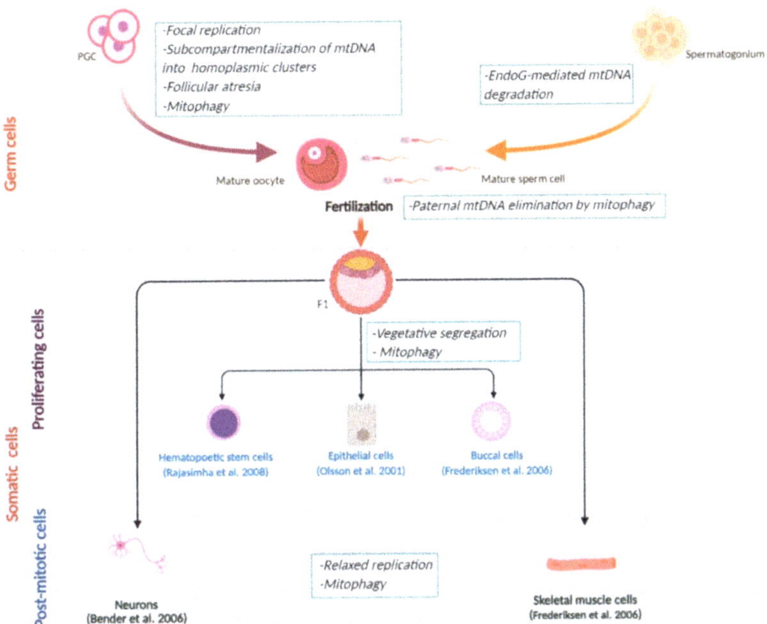

Figure 1. The genetic dynamics of mitochondrial DNA (mtDNA) is regulated via diverse mechanisms during mammalian development. In the maturation process, female oocytes undergo negative selection supposedly happening just before the developmental bottleneck. The oocytes with high mtDNA mutation load could be eliminated via follicular atresia. At the same time, during spermatogenesis, endonuclease G (EndoG)-mediated mtDNA degradation occurs. During fertilization, the mitophagy of male mitochondria takes place to prevent paternal mtDNA transmission to the next generation (F1). Since the early stages of embryogenesis, mtDNA is randomly distributed between daughter cells during mitosis due to vegetative segregation. In non-dividing cells, one of mtDNA variants might have a replicative advantage due to relaxed replication. These two mechanisms might proceed in parallel in different somatic tissues within one organism to regulate the genetic dynamics of mtDNA. Additionally, mtDNA heteroplasmy can shift toward healthy or mutant variant in development because of the mitophagy process (figure created in BioRender.com).

3.3.1. MtDNA Elimination during Spermatogenesis

DeLuca and O'Farrell analyzed all stages of male germ cells development of *D. Melanogaster* to track mtDNA fate during spermatogenesis [95]. Spermatids of Drosophila undergo a physical transformation during maturation. At the sperm individualization stage, spermatids get rid of almost all cytoplasm with the majority of organelles, including mitochondria. Mature spermatozoon still contains the remaining mitochondria but without mtDNA. A two-step mechanism is responsible for mitochondrial genome elimination. It includes primary nucleoid degradation with Endo G [95,96] combined with the elimination of remaining aggregates via their transport into caudal "waste bags", which are then extruded out of the spermatozoon [97]. In addition to Endo G, the molecular machinery to remove mtDNA in fruit fly spermatozoon contains a α-subunit of mitochondrial polymerase γ (Pol γ). Supposedly, the physical proximity of proteins responsible for the replication and the degradation of mtDNA underlies a balanced and stable nuclear control of mitochondrial genome copy number [98]. However, this hypothesis still needs further experimental confirmation.

Do the mammalian species have similar mechanisms of mtDNA degradation before the fertilization compared to Drosophila? Mouse sperm exhibits a three-fold decrease of mtDNA quantity during its maturation, leading to the impaired fertility in mice [99]. A reverse effect associated with the increase of mtDNA level in sperm is found in human males with oligozoospermia and asthenozoospermia [100–102]. Several studies suggest that the principal mechanism of mammalian mtDNA degradation is the poly-ubiquitination of defective mitochondria and its consequent proteolysis in the epididymis duct [103,104]. In mammals, including humans, mitochondria ubiquitination in sperm was confirmed by the co-localization of the ubiquitin labeling with mitochondria at all stages of spermatogenesis from spermatogenic cell to mature spermatozoon in fertilization [105,106].

3.3.2. Paternal mtDNA Elimination during Fertilization

A striking example of purifying selection is the elimination of mtDNA from the spermatozoon introduced to the oocyte during fertilization. It was initially considered that sperm mtDNA is not detected in the offspring due to a substantial dilution in the large amount of the oocyte mtDNA [8,97,107]. However, such model of a "passive dilution" has not been confirmed extensively due to the lack of a unified opinion on the number of mtDNA molecules in mammalian spermatozoon [108–110]. Studies in this field allowed formulating a hypothesis on the mechanism of a directed degradation ("active elimination") of paternal mtDNA in a zygote [111,112]. A convincing evidence from a time-lapse fluorescent microscopy from different stages of embryogenesis of *C. elegans* has demonstrated that sperm mitochondria easily get into the oocyte; however, two hours after fertilization, they are all eliminated via autophagy [113,114]. Paternal mitochondria become depolarized in the oocyte. Its inner mitochondrial membrane (IMM) permeabilizes, which allows Endo G transfer from the intermembrane space of male mitochondria into the mitochondrial matrix [97,115]. This leads to the cleavage of spermatozoon's mitochondrial DNA [116]. MtDNA degradation promotes the formation of autophagy initiation and proteasome degradation signals on the surface of mitochondria [104,117,118]. When the assembly of the autophagolysosome is inhibited, paternal mtDNA molecules are retained until the late embryonic stages and contribute to the formation of bioenergetics function of mitochondria in the next generation. Additionally, the co-localization of autophagy markers (LC3, GABARAP, and p62) with paternal mitochondria has been detected in zygotes of mice [113,119]. Most likely, mitophagy proceeds via the Parkin/Mitochondrial ubiquitin ligase 1 (MulI)-dependent pathway [120]. Despite the differences in the pathways of mitophagy activation caused by fertilization, it is safe to assume that the elimination of mtDNA via autophagy is a conserved mechanism among the absolute majority of animal species [121]. Sato et al. and Kaneda et al. review the information on the paternal mtDNA degradation in different animals in detail [121,122].

Despite the growing evidence of active elimination of paternal mtDNA in mammals, any accurate information about its fate in fertilization and embryo formation is still lacking [112]. However, there are already several well-documented cases of paternal mtDNA inheritance in human and the scientific community aims to explore such events. [10,123]. It is highly possible that a false impression of paternal mtDNA inheritance could be formed due to the presence of NUMT ("Nuclear copies of mitochondrial genes") pseudogenes, which are integrated into nuclear genome due to intergenomic recombination. However, there are no direct experiments proving this as for now; thus, one cannot exclude rare biparental mtDNA transmission events.

3.4. The Driving Forces of mtDNA Segregation

More than 20 years ago, Jenuth et al. have identified the competition between mtDNA haplogroups during the transfer from mother to offspring [37]. However, the molecular basis of this phenomenon has stayed unexplored for a long time. To reveal the factors influencing mtDNA segregation, special conplastic mice were produced [124,125]. Heteroplasmy preferentially shifted toward the C57 variant in the germline of conplastic mice possessing the C57 mouse line nuclear genome and different (C57 and NZB) mtDNA haplotypes [126]. Such a preference could be explained by the functional complementation of the nuclear and the mitochondrial genomes during OXPHOS system assembly. In mitochondria, around 70 proteins for the OXPHOS system encoded by the nuclear DNA and 13 structural proteins encoded by the mtDNA have to physically complement each other to form functional multiprotein respiratory complexes. The incompatibility between the products of the nuclear and the mitochondrial genomes promotes the reduction of mitochondrial activity and cellular adaptive reaction at the early embryonic stages [125]. That makes the heteroplasmic state unstable and promotes a selective segregation of cognate (to the respective nuclear DNA) mtDNA haplotype in the germline [94].

To understand the mito-nuclear heteroplasmy regulation, it is crucial to know the components of the retrograde and the anterograde signaling maintaining certain mtDNA haplotypes. After analyzing mtDNA segregation during the lifespan in 19 different tissues of heteroplasmic C57/NZB mice with C57 nuclear background, it was determined that there is an association between cellular metabolic program, oxidative phosphorylation, and mitochondrial genome segregation [77]. It is known that the ETC of any cell type is tuned to use of certain molecules as energetic substrates and, therefore, for a certain metabolic pathway (for instance, glucose or lipid metabolism). In the case of mtDNA homoplasmy, the function of the respiratory chain is not altered compared to the default metabolic program. In the case of heteroplasmy, there is a dose-dependent imbalance between cellular metabolism and the ETC activity, which in turn induces the respiratory chain rearrangement and intensive ROS production [127]. In response to the increase of ROS level, the proteins encoded by the nuclear genome SCAF1 (supercomplex assembly factor 1), NNT (mitochondrial NAD(P) transhydrogenase), and OMA1 (metalloendopeptidase, which is responsible for the control of the mitochondrial fusion) are imported into mitochondria to fine-tune the oxidative phosphorylation and to reduce the ROS level [128]. At the same time, the ETC dysfunction activates factors that coordinate the replication of mtDNA, mitochondria biogenesis, and mitophagy [129]. In the end, a suboptimal mitochondrial function due to heteroplasmy leads to the degradation of mitochondria, which have mtDNA variants generating excessive ROS.

Systematic studies of the driving forces of mtDNA heteroplasmy have considerably broadened the understanding of the mechanisms of mtDNA segregation and the interaction between the nuclear and the mitochondrial genomes. During the parallel evolution of these genomes, the pathways to coordinate a stable mitochondrial function were established. Along those lines, mtDNA segregation and elimination potentially maintain the mitochondrial homoplasmic state. Consequently, homoplasmy is more preferable than heteroplasmy, because it provides more stable energy input for the cell and reduces ROS production. The alteration of any component of this mechanism would potentially lead the cell, the tissue, or the organism to death.

4. Artificial Mechanisms of Mitochondrial Genome Elimination

Mitochondrial pathologies progressing with age do not currently have any common term and are classified into different subgroups of human diseases according to the International classification of human diseases (ICD-11) of the World Health Organization. The lack of a shared etiology and pathogenesis of mitochondrial diseases causes complications in patient's diagnostics and therapy. Modern clinical diagnostics of mitochondrial pathologies with traditional histochemical, immunohistochemical, and biochemical techniques in combination with high-throughput screening of mitochondrial and nuclear DNA considerably improves early diagnostics [112]. However, in many cases, after determining a type of mitochondrial pathology, a clinician is not able to find an adequate curing approach due to the lack of such disease-suppressing therapies in clinical practice. Sometimes, patients undergo neurotrophic and metabolic therapies aimed at temporal symptomatic suppression [112–115]. At the same time, modern reproductive and gene therapy techniques not only enable reducing the heteroplasmy level in a patient's somatic cells but also prevent a transfer of pathogenic mtDNA copies to the next generations.

4.1. Reproductive Technologies for the Prevention of Mutant mtDNA Transfer

Women with high heteroplasmy levels in oocytes are often diagnosed with infertility, because the reduced activity of the OXPHOS complexes leads to the suppression of embryonic development at early stages [63,130]. In vitro fertilization (IVF) with preimplantation genetic diagnostics (PGD) of embryos of mtDNA mutation load has become an effective instrument of clinical practice for mitochondrial disease prevention. During this procedure, one or two cells from embryos are taken for the analysis of heteroplasmy level to assess the risk of mitochondrial disease manifestation. For a birth of a healthy child, it is crucial to have a low (<5%) heteroplasmy level in the embryo [131]. It is assumed that this level is similar in all the cells and stable during the course of embryonic development. Considering the fact that mtDNA segregation mechanisms during embryogenesis are not fully studied, one cannot exclude the return to the initial heteroplasmy level in the whole embryo or in its specific tissues. In addition to that, some women have a significant number of mutant mtDNA molecules in the majority of oocytes; therefore, PGD does not assist well in such cases [132]. Instead, to prevent the transfer of mutant mtDNA to the offspring, mitochondrial replacement therapy (MRT) or mitochondrial donation techniques were established [133–135]. This technology was first tested on animals [136–138]. Thereafter, it has been used in humans to prevent the transmission of mitochondrial diseases. The procedure might include the transfer of a maternal spindle (MST) [138,139], of a pronucleus (PNT) [140], or of a first polar body [141–144] into enucleated oocytes or zygotes of a healthy donor. As a result, a child born with MRT would have nuclear DNA from both parents but mitochondria from a donor woman. However, despite the reports about successful births of children from "three parents" [140,145], the extensive application of mitochondrial donation still remains impractical or even illegal in many countries due to unresolved technical and ethical issues [146–148].

The reversal of embryos to a pathogenic mitochondrial type represents one of the encountered technical problems. It is considered to occur with low possibility because during the MRT, less than 3% of maternal mtDNA happen to be transferred [132,138]. However, it was shown that around 15% of embryonic stem cells produced from the embryos after MRT show a complete return to the initial, mutant mtDNA variant [149–152]. Such a phenomenon might be based on a reciprocal interaction between the nuclear and the mitochondrial genome [153]. Another crucial disadvantage of mitochondrial donation is that it not only reduces the level of heteroplasmy in the next generation but also does not block the transfer of mutant mtDNA completely. This issue could be resolved using the combination of mitochondrial donation with gene therapy tools [154,155].

4.2. Gene Therapeutic Approaches for the Prevention of Pathogenic mtDNA Transmission

In addition to the previously discussed mitochondrial genome characteristics, another critical trait is the retention of mtDNA copy number in a cell. MtDNA haplotypes could degrade due to multiple reasons leading to the overall reduction of mtDNA pool in mitochondria. Surpassing a critical threshold of mtDNA content could cause the death of not only mitochondria but the whole cell. To avoid cell death, mtDNA haplotypes unaffected by degradation factors actively replicate to restore the initial mtDNA copy number [156]. If one could direct natural elimination mechanisms against pathogenic mtDNA molecules, it is possible to repopulate the mitochondrial genome with wild-type molecules to restore the impaired mitochondrial function. Many actively developed gene therapeutic approaches for shifting heteroplasmy level are based on the principle mentioned above. These therapeutic instruments are being used in oocytes to prevent pathogenic mtDNA transmission to the next generation, as well as in somatic cells to treat mitochondrial diseases [154].

4.2.1. Anti-Replicative Approaches

One of several possible ways to shift heteroplasmy toward the wild-type variant is to suppress specifically the replication of mutant mtDNA. For the first time, this methodology was demonstrated using peptide nucleic acids (PNAs) [157]. PNAs are the artificial analogs of oligonucleotides, in which a pseudo-peptide scaffold replaces the sugar phosphate one. The latter one contains N-(2-aminoethyl) glycine as a monomer. The PNA structure makes it more resistant to nuclease and protease cleavage [158] and enables introducing modifications to directly import PNAs into the mitochondrial matrix [159,160]. Additionally, PNAs bind to the complementary single-stranded DNA with higher affinity than analogous complementary DNA [161]. PNA:DNA duplexes with single nucleotide mismatches are more stable than similar DNA:DNA duplexes [162]. Using in vitro replication in physiological conditions, Taylor et al. demonstrated that PNAs might selectively inhibit the replication of the respective single-stranded mtDNAs with deletions or single-nucleotide mutations [157]. Nevertheless, no striking results have been demonstrated on either cell culture or isolated mitochondria. Other than the rather effective import of PNAs into the mitochondrial matrix, the researchers did not observe any inhibition of mutant mtDNA replication [159,160]. That could be explained by the ineffectiveness or complete absence of PNA binding to mtDNA or by the displacement of PNA molecules away from the mtDNA during its replication or transcription. In any case, currently, PNAs could not be used for heteroplasmy shifting, and further experiments are necessary to search for potential PNA modifications to improve the efficiency of target DNA binding in living cells.

Another group has been developing a similar approach to shift the heteroplasmy level. For the selective inhibition of mutant mtDNA replication, they used short RNA molecules with a region complementary to a target mtDNA sequence. Additionally, they inserted F- and D-stem loops from yeast transport RNA (tRNALys(CUU)) or the α and γ domains of 5S human ribosomal RNA into the structure of their recombinant anti-replicative RNAs. As demonstrated earlier, these fragments could promote effective RNA import inside mitochondria [163–165]. The supposed mechanism of action of anti-replicative RNA molecules is associated with the incapability of helicase Twinkle in mitochondrial replisome to displace RNA in short RNA–DNA duplexes [166]. Using cultured cybrid cells, it was demonstrated that the anti-replicative RNA approach causes heteroplasmy level reduction for large deletion associated with Kearns–Sayer syndrome and for pathogenic point mutation A13514G in the *ND5* gene. Notably, only some of the tested RNA variants induced the suppression of mutant mtDNA replication [167–169]. Similar to PNA methodology, it is questionable whether a single-stranded mtDNA is available for the anti-replicative RNA during the replication and the effectiveness of their binding. Additionally, although many studies demonstrate the transport of different RNAs inside mitochondria (detailed review by [170]), currently, there are no common opinion on the molecular mechanisms of nucleic acid trafficking inside mitochondria and its possible function there [171].

4.2.2. Anti-Genomic Approaches

An alternative method for the reduction of heteroplasmy level toward wild-type mtDNA is a specific elimination of pathogenic mtDNA. Considering the nature of mitochondrial genome multicopy, mechanisms of mtDNA repopulation, and the apparent however debated inefficiency of active mechanisms of double-stranded break (DSB) repair of mtDNA [20,172], one could speculate about the high effectiveness of a directed DSB introduction into mutant mtDNA. Several studies illustrate the presence of some DSB repair mechanisms taking place in the mitochondrial genome—namely, homologous recombination (HR) and non-homologous end joining (NHEJ). One of the first indicators of HR in mitochondria was the finding of circular dimers and catenanes in mammalian mtDNA [173–176]. Moreover, human heart cells possess a unique configuration of complex mtDNA catenated networks [177]. Another evidence favoring the possibility of the HR activity in mammalian mitochondrial genome is the identification of maternal–paternal mtDNA hybrids in muscle cells of a unique patient [178]. This is an unprecedented case for humans, although the recombination of mtDNA variants from both parents has been detected in other vertebrate and non-vertebrate animal species [179–181]. Thirdly, studies on the artificial introduction of DSBs into the mouse mitochondrial genome have shown the occurrence of interspecific DNA exchange between different mtDNA haplotypes [182,183]. All this, plus the identification of several analogs of well-identified in yeast mtDNA HR protein participants in mammalian mitochondria suggest that HR is possible in mammalian mtDNA even in physiological conditions [184]. Another potential mechanism of DSB repair in the mitochondrial genome is one particular type of NHEJ—microhomology-mediated end joining (MMEJ), as opposed to classical NHEJ [185]. It was indirectly confirmed when Meiotic recombination 11 (Mre11) and Poly(ADP-Ribose) Polymerase 1 (PARP1)proteins were found to localize in mitochondria and interact with mtDNA [186,187], and then, it was directly demonstrated to be the principal DSB repair process in mitochondria [185].

Nevertheless, after a DSB is introduced, most of the linear mtDNA molecules rapidly degrade due to the activity of mitochondrial replicative proteins (mtDNA polymerase γ—POLG, DNA helicase Twinkle, exonuclease MGME1) [188], leading to heteroplasmy shifts. For DSB presentation into DNA structure, different nucleases could be used. The majority of proteins essential for mitochondria functioning are encoded in the nuclear DNA. Therefore, TIM (trafficking through inner membrane)/TOM (trafficking through outer membrane) machinery has been formed for the protein transport through the inner and the outer mitochondrial membranes, respectively. The presence of a Mitochondrial Targeting Sequence (MTS) at the protein N-terminus promotes its transport inside mitochondria [189]. Therefore, adding an MTS to the enzyme, which specifically recognizes mutant mtDNA and introduces a DSB there, reduces the heteroplasmy level. Almost all instruments for the directed elimination of pathogenic mtDNA are based on the above principle, and these anti-genomic approaches are currently considered as the most promising future therapeutic methodologies for diseases caused by mtDNA mutations.

For the first time, Srivastava and Moraes applied this methodology to manipulate the heteroplasmy level [190]. They combined restriction endonuclease PstI with the COX8A mitochondrial localization signal (MLS). By using cybrid cells with mouse and rat mtDNA (having different numbers of PstI cleavage sites), it was demonstrated that the application of a mitochondrial restriction endonuclease causes a considerable heteroplasmic shift. Mt8993T>G mutation in the *ATP6* gene leads to the formation of an SmaI endonuclease recognition site. This fact was employed for the specific degradation of a mutant mtDNA using the SmaI enzyme containing the MLS [191]. The transient expression of modified SmaI in cybrid cells reduced the heteroplasmy level for mt8993T>G mutation and restored ATP production and mitochondrial membrane potential. Therefore, the above-mentioned works confirmed the effectiveness of mitochondrial restriction endonucleases for heteroplasmy level reduction for both scientific and potential therapeutic purposes.

However, the number of mutations forming unique recognition sites for restriction endonucleases limits the application of this approach. Despite the identification of novel mtDNA polymorphisms, only a small fraction of them could become a target for mitochondrial restriction endonucleases. In addition to that, the nuclear genome contains a large number of restriction sites; therefore, one should always consider the off-target effects there, which potentially lead to unnecessary mutations.

Mitochondrially targeted designer nucleases enabled at least partly overcoming this issue. Initially developed for the nuclear genome editing, the zinc finger nucleases (ZFNs) [192] and transcription activator-like effector nucleases (TALENs) [193] have been adapted for the elimination of pathogenic mtDNA variants. Such nucleases consist of three parts (the MLS, the DNA-binding domain, the nuclease catalytic domain) fused in one protein [194,195]. As for mitochondrial restriction endonucleases, the MLS comprised a peptide of a protein naturally imported into mitochondria [196]. A nuclease domain was taken from FokI nuclease, which produces DNA double-stranded breaks (DSBs) [197]. The binding specificity of engineered nucleases with pathogenic mtDNA molecules was promoted by a DNA-binding domain from a bacterial transcription activator-like effector (TALE) protein in case of mitochondrially targeted TALE nucleases (mitoTALENs) [194] or by zinc finger DNA-binding domain for mitochondrially targeted zinc finger-nucleases (mtZFNs) [195].

Transcription factors often contain zinc finger type DNA-binding domains [198], which, in turn, include the nuclear localization signal (NLS) [199]. For the effective mtZFN import to mitochondria and the reduction of off-target effects in the nuclear DNA, NLS should be eliminated [200]. Importantly, for mtZFNs and mtTALENs, one could construct a binding domain for virtually any DNA sequence [201], considerably expanding the spectrum of targetable mtDNA pathogenic mutations. MtZFNs and mtTALENs were successfully utilized to shift the heteroplasmy level for both point mutations and deletions of mtDNA to restore the affected mitochondrial functions [155,194,195,202–205]. Disadvantages of mtZFNs and mtTALENs include complex and laborious process of DNA-binding domain construction, which are associated with the repetitive nature of these domains for mtTALENs and the complicated screening process for specific mtZFNs.

All of the above-described nucleases are artificially made proteins; therefore, the recognition and binding of a target DNA sequence is based on DNA–protein interactions. Clustered Regularly Interspaced Short Palindromic Repeats(CRISPR) and CRISPR-associated protein 9 (Cas9)—CRISPR/Cas9 is a novel and popular technology for nuclear genome editing that employs a single-guide RNA (sgRNA)-mediated approach to target DNA sequences [206]. Target DNA recognition proceeds based on DNA–RNA binding, which makes it more specific compared to other methods. Additionally, CRISPR/Cas9 is more effective in genome editing and more flexible due to a simple procedure of the sgRNA customization [207]. Therefore, many groups including ours are working on the optimization of this system for mutant mtDNA elimination and mitochondrial genome editing. However, still it is far from an unambiguous confirmation for CRISPR/Cas9 effectiveness in mitochondria. Moreover, some researchers question even the theoretical possibility of its effective use for pathogenic mtDNA elimination [171]. Analogous to anti-replicative RNAs, such skepticism arises due to the lack of information on the molecular mechanisms of RNA transport inside mitochondria, speculating that the whole concept of mitochondrial CRISPR/Cas9 is controversial.

4.2.3. MtDNA Base-Editing

MtDNA editing is an exciting concept because it is essential to correct pathogenic mtDNA mutations to treat mitochondrial diseases and also to introduce mutations into mtDNA to acquire disease models and study fundamental processes of mtDNA function. The existing nuclear genome editing methodology is based on a modified CRISPR/Cas9 system [208,209]; therefore, such direction for mtDNA editing seems controversial. However, very recently, a protein-only system (DddA-derived cytosine base editor, DdCBE) has been introduced. It is based on bacterial toxin DddA with deaminase activity [210]. This enzyme enables deaminating cytosine (C) preceded by a thymine (T) residue in a double-stranded DNA molecule. Deamination promotes C conversion to uracil (U), which pairs with adenine (A) residue. As a result, during mtDNA replication, A would be inserted complementary to U, and CG would be converted to a UA pair. Consequently, through DNA repair mechanisms, U is substituted by T residue, resulting in CG transition into TA. To reduce the level of unspecific deamination and cellular toxicity, the DddA domain was split in two halves, which were both fused to the programmable DNA-binding TALE domains, analogous to mtTALENs. Upon the binding of these proteins with two adjacent mtDNA motifs, two halves of DddA could associate to form a functional cytidine deaminase. Using human cell cultures, Mok et al. demonstrated that DdCBE effectively and specifically converts CG pairs into TA pairs. It does not show considerable off-target effects on both the mitochondrial and the nuclear DNA [210]. It is a promising technology for mitochondrial disease treatment because the majority of pathogenic mtDNA mutations are T > C transitions. Additionally, such an approach might be utilized to study the mechanisms of mitochondrial diseases associated with mtDNA mutations and aging.

5. Conclusions

Mitochondrial diseases develop due to inherited or spontaneous mtDNA mutations. During the evolution, negative selection mechanisms were established to discard defective copies of mtDNA. These mechanisms reduce but not eliminate the risk of the manifestation of mitochondrial pathologies. Considering that, it is crucial to develop and modify mtDNA editing tools. At the same time, the understanding of natural mechanisms of mtDNA heteroplasmy level would facilitate the invention of new methodologies of mitochondrial disease prevention and treatment.

Author Contributions: Conceptualization, E.G.Z. and K.E.O.; writing—original draft preparation, E.G.Z., K.E.O.; writing—review and editing, V.V.M., I.O.M.; visualization, E.G.Z.; supervision, K.E.O.; funding acquisition, K.E.O. All authors have read and agreed to the published version of the manuscript.

Funding: This work was supported by the Ministry of Education and Science of Russian Federation, grant number 2019-0546 (FSUS-2020-0040) and by the RFBR grant number 19-34-90087.

Conflicts of Interest: The authors declare no conflict of interest.

References

1. Nass, M.M.; Nass, S. Intramitochondrial fibers with DNA characteristics. I. Fixation and staining reactions. *J. Cell Biol.* **1963**, *19*, 593–611. [CrossRef]
2. Anderson, S.; Bankier, A.T.; Barrell, B.G.; de Bruijn, M.H.L.L.; Coulson, A.R.; Drouin, J.; Eperon, I.C.; Nierlich, D.P.; Roe, B.A.; Sanger, F.; et al. Sequence and organization of the human mitochondrial genome. *Nature* **1981**, *290*, 457–465. [CrossRef]
3. Schatz, G.; Haslbrunner, E.; Tuppy, H. Deoxyribonucleic acid associated with yeast mitochondria. *Biochem. Biophys. Res. Commun.* **1964**, *15*, 127–132. [CrossRef]
4. Nicholls, T.J.; Minczuk, M. In D-loop: 40 years of mitochondrial 7S DNA. *Exp. Gerontol.* **2014**, *56*, 175–181. [CrossRef]
5. Francisco, J.F.; Brown, G.G.; Simpson, M.V. Further studies on types A and B rat mtDNAs: Cleavage maps and evidence for cytoplasmic inheritance in mammals. *Plasmid* **1979**, *2*, 426–436. [CrossRef]
6. Tatarenkov, A.; Avise, J.C. Rapid concerted evolution in animal mitochondrial DNA. *Proc. Biol. Sci.* **2007**, *274*, 1795–1798. [CrossRef]
7. Giles, R.E.; Blanc, H.; Cann, H.M.; Wallace, D.C. Maternal inheritance of human mitochondrial DNA (genetic polymorphism/restriction endonuclease cleavage map/blood platelets). *Genetics* **1980**, *77*, 6715–6719.
8. Gyllensten, U.; Wharton, D.; Josefsson, A.; Wilson, A.C. Paternal inheritance of mitochondrial DNA in mice. *Nature* **1991**, *352*, 255–257. [CrossRef] [PubMed]
9. Kidgotko, O.V.; Kustova, M.Y.; Sokolova, V.A.; Bass, M.G.; Vasilyev, V.B. Transmission of human mitochondrial DNA along the paternal lineage in transmitochondrial mice. *Mitochondrion* **2013**, *13*, 330–336. [CrossRef] [PubMed]
10. Luo, S.; Valencia, C.A.; Zhang, J.; Lee, N.C.; Slone, J.; Gui, B.; Wang, X.; Li, Z.; Dell, S.; Brown, J.; et al. Biparental inheritance of mitochondrial DNA in humans. *Proc. Natl. Acad. Sci. USA* **2018**, *115*, 13039–13044. [CrossRef] [PubMed]
11. St John, J.; Sakkas, D.; Dimitriadi, K.; Barnes, A.; Maclin, V.; Ramey, J.; Barratt, C.; De Jonge, C. Failure of elimination of paternal mitochondrial DNA in abnormal embryos. *Lancet* **2000**, *355*, 200. [CrossRef]
12. Legros, F.; Malka, F.; Frachon, P.; Lombès, A.; Rojo, M. Organization and dynamics of human mitochondrial DNA. *J. Cell Sci.* **2004**, *117*, 2653–2662. [CrossRef]
13. Satoh, M.; Kuroiwa, T. Organization of multiple nucleoids and DNA molecules in mitochondria of a human cell. *Exp. Cell Res.* **1991**, *196*, 137–140. [CrossRef]
14. Cassano, A.G.; Anderson, V.E.; Harris, M.E. Evidence for direct attack by hydroxide in phosphodiester hydrolysis. *J. Am. Chem. Soc.* **2002**, *124*, 10964–10965. [CrossRef] [PubMed]
15. Kaufman, B.A.; Picard, M.; Sondheimer, N. Mitochondrial DNA, nuclear context, and the risk for carcinogenesis. *Environ. Mol. Mutagen.* **2019**, *60*, 455–462. [CrossRef] [PubMed]
16. Floros, V.I.; Pyle, A.; DIetmann, S.; Wei, W.; Tang, W.W.C.; Irie, N.; Payne, B.; Capalbo, A.; Noli, L.; Coxhead, J.; et al. Segregation of mitochondrial DNA heteroplasmy through a developmental genetic bottleneck in human embryos. *Nat. Cell Biol.* **2018**, *20*, 144–151. [CrossRef]
17. Brown, W.M.; Prager, E.M.; Wang, A.; Wilson, A.C. Mitochondrial DNA sequences of primates: Tempo and mode of evolution. *J. Mol. Evol.* **1982**, *18*, 225–239. [CrossRef]
18. Stewart, J.B.; Chinnery, P.F. The dynamics of mitochondrial DNA heteroplasmy: Implications for human health and disease. *Nat. Rev. Genet.* **2015**, *16*, 530–542. [CrossRef]
19. Gorman, G.S.; Schaefer, A.M.; Ng, Y.; Gomez, N.; Blakely, E.L.; Alston, C.L.; Feeney, C.; Horvath, R.; Yu-Wai-Man, P.; Chinnery, P.F.; et al. Prevalence of nuclear and mitochondrial DNA mutations related to adult mitochondrial disease. *Ann. Neurol.* **2015**, *77*, 753–759. [CrossRef]
20. Zinovkina, L.A. Mechanisms of Mitochondrial DNA Repair in Mammals. *Biochem.* **2018**, *83*, 233–249. [CrossRef]

21. Torregrosa-Muñumer, R.; Hangas, A.; Goffart, S.; Blei, D.; Zsurka, G.; Griffith, J.; Kunz, W.S.; Pohjoismäki, J.L.O. Replication fork rescue in mammalian mitochondria. *Sci. Rep.* **2019**, *9*. [CrossRef] [PubMed]
22. Spelbrink, J.N. Functional organization of mammalian mitochondrial DNA in nucleoids: History, recent developments, and future challenges. *IUBMB Life* **2010**, *62*, 19–32. [CrossRef] [PubMed]
23. Kauppila, J.H.K.; Bonekamp, N.A.; Mourier, A.; Isokallio, M.A.; Just, A.; Kauppila, T.E.S.; Stewart, J.B.; Larsson, N. Base-excision repair deficiency alone or combined with increased oxidative stress does not increase mtDNA point mutations in mice. *Nucleic Acids Res.* **2018**, *46*, 6642–6649. [CrossRef] [PubMed]
24. Wallace, D.C. Diseases of the mitochondrial DNA. *Annu. Rev. Biochem.* **1992**, *61*, 1175–1212. [CrossRef] [PubMed]
25. Rossignol, R.; Faustin, B.; Rocher, C.; Malgat, M.; Mazat, J.P.; Letellier, T. Mitochondrial threshold effects. *Biochem. J.* **2003**, *370*, 751–762. [CrossRef] [PubMed]
26. Bacman, S.R.; Gammage, P.A.; Minczuk, M.; Moraes, C.T. Manipulation of mitochondrial genes and mtDNA heteroplasmy. *Methods Cell Biol.* **2020**, *155*, 441–487. [CrossRef]
27. Tuppen, H.A.L.; Blakely, E.L.; Turnbull, D.M.; Taylor, R.W. Mitochondrial DNA mutations and human disease. *Biochim. Biophys. Acta Bioenerg.* **2010**, *1797*, 113–128. [CrossRef] [PubMed]
28. Rai, P.K.; Craven, L.; Hoogewijs, K.; Russell, O.M.; Lightowlers, R.N. Advances in methods for reducing mitochondrial DNA disease by replacing or manipulating the mitochondrial genome. *Essays Biochem.* **2018**, *62*, 455–465. [CrossRef]
29. Wei, W.; Chinnery, P.F. Inheritance of mitochondrial DNA in humans: Implications for rare and common diseases. *J. Intern. Med.* **2020**, *287*, 634–644. [CrossRef]
30. Payne, B.A.I.; Wilson, I.J.; Yu-Wai-Man, P.; Coxhead, J.; Deehan, D.; Horvath, R.; Taylor, R.W.; Samuels, D.C.; Santibanez-Koref, M.; Chinnery, P.F. Universal heteroplasmy of human mitochondrial DNA. *Hum. Mol. Genet.* **2013**, *22*, 384–390. [CrossRef]
31. Li, M.; Schönberg, A.; Schaefer, M.; Schroeder, R.; Nasidze, I.; Stoneking, M. Detecting heteroplasmy from high-throughput sequencing of complete human mitochondrial DNA genomes. *Am. J. Hum. Genet.* **2010**, *87*, 237–249. [CrossRef] [PubMed]
32. Giuliani, C.; Barbieri, C.; Li, M.; Bucci, L.; Monti, D.; Passarino, G.; Luiselli, D.; Franceschi, C.; Stoneking, M.; Garagnani, P. Transmission from centenarians to their offspring of mtDNA heteroplasmy revealed by ultra-deep sequencing. *Aging (Albany. NY)* **2014**, *6*, 454–467. [CrossRef] [PubMed]
33. 3van den Ameele, J.; Li, A.Y.Z.; Ma, H.; Chinnery, P.F. Mitochondrial heteroplasmy beyond the oocyte bottleneck. *Semin. Cell Dev. Biol.* **2020**, *97*, 156–166. [CrossRef] [PubMed]
34. Hauswirth, W.W.; Laipis, P.J. Mitochondrial DNA polymorphism in a maternal lineage of Holstein cows. *Proc. Natl. Acad. Sci. USA* **1982**, *79*, 4686–4690. [CrossRef]
35. Cao, L.; Shitara, H.; Sugimoto, M.; Hayashi, J.I.; Abe, K.; Yonekawa, H. New evidence confirms that the mitochondrial bottleneck is generated without reduction of mitochondrial DNA content in early primordial germ cells of mice. *PLoS Genet.* **2009**, *5*, e1000756. [CrossRef]
36. Cao, L.; Shitara, H.; Horii, T.; Nagao, Y.; Imai, H.; Abe, K.; Hara, T.; Hayashi, J.I.; Yonekawa, H. The mitochondrial bottleneck occurs without reduction of mtDNA content in female mouse germ cells. *Nat. Genet.* **2007**, *39*, 386–390. [CrossRef]
37. Jenuth, J.P.; Peterson, A.C.; Fu, K.; Shoubridge, E.A. Random genetic drift in the female germline explains the rapid segregation of mammalian mitochondrial DNA. *Nat. Genet.* **1996**, *14*, 146–151. [CrossRef]
38. Chen, X.; Prosser, R.; Simonetti, S.; Sadlock, J.; Jagiello, G.; Schon, E.A. Rearranged mitochondrial genomes are present in human oocytes. *Am. J. Hum. Genet.* **1995**, *57*, 239–247. [CrossRef] [PubMed]
39. Freyer, C.; Cree, L.M.; Mourier, A.; Stewart, J.B.; Koolmeister, C.; Milenkovic, D.; Wai, T.; Floros, V.I.; Hagström, E.; Chatzidaki, E.E.; et al. Variation in germline mtDNA heteroplasmy is determined prenatally but modified during subsequent transmission. *Nat. Genet.* **2012**, *44*, 1282–1285. [CrossRef] [PubMed]
40. Cree, L.M.; Samuels, D.C.; De Sousa Lopes, S.C.; Rajasimha, H.K.; Wonnapinij, P.; Mann, J.R.; Dahl, H.H.M.; Chinnery, P.F. A reduction of mitochondrial DNA molecules during embryogenesis explains the rapid segregation of genotypes. *Nat. Genet.* **2008**, *40*, 249–254. [CrossRef] [PubMed]
41. Wonnapinij, P.; Chinnery, P.F.; Samuels, D.C. Previous Estimates of Mitochondrial DNA Mutation Level Variance Did Not Account for Sampling Error: Comparing the mtDNA Genetic Bottleneck in Mice and Humans. *Am. J. Hum. Genet.* **2010**, *86*, 540–550. [CrossRef] [PubMed]
42. Johnston, I.G.; Burgstaller, J.P.; Havlicek, V.; Kolbe, T.; Rülicke, T.; Brem, G.; Poulton, J.; Jones, N.S. Stochastic modelling, bayesian inference, and new in vivo measurements elucidate the debated mtDNA bottleneck mechanism. *Elife* **2015**, *4*, e07464. [CrossRef] [PubMed]
43. Cree, L.M.; Samuels, D.C.; Chinnery, P.F. The inheritance of pathogenic mitochondrial DNA mutations. *Biochim. Biophys. Acta Mol. Basis Dis.* **2009**, *1792*, 1097–1102. [CrossRef] [PubMed]
44. Ling, F.; Niu, R.; Hatakeyama, H.; Goto, Y.I.; Shibata, T.; Yoshida, M. Reactive oxygen species stimulate mitochondrial allele segregation toward homoplasmy in human cells. *Mol. Biol. Cell* **2016**, *27*, 1684–1693. [CrossRef]
45. Chinnery, P.F.; Thorburn, D.R.; Samuels, D.C.; White, S.L.; Dahl, H.H.M.; Turnbull, D.M.; Lightowlers, R.N.; Howell, N. The inheritance of mitochondrial DNA heteroplasmy: Random drift, selection or both? *Trends Genet.* **2000**, *16*, 500–505. [CrossRef]
46. Ma, H.; Xu, H.; O'Farrell, P.H. Transmission of mitochondrial mutations and action of purifying selection in Drosophila melanogaster. *Nat. Genet.* **2014**, *46*, 393–397. [CrossRef]

47. Stewart, J.B.; Freyer, C.; Elson, J.L.; Wredenberg, A.; Cansu, Z.; Trifunovic, A.; Larsson, N.-G. Strong purifying selection in transmission of mammalian mitochondrial DNA. *PLoS Biol.* **2008**, *6*, e10. [CrossRef]
48. Hill, J.H.; Chen, Z.; Xu, H. Selective propagation of functional mitochondrial DNA during oogenesis restricts the transmission of a deleterious mitochondrial variant. *Nat. Genet.* **2014**, *46*, 389–392. [CrossRef]
49. Fan, W.; Waymire, K.G.; Narula, N.; Li, P.; Rocher, C.; Coskun, P.E.; Vannan, M.A.; Narula, J.; MacGregor, G.R.; Wallace, D.C. A mouse model of mitochondrial disease reveals germline selection against severe mtDNA mutations. *Science* **2008**, *319*, 958–962. [CrossRef]
50. Wei, W.; Tuna, S.; Keogh, M.J.; Smith, K.R.; Aitman, T.J.; Beales, P.L.; Bennett, D.L.; Gale, D.P.; Bitner-Glindzicz, M.A.K.; Black, G.C.; et al. Germline selection shapes human mitochondrial DNA diversity. *Science* **2019**, *364*. [CrossRef]
51. Wai, T.; Teoli, D.; Shoubridge, E.A. The mitochondrial DNA genetic bottleneck results from replication of a subpopulation of genomes. *Nat. Genet.* **2008**, *40*, 1484–1488. [CrossRef] [PubMed]
52. Mishra, P.; Chan, D.C. Mitochondrial dynamics and inheritance during cell division, development and disease. *Nat. Rev. Mol. Cell Biol.* **2014**, *15*, 634–646. [CrossRef] [PubMed]
53. Suen, D.F.; Narendra, D.P.; Tanaka, A.; Manfredi, G.; Youle, R.J. Parkin overexpression selects against a deleterious mtDNA mutation in heteroplasmic cybrid cells. *Proc. Natl. Acad. Sci. USA* **2010**, *107*, 11835–11840. [CrossRef] [PubMed]
54. Chen, Y.; Dorn, G.W. PINK1-phosphorylated mitofusin 2 is a parkin receptor for culling damaged mitochondria. *Science* **2013**, *340*, 471–475. [CrossRef]
55. Ziviani, E.; Tao, R.N.; Whitworth, A.J. Drosophila Parkin requires PINK1 for mitochondrial translocation and ubiquitinates Mitofusin. *Proc. Natl. Acad. Sci. USA* **2010**, *107*, 5018–5023. [CrossRef]
56. Narendra, D.P.; Jin, S.M.; Tanaka, A.; Suen, D.-F.; Gautier, C.A.; Shen, J.; Cookson, M.R.; Youle, R.J. PINK1 is selectively stabilized on impaired mitochondria to activate Parkin. *PLoS Biol.* **2010**, *8*, e1000298. [CrossRef]
57. Lazarou, M.; Sliter, D.A.; Kane, L.A.; Sarraf, S.A.; Wang, C.; Burman, J.L.; Sideris, D.P.; Fogel, A.I.; Youle, R.J. The ubiquitin kinase PINK1 recruits autophagy receptors to induce mitophagy. *Nature* **2015**, *524*, 309–314. [CrossRef]
58. Chan, D.C. Fusion and fission: Interlinked processes critical for mitochondrial health. *Annu. Rev. Genet.* **2012**, *46*, 265–287. [CrossRef]
59. Meeusen, S.; McCaffery, J.M.; Nunnari, J. Mitochondrial fusion intermediates revealed in vitro. *Science* **2004**, *305*, 1747–1752. [CrossRef]
60. Kandul, N.P.; Zhang, T.; Hay, B.A.; Guo, M. Selective removal of deletion-bearing mitochondrial DNA in heteroplasmic Drosophila. *Nat. Commun.* **2016**, *7*, 1–11. [CrossRef]
61. Liau, W.S.; Gonzalez-Serricchio, A.S.; Deshommes, C.; Chin, K.; LaMunyon, C.W. A persistent mitochondrial deletion reduces fitness and sperm performance in heteroplasmic populations of C. elegans. *BMC Genet.* **2007**, *8*, 1–11. [CrossRef]
62. Valenci, I.; Yonai, L.; Bar-Yaacov, D.; Mishmar, D.; Ben-Zvi, A. Parkin modulates heteroplasmy of truncated mtDNA in Caenorhabditis elegans. *Mitochondrion* **2015**, *20*, 64–70. [CrossRef]
63. Shoubridge, E.A.; Wai, T. Mitochondrial DNA and the mammalian oocyte. *Curr. Top. Dev. Biol.* **2007**, *77*, 87–111. [CrossRef] [PubMed]
64. Raap, A.K.; Jahangir Tafrechi, R.S.; van de Rijke, F.M.; Pyle, A.; Wählby, C.; Szuhai, K.; Ravelli, R.B.G.; de Coo, R.F.M.; Rajasimha, H.K.; Nilsson, M.; et al. Non-random mtDNA segregation patterns indicate a metastable heteroplasmic segregation unit in m.3243A>G cybrid cells. *PLoS ONE* **2012**, *7*, e52080. [CrossRef] [PubMed]
65. Zhang, H.; Burr, S.P.; Chinnery, P.F. The mitochondrial DNA genetic bottleneck: Inheritance and beyond. *Essays Biochem.* **2018**, *62*, 225–234. [CrossRef] [PubMed]
66. Chacinska, A.; Koehler, C.M.; Milenkovic, D.; Lithgow, T.; Pfanner, N. Importing mitochondrial proteins: Machineries and mechanisms. *Cell* **2009**, *138*, 628–644. [CrossRef] [PubMed]
67. Kukat, C.; Wurm, C.A.; Spahr, H.; Falkenberg, M.; Larsson, N.-G.; Jakobs, S. Super-resolution microscopy reveals that mammalian mitochondrial nucleoids have a uniform size and frequently contain a single copy of mtDNA. *Proc. Natl. Acad. Sci. USA* **2011**, *108*, 13534–13539. [CrossRef]
68. Krakauer, D.C.; Mira, A. Mitochondria and germ-cell death. *Nature* **1999**, *400*, 125–126. [CrossRef]
69. Palozzi, J.M.; Jeedigunta, S.P.; Hurd, T.R. Mitochondrial DNA Purifying Selection in Mammals and Invertebrates. *J. Mol. Biol.* **2018**, *430*, 4834–4848. [CrossRef]
70. Rajasimha, H.K.; Chinnery, P.F.; Samuels, D.C. Selection against pathogenic mtDNA mutations in a stem cell population leads to the loss of the 3243A→G mutation in blood. *Am. J. Hum. Genet.* **2008**, *82*, 333–343. [CrossRef]
71. Olsson, C.; Johnsen, E.; Nilsson, M.; Wilander, E.; Syvänen, A.C.; Lagerström-Fermér, M. The level of the mitochondrial mutation A3243G decreases upon ageing in epithelial cells from individuals with diabetes and deafness. *Eur. J. Hum. Genet.* **2001**, *9*, 917–921. [CrossRef]
72. Diaz, F.; Bayona-Bafaluy, M.P.; Rana, M.; Mora, M.; Hao, H.; Moraes, C.T. Human mitochondrial DNA with large deletions repopulates organelles faster than full-length genomes under relaxed copy number control. *Nucleic Acids Res.* **2002**, *30*, 4626–4633. [CrossRef] [PubMed]
73. Hart, L.M.T.; Jansen, J.J.; Lemkes, H.H.; de Knijff, P.; Maassen, J.A. Heteroplasmy levels of a mitochondrial gene mutation associated with diabetes mellitus decrease in leucocyte DNA upon aging. *Hum. Mutat.* **1996**, *7*, 193–197. [CrossRef]

74. Frederiksen, A.L.; Andersen, P.H.; Kyvik, K.O.; Jeppesen, T.D.; Vissing, J.; Schwartz, M. Tissue specific distribution of the 3243A→G mtDNA mutation. *J. Med. Genet.* **2006**, *43*, 671–677. [CrossRef] [PubMed]
75. Hämäläinen, R.H.; Manninen, T.; Koivumäki, H.; Kislin, M.; Otonkoski, T.; Suomalainen, A. Tissue-and cell-type-specific manifestations of heteroplasmic mtDNA 3243A>G mutation in human induced pluripotent stem cell-derived disease model. *Proc. Natl. Acad. Sci. USA* **2013**, *110*. [CrossRef]
76. Lorenz, C.; Lesimple, P.; Bukowiecki, R.; Zink, A.; Inak, G.; Mlody, B.; Singh, M.; Semtner, M.; Mah, N.; Auré, K.; et al. Human iPSC-Derived Neural Progenitors Are an Effective Drug Discovery Model for Neurological mtDNA Disorders. *Cell Stem Cell* **2017**, *20*, 659–674. [CrossRef]
77. Lechuga-Vieco, A.V.; Latorre-Pellicer, A.; Johnston, I.G.; Prota, G.; Gileadi, U.; Justo-Méndez, R.; Acín-Pérez, R.; Martínez-de-Mena, R.; Fernández-Toro, J.M.; Jimenez-Blasco, D.; et al. Cell identity and nucleo-mitochondrial genetic context modulate OXPHOS performance and determine somatic heteroplasmy dynamics. *Sci. Adv.* **2020**, *6*, eaba5345. [CrossRef] [PubMed]
78. Elson, J.L.; Samuels, D.C.; Turnbull, D.M.; Chinnery, P.F. Random intracellular drift explains the clonal expansion of mitochondrial DNA mutations with age. *Am. J. Hum. Genet.* **2001**, *68*, 802–806. [CrossRef]
79. Clark, K.A.; Howe, D.K.; Gafner, K.; Kusuma, D.; Ping, S.; Estes, S.; Denver, D.R. Selfish little circles: Transmission bias and evolution of large deletion-bearing mitochondrial DNA in Caenorhabditis briggsae nematodes. *PLoS ONE* **2012**, *7*, e41433. [CrossRef]
80. Dasgupta, S. Mitochondrion: I am more than a fuel server. *Ann. Transl. Med.* **2019**, *7*, 594. [CrossRef]
81. Sharma, L.K.; Fang, H.; Liu, J.; Vartak, R.; Deng, J.; Bai, Y. Mitochondrial respiratory complex I dysfunction promotes tumorigenesis through ROS alteration and AKT activation. *Hum. Mol. Genet.* **2011**, *20*, 4605–4616. [CrossRef]
82. Mayr, J.A.; Meierhofer, D.; Zimmermann, F.; Feichtinger, R.; Kögler, C.; Ratschek, M.; Schmeller, N.; Sperl, W.; Kofler, B. Loss of complex I due to mitochondrial DNA mutations in renal oncocytoma. *Clin. Cancer Res.* **2008**, *14*, 2270–2275. [CrossRef] [PubMed]
83. Park, J.S.; Sharma, L.K.; Li, H.; Xiang, R.; Holstein, D.; Wu, J.; Lechleiter, J.; Naylor, S.L.; Deng, J.J.; Lu, J.; et al. A heteroplasmic, not homoplasmic, mitochondrial DNA mutation promotes tumorigenesis via alteration in reactive oxygen species generation and apoptosis. *Hum. Mol. Genet.* **2009**, *18*, 1578–1589. [CrossRef] [PubMed]
84. Calabrese, C.; Iommarini, L.; Kurelac, I.; Calvaruso, M.A.; Capristo, M.; Lollini, P.-L.; Nanni, P.; Bergamini, C.; Nicoletti, G.; De Giovanni, C.; et al. Respiratory complex I is essential to induce a Warburg profile in mitochondria-defective tumor cells. *Cancer Metab.* **2013**, *1*, 11. [CrossRef] [PubMed]
85. Leone, G.; Abla, H.; Gasparre, G.; Porcelli, A.M.; Iommarini, L. The oncojanus paradigm of respiratory complex I. *Genes (Basel)* **2018**, *9*, 243. [CrossRef]
86. Gasparre, G.; Kurelac, I.; Capristo, M.; Iommarini, L.; Ghelli, A.; Ceccarelli, C.; Nicoletti, G.; Nanni, P.; De Giovanni, C.; Scotlandi, K.; et al. A mutation threshold distinguishes the antitumorigenic effects of the mitochondrial gene MTND1, an Oncojanus function. *Cancer Res.* **2011**, *71*, 6220–6229. [CrossRef]
87. Damm, F.; Bunke, T.; Thol, F.; Markus, B.; Wagner, K.; Göhring, G.; Schlegelberger, B.; Heil, G.; Reuter, C.W.M.; Püllmann, K.; et al. Prognostic implications and molecular associations of NADH dehydrogenase subunit 4 (ND4) mutations in acute myeloid leukemia. *Leukemia* **2012**, *26*, 289–295. [CrossRef]
88. Mcmahon, S.; Laframboise, T. Mutational patterns in the breast cancer mitochondrial genome, with clinical correlates. *Carcinogenesis* **2014**, *35*, 1046–1054. [CrossRef]
89. Birky, C.W. Uniparental inheritance of mitochondrial and chloroplast genes: Mechanisms and evolution. *Proc. Natl. Acad. Sci. USA* **1995**, *92*, 11331–11338. [CrossRef]
90. Boynton, J.E.; Harris, E.H.; Burkhart, B.D.; Lamerson, P.M.; Gillham, N.W. Transmission of mitochondrial and chloroplast genomes in crosses of Chlamydomonas. *Proc. Natl. Acad. Sci. USA* **1987**, *84*, 2391–2395. [CrossRef]
91. Hurst, L.D. Selfish genetic elements and their role in evolution: The evolution of sex and some of what that entails. *Philos. Trans. R. Soc. London. Ser. B Biol. Sci.* **1995**, *349*, 321–332. [CrossRef]
92. Ladoukakis, E.D. Evolutionary genetics: Direct evidence of recombination in human mitochondrial DNA. *Heredity (Edinb)* **2004**, *93*, 321. [CrossRef]
93. Greiner, S.; Sobanski, J.; Bock, R. Why are most organelle genomes transmitted maternally? *BioEssays* **2015**, *37*, 80–94. [CrossRef] [PubMed]
94. Sharpley, M.S.; Marciniak, C.; Eckel-Mahan, K.; McManus, M.; Crimi, M.; Waymire, K.; Lin, C.S.; Masubuchi, S.; Friend, N.; Koike, M.; et al. Heteroplasmy of mouse mtDNA is genetically unstable and results in altered behavior and cognition. *Cell* **2012**, *151*, 333–343. [CrossRef] [PubMed]
95. DeLuca, S.Z.; O'Farrell, P.H. Barriers to male transmission of mitochondrial DNA in sperm development. *Dev. Cell* **2012**, *22*, 660–668. [CrossRef]
96. Chan, D.C.; Schon, E.A. Eliminating Mitochondrial DNA from Sperm. *Dev. Cell* **2012**, *22*, 469–470. [CrossRef] [PubMed]
97. Yan, C.; Duanmu, X.; Zeng, L.; Liu, B.; Song, Z. Mitochondrial DNA: Distribution, mutations, and elimination. *Cells* **2019**, *8*, 379. [CrossRef]
98. Yu, Z.; O'Farrell, P.H.; Yakubovich, N.; DeLuca, S.Z. The mitochondrial DNA polymerase promotes elimination of paternal mitochondrial genomes. *Curr. Biol.* **2017**, *27*, 1033–1039. [CrossRef]
99. Wai, T.; Ao, A.; Zhang, X.; Cyr, D.; Dufort, D.; Shoubridge, E.A. The Role of Mitochondrial DNA Copy Number in Mammalian Fertility. *Biol. Reprod.* **2010**, *83*, 52–62. [CrossRef]

100. Pereira, L.; Gonçalves, J.; Franco-Duarte, R.; Silva, J.; Rocha, T.; Arnold, C.; Richards, M.; Macaulay, V. No evidence for an mtDNA role in sperm motility: Data from complete sequencing of asthenozoospermic males. *Mol. Biol. Evol.* **2007**, *24*, 868–874. [CrossRef]
101. Ruiz-Pesini, E.; Díez-Sánchez, C.; López-Pérez, M.J.; Enríquez, J.A. The role of the mitochondrion in sperm function: Is there a place for oxidative phosphorylation or is this a purely glycolytic process? *Curr. Top. Dev. Biol.* **2007**, *77*, 3–19. [CrossRef] [PubMed]
102. Tremellen, K. Oxidative stress and male infertility—A clinical perspective. *Hum. Reprod. Update* **2008**, *14*, 243–258. [CrossRef] [PubMed]
103. Thompson, W.E.; Ramalho-Santos, J.; Sutovsky, P. Ubiquitination of Prohibitin in mammalian sperm mitochondria: Possible roles in the regulation of mitochondrial inheritance and sperm quality control. *Biol. Reprod.* **2003**, *69*, 254–260. [CrossRef]
104. Sutovsky, P.; Moreno, R.D.; Ramalho-Santos, J.; Dominko, T.; Simerly, C.; Schatten, G. Ubiquitin tag for sperm mitochondria. *Nature* **1999**, *402*, 371–372. [CrossRef] [PubMed]
105. Sutovsky, P.; Moreno, R.D.; Ramalho-Santos, J.; Dominko, T.; Simerly, C.; Schatten, G. Ubiquitinated sperm mitochondria, selective proteolysis, and the regulation of mitochondrial inheritance in mammalian embryos. *Biol. Reprod.* **2000**, *63*, 582–590. [CrossRef] [PubMed]
106. Tipler, C. Purification and characterization of 26S proteasomes from human and mouse spermatozoa. *Mol. Hum. Reprod.* **1997**, *3*, 1053–1060. [CrossRef] [PubMed]
107. Pyle, A.; Hudson, G.; Wilson, I.J.; Coxhead, J.; Smertenko, T.; Herbert, M.; Santibanez-Koref, M.; Chinnery, P.F. Extreme-Depth Re-sequencing of mitochondrial DNA finds no evidence of paternal transmission in humans. *PLoS Genet.* **2015**, *11*, e1005040. [CrossRef] [PubMed]
108. May-Panloup, P.; Chrétien, M.F.; Savagner, F.; Vasseur, C.; Jean, M.; Malthièry, Y.; Reynier, P. Increased sperm mitochondrial DNA content in male infertility. *Hum. Reprod.* **2003**, *18*, 550–556. [CrossRef]
109. Manfredi, G.; Thyagarajan, D.; Papadopoulou, L.C.; Pallotti, F.; Schon, E.A.; Houston, H. The fate of human sperm-derived mtDNA in somatic cells. *Am. J. Hum. Genet* **1997**, *61*, 953–960. [CrossRef]
110. Orsztynowicz, M.; Pawlak, P.; Podstawski, Z.; Nizanski, W.; Partyka, A.; Gotowiecka, M.; Kosiniak-Kamysz, K.; Lechniak, D. Mitochondrial DNA copy number in spermatozoa of fertile stallions. *Reprod. Domest. Anim.* **2016**, *51*, 378–385. [CrossRef]
111. Carelli, V. Keeping in shape the dogma of mitochondrial DNA maternal inheritance. *PLoS Genet.* **2015**, *11*, e1005179. [CrossRef] [PubMed]
112. Burr, S.P.; Chinnery, P.F. Heredity and segregation of mtDNA. In *The Human Mitochondrial Genome*; Academic Press: New York, NY, USA, 2020; pp. 87–107. ISBN 9780128196564.
113. Al Rawi, S.; Louvet-Vallée, S.; Djeddi, A.; Sachse, M.; Culetto, E.; Hajjar, C.; Boyd, L.; Legouis, R.; Galy, V. Postfertilization autophagy of sperm organelles prevents paternal mitochondrial DNA transmission. *Science* **2011**, *334*, 1144–1147. [CrossRef] [PubMed]
114. Zhou, Q.; Li, H.; Xue, D. Elimination of paternal mitochondria through the lysosomal degradation pathway in C. elegans. *Cell Res.* **2011**, *21*, 1662–1669. [CrossRef] [PubMed]
115. Parrish, J.; Li, L.; Klotz, K.; Ledwich, D.; Wang, X.; Xue, D. Mitochondrial endonuclease G is important for apoptosis in C. elegans. *Nature* **2001**, *412*, 90–94. [CrossRef] [PubMed]
116. Wiehe, R.S.; Gole, B.; Chatre, L.; Walther, P.; Calzia, E.; Ricchetti, M.; Wiesmüller, L. Endonuclease G promotes mitochondrial genome cleavage and replication. *Oncotarget* **2018**, *9*, 18309–18326. [CrossRef]
117. Ashrafi, G.; Schwarz, T.L. The pathways of mitophagy for quality control and clearance of mitochondria. *Cell Death Differ.* **2013**, *20*, 31–42. [CrossRef]
118. Wei, Y.; Chiang, W.C.; Sumpter, R.; Mishra, P.; Levine, B. Prohibitin 2 Is an Inner Mitochondrial Membrane Mitophagy Receptor. *Cell* **2017**, *168*, 224–238.e10. [CrossRef]
119. Tsukamoto, S.; Kuma, A.; Murakami, M.; Kishi, C.; Yamamoto, A.; Mizushima, N. Autophagy is essential for preimplantation development of mouse embryos. *Science* **2008**, *321*, 117–120. [CrossRef]
120. Rojansky, R.; Cha, M.Y.; Chan, D.C. Elimination of paternal mitochondria in mouse embryos occurs through autophagic degradation dependent on PARKIN and MUL1. *Elife* **2016**, *5*, e17896. [CrossRef]
121. Sato, M.; Sato, K. Maternal inheritance of mitochondrial DNA by diverse mechanisms to eliminate paternal mitochondrial DNA. *Biochim. Biophys. Acta Mol. Cell Res.* **2013**, *1833*, 1979–1984. [CrossRef]
122. Kaneda, H.; Hayashi, J.I.; Takahama, S.; Taya, C.; Lindahl, K.F.; Yonekawa, H. Elimination of paternal mitochondrial DNA in intraspecific crosses during early mouse embryogenesis. *Proc. Natl. Acad. Sci. USA* **1995**, *92*, 4542–4546. [CrossRef] [PubMed]
123. Schwartz, M.; Vissing, J. Paternal Inheritance of Mitochondrial DNA. *N. Engl. J. Med.* **2002**, *347*, 576–580. [CrossRef] [PubMed]
124. Yu, X.; Gimsa, U.; Wester-Rosenlöf, L.; Kanitz, E.; Otten, W.; Kunz, M.; Ibrahim, S.M. Dissecting the effects of mtDNA variations on complex traits using mouse conplastic strains. *Genome Res.* **2009**, *19*, 159–165. [CrossRef] [PubMed]
125. Latorre-Pellicer, A.; Moreno-Loshuertos, R.; Lechuga-Vieco, A.V.; Sánchez-Cabo, F.; Torroja, C.; Acín-Pérez, R.; Calvo, E.; Aix, E.; González-Guerra, A.; Logan, A.; et al. Mitochondrial and nuclear DNA matching shapes metabolism and healthy ageing. *Nature* **2016**, *535*, 561–565. [CrossRef]
126. Moraes, C.T. Sorting mtDNA Species—The Role of nDNA-mtDNA Co-evolution. *Cell Metab.* **2019**, *30*, 1002–1004. [CrossRef]
127. Guarás, A.; Perales-Clemente, E.; Calvo, E.; Acín-Pérez, R.; Loureiro-Lopez, M.; Pujol, C.; Martínez-Carrascoso, I.; Nuñez, E.; García-Marqués, F.; Rodríguez-Hernández, M.A.; et al. The CoQH2/CoQ Ratio Serves as a Sensor of Respiratory Chain Efficiency. *Cell Rep.* **2016**, *15*, 197–209. [CrossRef]

128. Latorre-Pellicer, A.; Lechuga-Vieco, A.V.; Johnston, I.G.; Hämäläinen, R.H.; Pellico, J.; Justo-Méndez, R.; Fernández-Toro, J.M.; Clavería, C.; Guaras, A.; Sierra, R.; et al. Regulation of Mother-to-Offspring Transmission of mtDNA Heteroplasmy. *Cell Metab.* **2019**, *30*, 1120–1130.e5. [CrossRef]
129. Hock, M.B.; Kralli, A. Transcriptional control of mitochondrial biogenesis and function. *Annu. Rev. Physiol.* **2009**, *71*, 177–203. [CrossRef]
130. Russell, O.M.; Gorman, G.S.; Lightowlers, R.N.; Turnbull, D.M. Mitochondrial Diseases: Hope for the Future. *Cell* **2020**, *181*, 168–188. [CrossRef]
131. Samuels, D.C.; Wonnapinij, P.; Chinnery, P.F. Preventing the transmission of pathogenic mitochondrial DNA mutations: Can we achieve long-term benefits from germ-line gene transfer? *Hum. Reprod.* **2013**, *28*, 554–559. [CrossRef]
132. Mitalipov, S.; Amato, P.; Parry, S.; Falk, M.J. Limitations of preimplantation genetic diagnosis for mitochondrial DNA diseases. *Cell Rep.* **2014**, *7*, 935–937. [CrossRef] [PubMed]
133. Wolf, D.P.; Mitalipov, N.; Mitalipov, S. Mitochondrial replacement therapy in reproductive medicine. *Trends Mol. Med.* **2015**, *21*, 68–76. [CrossRef] [PubMed]
134. Gorman, G.S.; Grady, J.P.; Turnbull, D.M. Mitochondrial donation—How many women could benefit? *N. Engl. J. Med.* **2015**, *372*, 885–887. [CrossRef]
135. Cree, L.; Loi, P. Mitochondrial replacement: From basic research to assisted reproductive technology portfolio tool-technicalities and possible risks. *Mol. Hum. Reprod.* **2015**, *21*, 3–10. [CrossRef] [PubMed]
136. Byrne, J.A.; Pedersen, D.A.; Clepper, L.L.; Nelson, M.; Sanger, W.G.; Gokhale, S.; Wolf, D.P.; Mitalipov, S. Producing primate embryonic stem cells by somatic cell nuclear transfer. *Nature* **2007**, *450*, 497–502. [CrossRef] [PubMed]
137. Sato, A.; Kono, T.; Nakada, K.; Ishikawa, K.; Inoue, S.I.; Yonekawa, H.; Hayashi, J.I. Gene therapy for progeny of mito-mice carrying pathogenic mtDNA by nuclear transplantation. *Proc. Natl. Acad. Sci. USA* **2005**, *102*, 16765–16770. [CrossRef]
138. Tachibana, M.; Sparman, M.; Sritanaudomchai, H.; Ma, H.; Clepper, L.; Woodward, J.; Li, Y.; Ramsey, C.; Kolotushkina, O.; Mitalipov, S. Mitochondrial gene replacement in primate offspring and embryonic stem cells. *Nature* **2009**, *461*, 367–372. [CrossRef]
139. Alikani, M.; Fauser, B.C.J.; García-Valesco, J.A.; Simpson, J.L.; Johnson, M.H. First birth following spindle transfer for mitochondrial replacement therapy: Hope and trepidation. *Reprod. Biomed. Online* **2017**, *34*, 333–336. [CrossRef]
140. Zhang, J.; Liu, H.; Luo, S.; Lu, Z.; Chávez-Badiola, A.; Liu, Z.; Yang, M.; Merhi, Z.; Silber, S.J.; Munné, S.; et al. Live birth derived from oocyte spindle transfer to prevent mitochondrial disease. *Reprod. Biomed. Online* **2017**, *34*, 361–368. [CrossRef]
141. Wakayama, T.; Yanagimachi, R. The First Polar Body Can Be Used for the Production of Normal Offspring in Mice1. *Biol. Reprod.* **1998**, *59*, 100–104. [CrossRef]
142. Ma, H.; O'Neil, R.C.; Marti Gutierrez, N.; Hariharan, M.; Zhang, Z.Z.; He, Y.; Cinnioglu, C.; Kayali, R.; Kang, E.; Lee, Y.; et al. Functional Human Oocytes Generated by Transfer of Polar Body Genomes. *Cell Stem Cell* **2017**, *20*, 112–119. [CrossRef] [PubMed]
143. Wang, T.; Sha, H.; Ji, D.; Zhang, H.L.; Chen, D.; Cao, Y.; Zhu, J. Polar body genome transfer for preventing the transmission of inherited mitochondrial diseases. *Cell* **2014**, *157*, 1591–1604. [CrossRef] [PubMed]
144. Greenfield, A.; Braude, P.; Flinter, F.; Lovell-Badge, R.; Ogilvie, C.; Perry, A.C.F. Assisted reproductive technologies to prevent human mitochondrial disease transmission. *Nat. Biotechnol.* **2017**, *35*, 1059–1068. [CrossRef] [PubMed]
145. Zhang, J.; Zhuang, G.; Zeng, Y.; Grifo, J.; Acosta, C.; Shu, Y.; Liu, H. Pregnancy derived from human zygote pronuclear transfer in a patient who had arrested embryos after IVF. *Reprod. Biomed. Online* **2016**, *33*, 529–533. [CrossRef]
146. Palacios-González, C.; de Medina-Arellano, M.J. Mitochondrial replacement techniques and Mexico's rule of law: On the legality of the first maternal spindle transfer case. *J. Law Biosci.* **2017**, *4*, 50–69. [CrossRef]
147. Ishii, T. Potential impact of human mitochondrial replacement on global policy regarding germline gene modification. *Reprod. Biomed. Online* **2014**, *29*, 150–155. [CrossRef]
148. Hamilton, G. The mitochondria mystery. *Nature* **2015**, *525*, 444–446. [CrossRef]
149. Hyslop, L.A.; Blakeley, P.; Craven, L.; Richardson, J.; Fogarty, N.M.E.; Fragouli, E.; Lamb, M.; Wamaitha, S.E.; Prathalingam, N.; Zhang, Q.; et al. Towards clinical application of pronuclear transfer to prevent mitochondrial DNA disease. *Nature* **2016**, *534*, 383–386. [CrossRef]
150. Hudson, G.; Takeda, Y.; Herbert, M. Reversion after replacement of mitochondrial DNA. *Nature* **2019**, *574*, E8–E11. [CrossRef]
151. Kang, E.; Wu, J.; Gutierrez, N.M.; Koski, A.; Tippner-Hedges, R.; Agaronyan, K.; Platero-Luengo, A.; Martinez-Redondo, P.; Ma, H.; Lee, Y.; et al. Mitochondrial replacement in human oocytes carrying pathogenic mitochondrial DNA mutations. *Nature* **2016**, *540*, 270–275. [CrossRef]
152. Yamada, M.; Emmanuele, V.; Sanchez-Quintero, M.J.; Sun, B.; Lallos, G.; Paull, D.; Zimmer, M.; Pagett, S.; Prosser, R.W.; Sauer, M.V.; et al. Genetic Drift Can Compromise Mitochondrial Replacement by Nuclear Transfer in Human Oocytes. *Cell Stem Cell* **2016**, *18*, 749–754. [CrossRef]
153. Ramalho-Santos, J.; Varum, S.; Amaral, S.; Mota, P.C.; Sousa, A.P.; Amaral, A. Mitochondrial functionality in reproduction: From gonads and gametes to embryos and embryonic stem cells. *Hum. Reprod. Update* **2009**, *15*, 553–572. [CrossRef] [PubMed]
154. Nissanka, N.; Moraes, C.T. Mitochondrial DNA heteroplasmy in disease and targeted nuclease-based therapeutic approaches. *EMBO Rep.* **2020**, *21*, e49612. [CrossRef] [PubMed]
155. Reddy, P.; Ocampo, A.; Suzuki, K.; Luo, J.; Bacman, S.R.; Williams, S.L.; Sugawara, A.; Okamura, D.; Tsunekawa, Y.; Wu, J.; et al. Selective elimination of mitochondrial mutations in the germline by genome editing. *Cell* **2015**, *161*, 459–469. [CrossRef] [PubMed]

156. King, M.; Attardi, G. Human cells lacking mtDNA: Repopulation with exogenous mitochondria by complementation. *Science* **1989**, *246*, 500–503. [CrossRef] [PubMed]
157. Taylor, R.W.; Chinnery, P.F.; Turnbull, D.M.; Lightowlers, R.N. Selective inhibition of mutant human mitochondrial DNA replication in vitro by peptide nucleic acids. *Nat. Genet.* **1997**, *15*, 212–215. [CrossRef]
158. Demidov, V.V.; Potaman, V.N.; Frank-Kamenetskil, M.D.; Egholm, M.; Buchard, O.; Sönnichsen, S.H.; Nlelsen, P.E. Stability of peptide nucleic acids in human serum and cellular extracts. *Biochem. Pharmacol.* **1994**, *48*, 1310–1313. [CrossRef]
159. Chinnery, P.F.; Taylor, R.W.; Diekert, K.; Lill, R.; Turnbull, D.M.; Lightowlers, R.N. Peptide nucleic acid delivery to human mitochondria. *Gene Ther.* **1999**, *6*, 1919–1928. [CrossRef]
160. Muratovska, A.; Lightowlers, R.N.; Taylor, R.W.; Turnbull, D.M.; Smith, R.A.J.; Wilce, J.A.; Martin, S.W.; Murphy, M.P. Targeting peptide nucleic acid (PNA) oligomers to mitochondria within cells by conjugation to lipophilic cations: Implications for mitochondrial DNA replication, expression and disease. *Nucleic Acids Res.* **2001**, *29*, 1852–1863. [CrossRef]
161. Egholm, M.; Buchardt, O.; Christensen, L.; Behrens, C.; Freier, S.M.; Driver, D.A.; Berg, R.H.; Kim, S.K.; Norden, B.; Nielsen, P.E. PNA hybridizes to complementary oligonucleotides obeying the Watson-Crick hydrogen-bonding rules. *Nature* **1993**, *365*, 566–568. [CrossRef]
162. Orum, H.; Nielsen, P.E.; Egholm, M.; Berg, R.H.; Buchardt, O.; Stanley, C. Single base pair mutation analysis by PNA directed PCR clamping. *Nucleic Acids Res.* **1993**, *21*, 5332–5336. [CrossRef] [PubMed]
163. Kolesnikova, O.A.; Kazakova, H.; Comte, C.; Steinberg, S.; Kamenski, P.; Martin, R.P.; Tarassov, I.A.; Entelis, N. Selection of RNA aptamers imported into yeast and human mitochondria. *Rna* **2010**, *16*, 926–941. [CrossRef] [PubMed]
164. Magalhães, P.J.; Andreu, A.L.; Schon, E.A. Evidence for the presence of 5S rRNA in mammalian mitochondria. *Mol. Biol. Cell* **1998**, *9*, 2375–2382. [CrossRef] [PubMed]
165. Entelis, N.S.; Kolesnikova, O.A.; Dogan, S.; Martin, R.P.; Tarassov, I.A. 5 S rRNA and tRNA import into human mitochondria: Comparison of in vitro requirements. *J. Biol. Chem.* **2001**, *276*, 45642–45653. [CrossRef] [PubMed]
166. Bowmaker, M.R.; Yang, M.Y.; Yasukawa, T.; Reyes, A.; Jacobs, H.T.; Huberman, J.A.; Holt, I.J. Mammalian Mitochondrial DNA Replicates Bidirectionally from an Initiation Zone. *J. Biol. Chem.* **2003**, *278*, 50961–50969. [CrossRef]
167. Comte, C.; Tonin, Y.; Heckel-Mager, A.-M.; Boucheham, A.; Smirnov, A.; Auré, K.; Lombès, A.; Martin, R.P.; Entelis, N.; Tarassov, I. Mitochondrial targeting of recombinant RNAs modulates the level of a heteroplasmic mutation in human mitochondrial DNA associated with Kearns Sayre Syndrome. *Nucleic Acids Res.* **2013**, *41*, 418–433. [CrossRef]
168. Tonin, Y.; Heckel, A.-M.; Vysokikh, M.; Dovydenko, I.; Meschaninova, M.; Rötig, A.; Munnich, A.; Venyaminova, A.; Tarassov, I.A.; Entelis, N. Modeling of antigenomic therapy of mitochondrial diseases by mitochondrially addressed RNA targeting a pathogenic point mutation in mitochondrial DNA. *J. Biol. Chem.* **2014**, *289*, 13323–13334. [CrossRef]
169. Loutre, R.; Heckel, A.-M.; Jeandard, D.; Tarassov, I.; Entelis, N. Anti-replicative recombinant 5S rRNA molecules can modulate the mtDNA heteroplasmy in a glucose-dependent manner. *PLoS ONE* **2018**, *13*, e0199258. [CrossRef]
170. Jeandard, D.; Smirnova, A.; Tarassov, I.; Barrey, E.; Smirnov, A.; Entelis, N. Import of Non-Coding RNAs into Human Mitochondria: A Critical Review and Emerging Approaches. *Cells* **2019**, *8*, 286. [CrossRef]
171. Gammage, P.A.; Moraes, C.T.; Minczuk, M. Mitochondrial Genome Engineering: The Revolution May Not Be CRISPR-Ized. *Trends Genet.* **2018**, *34*, 101–110. [CrossRef]
172. Moretton, A.; Morel, F.; Macao, B.; Lachaume, P.; Ishak, L.; Lefebvre, M.; Garreau-Balandier, I.; Vernet, P.; Falkenberg, M.; Farge, G. Selective mitochondrial DNA degradation following double-strand breaks. *PLoS ONE* **2017**, *12*, e0176795. [CrossRef] [PubMed]
173. Pohjoismäki, J.L.O.; Goffart, S. Of circles, forks and humanity: Topological organisation and replication of mammalian mitochondrial DNA. *BioEssays* **2011**, *33*, 290–299. [CrossRef] [PubMed]
174. Holt, I.J.; Dunbar, D.R.; Jacobs, H.T. Behaviour of a population of partially duplicated mitochondrial DNA molecules in cell culture: Segregation, maintenance and recombination dependent upon nuclear background. *Hum. Mol. Genet.* **1997**, *6*, 1251–1260. [CrossRef] [PubMed]
175. Clayton, D.A.; Vinograd, J. Complex mitochondrial DNA in leukemic and normal human myeloid cells. *Proc. Natl. Acad. Sci. USA* **1969**, *62*, 1077–1084. [CrossRef]
176. Clayton, D.A.; Vinograd, J. Circular dimer and catenate forms of mitochondrial DNA in human leukaemic leucocytes. *Nature* **1967**, *216*, 652–657. [CrossRef]
177. Pohjoismäki, J.L.O.; Goffart, S.; Tyynismaa, H.; Willcox, S.; Ide, T.; Kang, D.; Suomalainen, A.; Karhunen, P.J.; Griffith, J.D.; Holt, I.J.; et al. Human heart mitochondrial DNA is organized in complex catenated networks containing abundant four-way junctions and replication forks. *J. Biol. Chem.* **2009**, *284*, 21446–21457. [CrossRef]
178. Kraytsberg, Y.; Schwartz, M.; Brown, T.A.; Ebralidse, K.; Kunz, W.S.; Clayton, D.A.; Vissing, J.; Khrapko, K. Recombination of Human Mitochondrial DNA. *Science* **2004**, *304*, 981. [CrossRef]
179. Hoolahan, A.H.; Blok, V.C.; Gibson, T.; Dowton, M. Evidence of animal mtDNA recombination between divergent populations of the potato cyst nematode Globodera pallida. *Genetica* **2012**, *140*, 19–29. [CrossRef]
180. Ujvari, B.; Dowton, M.; Madsen, T. Mitochondrial DNA recombination in a free-ranging Australian lizard. *Biol. Lett.* **2007**, *3*, 189–192. [CrossRef]
181. Ciborowski, K.L.; Consuegra, S.; De Leániz, C.G.; Beaumont, M.A.; Wang, J.; Jordan, W.C. Rare and fleeting: An example of interspecific recombination in animal mitochondrial DNA. *Biol. Lett.* **2007**, *3*, 554–557. [CrossRef]

182. Srivastava, S.; Moraes, C.T. Double-strand breaks of mouse muscle mtDNA promote large deletions similar to multiple mtDNA deletions in humans. *Hum. Mol. Genet.* **2005**, *14*, 893–902. [CrossRef] [PubMed]
183. Bacman, S.R.; Williams, S.L.; Moraes, C.T. Intra-and inter-molecular recombination of mitochondrial DNA after in vivo induction of multiple double-strand breaks. *Nucleic Acids Res.* **2009**, *37*, 4218–4226. [CrossRef] [PubMed]
184. Chen, X.J. Mechanism of Homologous Recombination and Implications for Aging-Related Deletions in Mitochondrial DNA. *Microbiol. Mol. Biol. Rev.* **2013**, *77*, 476–496. [CrossRef] [PubMed]
185. Tadi, S.K.; Sebastian, R.; Dahal, S.; Babu, R.K.; Choudhary, B.; Raghavan, S.C. Microhomology-mediated end joining is the principal mediator of double-strand break repair during mitochondrial DNA lesions. *Mol. Biol. Cell* **2016**, *27*, 223–235. [CrossRef]
186. Dmitrieva, N.I.; Malide, D.; Burg, M.B. Mre11 is expressed in mammalian mitochondria where it binds to mitochondrial DNA. *Am. J. Physiol. Regul. Integr. Comp. Physiol.* **2011**, *301*, R632–R640. [CrossRef]
187. Brunyanszki, A.; Szczesny, B.; Virág, L.; Szabo, C. Mitochondrial poly(ADP-ribose) polymerase: The Wizard of Oz at work. *Free Radic. Biol. Med.* **2016**, *100*, 257–270. [CrossRef]
188. Peeva, V.; Blei, D.; Trombly, G.; Corsi, S.; Szukszto, M.J.; Rebelo-guiomar, P.; Gammage, P.A.; Kudin, A.P.; Becker, C.; Altmüller, J.; et al. Linear mitochondrial DNA is rapidly degraded by components of the replication machinery. *Nat. Commun.* **2018**, *9*, 1727. [CrossRef]
189. Schmidt, O.; Pfanner, N.; Meisinger, C. Mitochondrial protein import: From proteomics to functional mechanisms. *Nat. Rev. Mol. Cell Biol.* **2010**, *11*, 655–667. [CrossRef]
190. Srivastava, S.; Moraes, C.T. Manipulating mitochondrial DNA heteroplasmy by a mitochondrially targeted restriction endonuclease. *Hum. Mol. Genet.* **2001**, *10*, 3093–3099. [CrossRef]
191. Tanaka, M.; Borgeld, H.-J.; Zhang, J.; Muramatsu, S.; Gong, J.-S.; Yoneda, M.; Maruyama, W.; Naoi, M.; Ibi, T.; Sahashi, K.; et al. Gene therapy for mitochondrial disease by delivering restriction endonuclease SmaI into mitochondria. *J. Biomed. Sci.* **2002**, *9*, 534–541. [CrossRef]
192. Carroll, D. Genome engineering with zinc-finger nucleases. *Genetics* **2011**, *188*, 773–782. [CrossRef]
193. Joung, J.K.; Sander, J.D. TALENs: A widely applicable technology for targeted genome editing. *Nat. Rev. Mol. Cell Biol.* **2013**, *14*, 49–55. [CrossRef] [PubMed]
194. Bacman, S.R.; Williams, S.L.; Pinto, M.; Peralta, S.; Moraes, C.T. Specific elimination of mutant mitochondrial genomes in patient-derived cells by mitoTALENs. *Nat. Med.* **2013**, *19*, 1111–1113. [CrossRef] [PubMed]
195. Gammage, P.A.; Rorbach, J.; Vincent, A.I.; Rebar, E.J.; Minczuk, M. Mitochondrially targeted ZFNs for selective degradation of pathogenic mitochondrial genomes bearing large-scale deletions or point mutations. *EMBO Mol. Med.* **2014**, *6*, 458–466. [CrossRef] [PubMed]
196. Chin, R.M.; Panavas, T.; Brown, J.M.; Johnson, K.K. Optimized Mitochondrial Targeting of Proteins Encoded by Modified mRNAs Rescues Cells Harboring Mutations in mtATP6. *Cell Rep.* **2018**, *22*, 2818–2826. [CrossRef]
197. Hiroyuki, S.; Susumu, K. New restriction endonucleases from Flavobacterium okeanokoites (FokI) and Micrococcus luteus (MluI). *Gene* **1981**, *16*, 73–78. [CrossRef]
198. Ravasi, T.; Huber, T.; Zavolan, M.; Forrest, A.; Gaasterland, T.; Grimmond, S.; Arakawa, T.; Carninci, P.; Kawai, J.; Hayashizaki, Y. Systematic characterization of the zinc-finger-containing proteins in the mouse transcriptome. *Genome Res.* **2003**, *13*, 1430–1442. [CrossRef]
199. Macara, I.G. Transport into and out of the Nucleus. *Microbiol. Mol. Biol. Rev.* **2001**, *65*, 570–594. [CrossRef]
200. Minczuk, M.; Papworth, M.A.; Kolasinska, P.; Murphy, M.P.; Klug, A. Sequence-specific modification of mitochondrial DNA using a chimeric zinc finger methylase. *Proc. Natl. Acad. Sci. USA* **2006**, *103*, 19689–19694. [CrossRef]
201. Strauß, A.; Lahaye, T. Zinc fingers, TAL effectors, or Cas9-based DNA binding proteins: What's best for targeting desired genome loci? *Mol. Plant* **2013**, *6*, 1384–1387. [CrossRef]
202. Gammage, P.A.; Gaude, E.; Van Haute, L.; Rebelo-Guiomar, P.; Jackson, C.B.; Rorbach, J.; Pekalski, M.L.; Robinson, A.J.; Charpentier, M.; Concordet, J.P.; et al. Near-complete elimination of mutant mtDNA by iterative or dynamic dose-controlled treatment with mtZFNs. *Nucleic Acids Res.* **2016**, *44*, 7804–7816. [CrossRef] [PubMed]
203. Minczuk, M.; Papworth, M.A.; Miller, J.C.; Murphy, M.P.; Klug, A. Development of a single-chain, quasi-dimeric zinc-finger nuclease for the selective degradation of mutated human mitochondrial DNA. *Nucleic Acids Res.* **2008**, *36*, 3926–3938. [CrossRef] [PubMed]
204. Pereira, C.V.; Bacman, S.R.; Arguello, T.; Zekonyte, U.; Williams, S.L.; Edgell, D.R.; Moraes, C.T. mitoTev-TALE: A monomeric DNA editing enzyme to reduce mutant mitochondrial DNA levels. *EMBO Mol. Med.* **2018**, *10*, e8084. [CrossRef] [PubMed]
205. Phillips, A.F.; Millet, A.R.; Tigano, M.; Dubois, S.M.; Crimmins, H.; Babin, L.; Charpentier, M.; Piganeau, M.; Brunet, E.; Sfeir, A. Single-Molecule Analysis of mtDNA Replication Uncovers the Basis of the Common Deletion. *Mol. Cell* **2017**, *65*, 527–538.e6. [CrossRef] [PubMed]
206. Jinek, M.; Chylinski, K.; Fonfara, I.; Hauer, M.; Doudna, J.A.; Charpentier, E. A programmable dual-RNA-guided DNA endonuclease in adaptive bacterial immunity. *Science* **2012**, *337*, 816–821. [CrossRef] [PubMed]
207. Hsu, P.D.; Lander, E.S.; Zhang, F. Development and applications of CRISPR-Cas9 for genome engineering. *Cell* **2014**, *157*, 1262–1278. [CrossRef]
208. Gaudelli, N.M.; Komor, A.C.; Rees, H.A.; Packer, M.S.; Badran, A.H.; Bryson, D.I.; Liu, D.R. Programmable base editing of T to G C in genomic DNA without DNA cleavage. *Nature* **2017**, *551*, 464–471. [CrossRef]

209. Komor, A.C.; Kim, Y.B.; Packer, M.S.; Zuris, J.A.; Liu, D.R. Programmable editing of a target base in genomic DNA without double-stranded DNA cleavage. *Nature* **2016**, *533*, 420–424. [CrossRef]
210. Mok, B.Y.; de Moraes, M.H.; Zeng, J.; Bosch, D.E.; Kotrys, A.V.; Raguram, A.; Hsu, F.S.; Radey, M.C.; Peterson, S.B.; Mootha, V.K.; et al. A bacterial cytidine deaminase toxin enables CRISPR-free mitochondrial base editing. *Nature* **2020**, *583*, 631–637. [CrossRef]

life

MDPI

Review

Mitochondrial Kinases and the Role of Mitochondrial Protein Phosphorylation in Health and Disease

Veronika Kotrasová [1,†], Barbora Keresztesová [1,2,†], Gabriela Ondrovičová [1], Jacob A. Bauer [1], Henrieta Havalová [1], Vladimír Pevala [1], Eva Kutejová [1,*] and Nina Kunová [1,2,*]

1 Institute of Molecular Biology, Slovak Academy of Sciences, Dúbravská Cesta 21, 845 51 Bratislava, Slovakia; veronika.kotrasova@savba.sk (V.K.); umbibarb@savba.sk (B.K.); gabriela.ondrovicova@savba.sk (G.O.); jacob.bauer@savba.sk (J.A.B.); umbiheha@savba.sk (H.H.); vladimir.pevala@savba.sk (V.P.)

2 First Faculty of Medicine, Institute of Biology and Medical Genetics, Charles University, 128 00 Prague, Czech Republic

* Correspondence: eva.kutejova@savba.sk (E.K.); nina.kunova@savba.sk (N.K.)

† Authors contributed equally.

Abstract: The major role of mitochondria is to provide cells with energy, but no less important are their roles in responding to various stress factors and the metabolic changes and pathological processes that might occur inside and outside the cells. The post-translational modification of proteins is a fast and efficient way for cells to adapt to ever changing conditions. Phosphorylation is a post-translational modification that signals these changes and propagates these signals throughout the whole cell, but it also changes the structure, function and interaction of individual proteins. In this review, we summarize the influence of kinases, the proteins responsible for phosphorylation, on mitochondrial biogenesis under various cellular conditions. We focus on their role in keeping mitochondria fully functional in healthy cells and also on the changes in mitochondrial structure and function that occur in pathological processes arising from the phosphorylation of mitochondrial proteins.

Keywords: mitochondria; kinases; phosphorylation; disease

Citation: Kotrasová, V.; Keresztesová, B.; Ondrovičová, G.; Bauer, J.A.; Havalová, H.; Pevala, V.; Kutejová, E.; Kunová, N. Mitochondrial Kinases and the Role of Mitochondrial Protein Phosphorylation in Health and Disease. *Life* **2021**, *11*, 82. https://doi.org/10.3390/life11020082

Received: 29 December 2020
Accepted: 20 January 2021
Published: 23 January 2021

1. Introduction

Cells continuously undergo numerous changes as a result of metabolic conditions, stress and pathological situations. Mitochondria are organelles that contribute substantially to managing all of these challenges. Generally, cells respond to varying conditions in two ways. One way is by synthesizing newly demanded proteins and degrading those whose level must be reduced. For mitochondria, a large amount of energy and careful signaling is needed to manage all the messages continuously flowing into and out of the organelle. The second way is to modulate protein functionality through post-translational modifications (PTMs), which enables a rapid adaptation by precise modulation of stability, structure and function of particular proteins. Proteins can be phosphorylated, succinylated, ubiquitinated, sumoylated, acetylated, glycosylated, nitrated and malonylated. PTM is a reversible process which allows modified proteins to be quickly returned to their original state when conditions are altered. Phosphorylation is a PTM that is involved both in signaling and in modifying protein functions and is frequently employed in mitochondria. Accordingly, proteomic studies have found that ~40% of the organellar proteome is phosphorylated [1–3]. Moreover, to date, at least 25 protein kinases have been reported to either associate with mitochondria or have mitochondrial substrates (for rev. see [4–7]). Both Ser/Thr kinases, such as AKT [8,9], CK2 [10,11], GSK-3β [12], PKA [10,11], isoforms of PKC [13,14], PINK1 [15–17], components of the MAPK signaling pathway (JKN, p38) [18,19], and LRRK2 [20], and tyrosine kinases, such as ABL [21,22] and members of the SRC family (FYN [23], SRC [24–26], LYN [27], FGR [28,29], CSK [30] and EGFR [31]) have been identified to function within mitochondria. The phosphorylation of mitochondrial proteins has been connected with the dysregulation of mitochondrial functions that is found in neurodegenerative diseases, cancer, diabetes, heart dysfunction and aging. Here, we provide an overview of mitochondrial kinases, their substrates and functions and the diseases associated with their dysregulation.

2. Kinases on the Outer Mitochondrial Membrane

Protein kinases selectively modify other proteins by covalently attaching a phosphate group to a specific amino-acid residue, thus allowing cells to rapidly and reversibly adapt to changing cellular conditions. The human kinome consists of more than 500 kinases [32], which potentially affect nearly every cellular process, including those taking place in mitochondria. A growing number of mitochondrial proteins have been found to be phosphorylated, but only a fraction of the known protein kinases has been found either within the organelle itself or associated with its outer (OMM) or inner mitochondrial membranes (IMM) [4,5,33].

Almost no protein kinase has a known mitochondrial targeting signal, so their transport into mitochondria is still poorly understood. Several studies have shown that some kinases originally present in the cytosol use proteins embedded in the OMM as a scaffold [34,35]. For example, the cytosolic Jun N-terminal kinase (JNK) interacts with the membrane-bound protein Sab (SH3-binding protein 5), and disrupting this interaction has a neuroprotective effect in rat Parkinson's disease (PD) models and cerebral ischemia [36,37]. Interestingly, an anchoring protein for one kinase might be a substrate for another. In this case, Sab was shown to be phosphorylated by p38 (a mitogen-activated protein kinase, MAPK) [38]. A second example is the cAMP-dependent protein kinase A (PKA), which can be docked to the OMM by the A-kinase anchoring proteins (AKAPs), for example AKAP1 or the AKAP84 found in sperm [34,39]. AKAPs usually bind specifically to PKA regulatory subunit II, although several dual specificity AKAPs, which bind both regulatory subunits I and II, have been reported [40]. AKAP1 was shown to be phosphorylated by AMPK (5′-AMP-activated protein kinase) [41]. AMPK, in addition to AKAP1, has only two other currently known mitochondrial substrates, ACC2 (acetyl-CoA carboxylase 2), a lipid metabolism protein, and the mitochondrial fission factor (MFF), the receptor of mitochondrial fission [42,43].

PKA is also associated with other mitochondrial compartments, though the process by which PKA moves across the OMM into the matrix is still not fully understood. PKA-mediated phosphorylation affects a wide variety of mitochondrial functions, including protein import, oxidative phosphorylation, steroid hormone metabolism, mtDNA maintenance and apoptosis (see below for these). It also plays a role in mitophagy, a selective mitochondrial degradation process. PKA regulates mitophagy by modifying DRP1 (dynamin-related protein 1), a protein belonging to the dynamin family of large GTPases. Phosphorylation of DRP1 at Ser637 inhibits mitophagy and promotes mitochondrial fusion, most likely by increasing its binding to mitochondrial elongation factor 1/2 (MIEF1/2) [44,45]. A decrease in mitochondrial fission is also caused by the phosphorylation of DRP1 by AMPK, which plays a protective role in nervous system mitochondria, as well as by DRP1 phosphorylation by the Ca^{2+}/calmodulin-dependent protein kinase (CAMK) that acts in cardiomyocytes [46,47]. Mitochondrial fission is promoted, on the other hand, by DRP1 phosphorylation at Ser585 by CDK5 (cyclin-dependent kinase 5) [48], by modification of Tyr266 by the Abelson tyrosine kinase (ABL) [22], and by phosphorylation at the highly conserved Ser616 by ERK1/2 (extracellular receptor kinase 1/2), which increased mitochondrial fragmentation in human pancreatic adenocarcinoma [49]. In an Alzheimer's disease (AD) mouse model, GSK-3β (glycogen synthase kinase 3β) causes phosphorylation of DRP1 at Ser40 and Ser44, which promotes its GTPase activity, thereby increasing mitochondrial division and neuronal apoptosis [50]. In dopaminergic mouse neuronal cells (SN4741), DRP1 Ser616 is phosphorylated by p38 MAPK, which increases its mitochondrial translocation, resulting in mitochondrial dysfunction and neuronal loss [51].

The phosphorylation of DRP1 associated with mitophagy inhibition is abolished by the activity of *PTEN*-induced putative kinase 1 (PINK1) [52]. PINK1, when targeted into the mitochondria, disrupts the anchoring of PKA to the OMM by inhibiting its binding to AKAP1 and thus ensuring the necessary fission of damaged mitochondria for organellar degradation [53].

PINK1 is one of the few mitochondrial kinases to contain a true mitochondrial targeting sequence (others are creatine kinase [54] and PAK5 (p21-activated kinase 5) [55]). In healthy mitochondria, PINK1 translocates into the matrix and, after phosphorylating its substrates, immediately undergoes rapid cleavage by the mitochondrial proteases MPP (mitochondrial processing peptidase) and PARL (presenilins-associated rhomboid-like protein) with further targeting to the proteasome [56–59]. When mitochondria lose their membrane potential, as a result of aging or chemical exposure, auto-phosphorylated PINK1 accumulates in the OMM, forming a complex with proteins of the TOM (translocase of the outer membrane) machinery [60,61]. The kinase domain of PINK1 also phosphorylates the E3-ubiquitin ligase Parkin, which activates its enzymatic functions and triggers the processes leading to mitophagy [62].

Mutations in genes encoding either PINK1 or Parkin are implicated in the pathogenesis connected with Parkinson's disease [63–66] along with mutations to LRRK2 (leucine-rich repeat kinase 2), which is also found in the OMM [67,68]. Mutated LRRK2 confers aberrant kinase activity, which results in the increased phosphorylation of its substrates, leading to malfunctions in various mitochondrial processes including increased reactive-oxygen species (ROS) production [69], mitochondrial dynamics [70], and mitophagy [71] and also dysregulated calcium homeostasis [72].

Two other proteins involved in the mitochondrial fusion/fission cycle are also protein kinase targets. Mitofusin 1 (Mfn1) when phosphorylated by ERK1/2 inhibits mitochondrial fusion [73] while mitofusin 2 (Mfn2) when phosphorylated by JNK becomes a substrate for the proteasome and promotes the fragmentation of mitochondria in human osteosarcoma cells [74]. Phosphorylation of Mfn2 by PKA at Ser442 inhibits the Ras signaling pathway in rat vascular smooth muscle (VSM) cells, suppressing VSM cell growth in vascular disorders such as post-angioplasty restenosis and atherosclerosis [75].

Additionally, two less well-known protein kinases have also been found associated with the OMM. Mammalian target of rapamycin (mTOR) is an important regulator of cell growth, protein and lipid synthesis, autophagy and other mitochondrial functions (for rev. see [76]); it was found in the outer membrane of human T-cell mitochondria [77,78]. A direct substrate for this membrane-associated protein is still unknown, however it has been proposed to have a role in apoptosis and leukemia [79]. At least in three cancer cell lines, mTOR phosphorylates hexokinase II, which catalyzes one of the first steps of glucose metabolism in mitochondria. This phosphorylation may have a decisive role in switching between glycolytic pathway and oxidative phosphorylation, thus increasing tumor cell resistance [80,81]. In neural cells, including those in neuroblastoma, PAK5 was shown to be localized to the OMM [55,82] where it phosphorylates the pro-apoptotic protein Bad, thereby inhibiting apoptosis [55,82].

To date, protein kinases confirmed to be localized to the OMM are Ser/Thr kinases. In addition to PKA [83,84], JNK [85], ERK1/2 [86], p38 MAPK [87], PINK1 [88], CDK5 [48], LRRK2 [89], mTOR [77], PAK5 [82], CAMK [47] and AMPK [90], they also include cyclin-dependent kinase 11 (CDK11) [91], casein kinases 1 and 2 (CK1, CK2) [92,93], and three isoforms of protein kinase C (PKC α, PKC δ and PKC ε) [94–96]. All these kinases phosphorylate diverse mitochondrial substrates, resulting in many different alterations to organellar dynamics, protein import, metabolism, respiratory complex activity and apoptosis, all of which will be more extensively discussed in the following chapters.

3. Phosphorylation in Mitochondrial Import Machinery

Mitochondria are semiautonomous organelles with their own transcription-translation system. Although they are able to synthesize some of their functionally important components, most of the essential ions, small molecules, RNAs [97,98] and proteins [99] need to be imported from the cytoplasm.

Indeed, most mitochondrial proteins are synthesized in the cytosol and only afterwards transported to their designated location within the organelle. Protein kinases alter these processes in two different ways: they either phosphorylate the mitochondrial signaling pre-sequences of these proteins [100] or they modify the components of the import machinery [99,101]. These components include the outer-membrane complexes translocase of the outer membrane (TOM), mitochondrial import complex (MIM) and sorting and assembly machinery (SAM) and the inter- and inner-membrane systems translocase of the inner membrane (TIM) and mitochondrial intermembrane space assembly (MIA) (Figure 1). Moreover, the outer-membrane TOM complex was shown to directly cooperate with both MIA and TIM in the inner mitochondrial membrane, as well as the SAM complex through the activity of the inter-membrane proteins TIM9 and TIM10 [102].

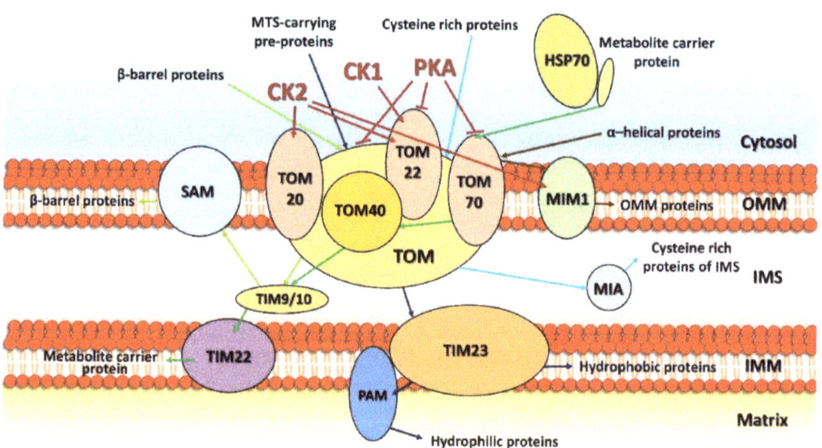

Figure 1. The major mitochondrial protein import pathways and the influence of protein kinases in *Saccharomyces cerevisiae*. MTS (mitochondrial targeting sequence)-carrying pre-proteins are imported through the TOM and TIM23 complexes. Proteins containing a hydrophobic sorting signal are embedded into the IMM while hydrophilic proteins are sent into the mitochondrial matrix through the PAM (protein import motor) complex. Cysteine-rich proteins are imported by the TOM and MIA protein translocation systems. The precursors of β-barrel proteins are translocated through the TOM and TIM9/10 complexes and are sorted and assembled by the SAM complex. The precursors of metabolite carriers are imported through TOM, TIM9/10 and TIM22, and several α-helical OMM proteins are imported by the MIM complex. Protein kinase A (PKA) phosphorylates the TOM40 and TOM22 precursors in the cytosol, thereby impairing their translocation and assembly into the mature TOM complex. PKA also phosphorylates TOM70 which abolishes its interaction with the metabolite carrier/chaperone HSP70 complex. CK1 phosphorylates TOM22, promoting its assembly, whereas CK2 phosphorylates both, TOM20 and TOM22. Phosphorylation of TOM22 facilitates its interaction with TOM20 and stimulates its assembly into the mature TOM complex. CK2 further phosphorylates MIM1 enhancing its stability and promoting the MIM1-dependent import of the TOM20 and TOM70 precursors. OMM, mitochondrial outer membrane; IMM, mitochondrial inner membrane.; IMS, intermembrane space.

The phosphorylation of mitochondrial pre-sequences was studied in the plant *Arabidopsis thaliana* [100]. By searching the PhosPhAt 4.0 Database [103], 103 mitochondrial pre-proteins with one or more experimentally determined phosphorylation sites were found. These included subunits of respiratory complexes I–V and proteins with functions in the tricarboxylic acid (TCA) cycle, protein translocation and degradation, RNA editing, transport and several others. More detailed studies of the respiratory complex subunits CAL2, SDH1-2, COX5b1, ATP17, F1β-1, F1β-2, and F1β-3 have shown that the N-terminal pre-sequences could be phosphorylated by several kinases with different but overlapping specificities, which might influence the protein import into plant mitochondria [100].

The effect of phosphorylation on the *Saccharomyces cerevisiae* mitochondrial import complexes has been well described [10,11,104,105] (Figure 1). A mass spectrometry analysis of purified *S. cerevisiae* mitochondria allowed the different phosphorylation sites of proteins in the TOM and MIM complexes to be mapped [105]. Several serine and threonine residues were found to be phosphorylated, including Ser84, Ser87, Thr92, Ser168 and Ser172 in TOM20; Thr5, Ser20, Ser44, Ser46, Thr76, and Thr129 in TOM22; Ser2,Thr5, Ser11, Ser50, Ser54, Ser77, and Thr220 in TOM40; Ser10 in TOM5; Ser16 in TOM6; Ser6 in TOM7; Ser65, Thr66, Ser69, Ser78, Ser166, Ser174, Thr228, Thr234, Thr232, and Thr520 in TOM70; Ser55, Ser73, and Ser96 in TOM 71; and Ser12 and Ser16 in MIM1 (all from [106]).

TOM20, TOM22 and TOM70 function as receptors for the import of nuclear-encoded precursor proteins into the mitochondria. TOM20 and TOM70 are responsible for the initial recognition of mitochondrial precursors that are later transferred to the central receptor, TOM22, and, through the action of TOM40, transported across the outer mitochondrial membrane. TOM20 is mainly phosphorylated by casein kinase 2 at Ser172, and inactivation of CK2 was shown to reduce the activity of both TOM20 and TOM22 [105]. Interestingly, however, the amino-acid substitution mutants TOM20^{S172A} (no phosphorylation at Ser172) and TOM20^{S172E} (mimicking phosphorylation at Ser172) showed no abnormalities in mitochondrial protein composition, TOM assembly or protein import [105,107]. In addition to being a CK2 substrate, TOM22 is also phosphorylated by casein kinase 1 [107]. CK2 modifies TOM22 at two proximal serines, Ser44 and Ser46, both of which are present in the active TOM complex and are already phosphorylated in the cytosol. On the other hand, the CK1-mediated phosphorylation of Thr57, which is required for the assembly of the TOM complex, takes place only after it has been imported into the mitochondria. Phosphorylation and TOM22 assembly are significantly increased in mitochondria obtained from yeast grown on glucose (a fermentable carbon source) compared to those from cells grown on glycerol (non-fermentable). Further observation showed that phosphorylations to TOM22 mediated by CK1 and CK2 are important for the recruitment and assembly of TOM20 into the TOM complex, thereby emphasizing the complex effects of phosphorylation on the formation and stability of the OMM import machinery and the biogenesis and dynamics of mitochondria. This is even more apparent for the phosphorylation of TOM22, TOM70 and TOM40 by PKA. PKA phosphorylates TOM22 at Thr76 in the cytosol, which decreases its binding to mitochondria and thus considerably reduces its membrane integration. Both CK1 and PKA, therefore, have opposites effects on TOM22 under fermentable growth conditions. Whereas CK1-mediated phosphorylation stimulates TOM22 function, PKA strongly inhibits it; thus, CK1 moderates the negative effect of PKA. In cells grown on a fermentable carbon source, TOM70 is phosphorylated at Ser174 in a PKA-dependent manner [105] and the PKA treatment of mitochondria isolated from such cells showed that only a fraction of TOM70 is phosphorylated in vivo. Here, it seems that phosphorylation does not affect either the targeting or the oligomerization of TOM70 but does influence its protein-transport function. TOM70 is the main receptor of non-cleavable hydrophobic protein carriers (like AAC, the phosphate carrier and the dicarboxylate carrier) [108,109], which are delivered to TOM70 by the cytosolic HSP70 complex. TOM70 phosphorylation at Ser174 substantially reduces the interaction between TOM70 and the precursor carrier protein complex with Hsp70, thereby leading to a great impairment in the import of metabolic carriers into the mitochondria [11,105].

TOM40, a hydrophobic channel that enables the majority of mitochondrial proteins to be imported [110], is phosphorylated by PKA at Ser54 in the cytosol [104]. This phosphorylation blocks the formation of the SAM–TOM40 complex intermediate important for the assembly of TOM40 into the mature TOM complex and inhibits its import into the OMM.

The whole process is negatively regulated by the presence of a fermentable carbon source like glucose or sucrose when PKA is rapidly activated [111]. Under these conditions, the assembly of a functional TOM complex is impaired, leading to lower mitochondrial activity. When switched to nonfermentable growth conditions, PKA activity lowers, and TOM40 is imported into the OMM in response, forming the channel that allows the effective protein transport and increased mitochondrial activity [104].

MIM1 is located in the OMM [112,113]. Although it is not a structural part of the TOM complex, it is required for the membrane insertion of the TOM complex components, including TOM20, TOM70 and TOM40. MIM1 is phosphorylated at Ser12 and Ser16 by CK2, partially in the cytosol and fully after its transport into the mitochondria [105], which strongly influences the assembly of TOM70 and TOM20 into the functional TOM complex.

Several MS analyses have shown that TOM complex components of human mitochondria are also phosphorylated [114–118], although more detailed functional studies are still lacking. Phosphorylated serine, threonine and tyrosine residues were found in TOM20 (Ser55, Ser135, Ser138), TOM22 (Ser15, Thr43, Thr45, Thr125), TOM40 (Tyr129, Ser142, Thr144, Ser157, Tyr194, Thr254, Thr255, Thr273, Ser317, Ser320, Thr338) and TOM70 (Ser19, Thr33, Thr85, Ser91, Ser96, Ser110, Tyr153, Tyr156, Ser253, Tyr260, Ser279, Tyr310, Tyr327, Tyr339, Ser434, Tyr601, Tyr607) [106].

Presently, it can only be assumed that some of these modifications may play roles in human diseases, especially cancer as, to the best of our knowledge, no individual studies have been done to date. In humans as in yeast, glucose repression substantially inhibits mitochondrial protein import and thereby reduces mitochondrial biogenesis. In summary, phosphorylation of the import machinery components regulates the translocation of proteins at multiple levels and allows for very precise cellular responses to metabolic and energetic demands.

4. Kinases and Metabolism

Mitochondria are the centers of energy production for most eukaryotic organisms. They integrate the metabolism of carbohydrates and amino acids and fatty acids to generate energy in the form of ATP. The key enzyme is the mitochondrial pyruvate dehydrogenase complex (PDC), which catalyzes the oxidative decarboxylation of pyruvate to acetyl-CoA [119]. This irreversible conversion links the glycolysis occurring in the cytosol to the Krebs cycle (citric or tricarboxylic acid (TCA) cycle) in the mitochondria and the synthesis of fatty acids and steroids [120,121]. PDC is one of the largest and most complex multi-enzyme systems known [122]; its activity is controlled by multiple reversible phosphorylations. In fact, PDC was one of the first phosphorylation-regulated enzymes to be described [123].

The mammalian PDC is composed of four major components: a thiamine diphosphate-dependent pyruvate dehydrogenase (E1), a dihydrolipoamide transacetylase (E2) containing covalently bound lipoyl groups, the flavoenzyme lipoamide dehydrogenase (E3), and an E3-binding protein (E3BP) [120]. It also contains a family of four pyruvate dehydrogenase kinases (PDKs 1–4), which inactive it through phosphorylation, and two pyruvate dehydrogenase phosphatases (PDPs 1 and 2), which dephosphorylates the enzyme to its active form [121]. The PDK isoenzymes belong to the ATPase/kinase superfamily [124–126] and phosphorylate three serine residues in the E1 α subunit: Ser293 (phosphorylation site 1), Ser300 (phosphorylation site 2) and Ser232 (phosphorylation site 3) [120,121,127,128]. Phosphorylation of Ser264 completely inactivates PDC, while phosphorylation of sites 2 and 3 maintains a low catalytic activity [129]. All four PDKs modify site 1 and 2; site 3 is phosphorylated only by PDK1 [129,130]. The activity of the PDKs has to be tightly regulated. Short-term regulation is carried out by metabolites; acetyl-CoA and NADH (both are formed by the pyruvate dehydrogenase (PDH) reaction and fatty acid β-oxidation), activate the complex while the pyruvate generated by glycolysis or circulating lactate inhibit it [131].

The PDC can also be modulated by cytosolic JNK, one of the MAPKs [119]. Phosphorylation-activated JNK was found in the OMM of primary cortical neurons where it was translocated in a response to H_2O_2 or anisomycin treatment (an activator of JNK/p38) and the PDC was shown to be its specific phosphorylation target [119]. Phosphorylated PDC reduces its pyruvate dehydrogenase activity, causing the cell to shift to anaerobic pyruvate metabolism, which is characterized by increased levels of lactate in the cytosol [119]. In aging rat heart, the amount of the phosphorylated, and therefore inactive, form of PDH was significantly increased, however the PDK activity was only slightly lower than that in younger rat cardiomyocytes [132]. The JNK-mediated inhibition of PDH and concomitant reduction of energy production seems to be especially important in tissues with high energy uptake like heart and brain or during brain aging [133]. Deficiencies in PDH regulation caused by malfunctions of JNK were shown in the pathogenesis of neurodegenerative disorders including Parkinson's and Alzheimer's diseases [134,135].

Hypoxia-induced glycolysis in the cytosol is coupled to suppressed mitochondrial pyruvate metabolism and oxidative phosphorylation [136]. Thus, pyruvate accumulated in the cytosol is converted to lactate (the so-called Warburg effect) [137] and the subsequent metabolic changes associated with acidosis promote cell proliferation and tumorigenesis [136,138,139]. Hypoxic conditions induce a signaling pathway leading to the translocation of cytosolic PGK1 (phosphoglycerate kinase 1), the first ATP-generating enzyme in the glycolytic pathway [140], into the mitochondria [136]. Mitochondrial PGK1, whose expression is upregulated in breast [141], pancreatic [142], ovarian [143], metastatic gastric [144] and colon cancer [145] and hepatocellular carcinomas [146], phosphorylates, and thereby activates, PDK1 at Thr338 [136]. The enhanced activity of PDK1 results in the inactivation of PDH by phosphorylation of Ser293, leading to the suppression of pyruvate utilization and ROS production in mitochondria, which is a poor prognosis for patients with glioblastoma [136].

PDC was further shown to be phosphorylated by the Ser/Thr kinase TPKI/GSK-3β (Tau protein kinase I/glycogen synthase kinase 3β) [147], which was originally identified as a tau-kinase associated with brain microtubules, possibly involved in phosphorylation of the Tau protein that occurs in Alzheimer's disease [148]; this kinase is presently known to play a key role in the molecular pathophysiology of many neurodegenerative diseases [12]. TPKI/GSK-3β translocates into the mitochondria upon stress-induction, where it phosphorylates and inhibits PDH in rat hippocampal cell cultures, again resulting in dysfunctional mitochondria and neuronal death [12,147].

A phosphoproteomic study of mitochondria isolated from potato tubers confirmed that the PDC is massively phosphorylated. The phosphorylation of formate dehydrogenase (FDH, which catalyzes the oxidation of formate to CO_2, reducing NAD^+ to NADH in the process) at Thr76 and Thr333 was also observed [149]. Phosphorylation of FDH decreased in response to higher levels of pyruvate, formate and NAD^+, whereas its activity was strongly increased by low mitochondrial oxygen concentrations, suggesting that phosphorylated FDH has a role in hypoxia [149].

In addition to the pyruvate and formate dehydrogenases, several other mitochondrial metabolic enzymes are phosphorylated by kinases. Aconitase, the enzyme catalyzing the second step of the TCA cycle (and also responsible for the metabolic regulation of iron and the stabilization of mtDNA) [150,151], is modified by the FGR tyrosine kinase at Tyr71, Tyr544 and Tyr665 in rat brain mitochondria [29,152]. All three tyrosines reside in a region that is highly conserved in all eukaryotes, which indicates that these phosphorylations are likely important for the stability and enzymatic activity of the protein [29]. FGR phosphorylates succinate dehydrogenase complex subunit A (SDHA) at Tyr535 and Tyr596 [29,152]. In *S. cerevisiae*, phosphorylation of citrate synthase, which mediates the first step of TCA cycle, at Ser462 completely abolishes its catalytic activity by inhibiting homodimer formation [153]. A mass spectrometry analysis of *Solanum tuberosum* mitochondria found that NAD-isocitrate dehydrogenase, succinyl-CoA ligase subunits α/β, NAD-malate dehydrogenase and 2 isoforms of NAD-malic enzyme were phosphorylated [149] (the physiological significance of these modifications remains largely unknown) [154].

Steroid hormone biosynthesis in mitochondria is also controlled through modifications mediated by protein kinases. The first and rate-limiting step in steroidogenesis is the transfer of cholesterol from the OMM to the IMM, which is facilitated by the Steroidogenic Acute Regulatory protein (StAR) [155]. Its inactive form resides within the IMM or matrix. Upon hormone stimulation, StAR is translocated to the OMM in a mitochondria-fusion-dependent manner and activated [156]. Here, phosphorylation mediated by PKA followed by ERK1/2 occurs, both of which are necessary for maximal steroid production [155,157]. StAR contains two consensus PKA-phosphorylation sites, Ser55/56/57 and Ser194/195, which are conserved across all eukaryotes [158,159]. PKA phosphorylation allows the binding of a cholesterol molecule to the hydrophobic pocket of StAR [157]. ERK1/2 phosphorylates StAR at Ser232, which elevates its activity and increases cholesterol transport [155,157].

The StAR activity cycle enables many cholesterol molecules to be metabolized per unit of StAR protein. When unnecessary, StAR may be dephosphorylated, and thereby deactivated, by mitochondrial phosphatases, but left intact to become available later to bind another cholesterol molecule or it may be translocated into the mitochondrial matrix and degraded by the ATP-dependent protease Lon [160].

5. OXPHOS System Modification

NADH and $FADH_2$, the electron carrier molecules produced by the TCA cycle, transfer their electrons to oxygen in several steps through respiratory chain complexes embedded in the inner mitochondrial membrane. This metabolic pathway, called oxidative phosphorylation (OXPHOS), uses a series of enzymes to transport electrons and generate the proton gradient used in ATP production. The OXPHOS system consists of five protein complexes (NADH dehydrogenase, succinate dehydrogenase, cytochrome *c* reductase, cytochrome *c* oxidase and ATP synthase) each of which is made up of numerous subunits, some of which are encoded by the mitochondrial DNA, and regulated by various PTMs including phosphorylation [161] (Figure 2).

NADH dehydrogenase (complex I), the largest of the respiratory chain complexes, is conserved in all eukaryotes except several yeast species (e.g., *S. cerevisiae, Schizosaccharomyces pombe*) [162]. It catalyzes the oxidation of NADH by reducing coenzyme Q, and this process is associated with the translocation of protons across the IMM. NADH dehydrogenase is an L-shaped protein complex consisting of several cofactors and as many as 45 subunits, 14 of which form the evolutionarily widely conserved core of the enzyme [163,164]. A bioinformatics analysis identified several potential phosphorylation sites, indicating that phosphorylation is prevalent in various subunits of complex I and may play a role in its assembly and regulation [165].

In mouse fibroblasts, a nuclear-encoded 18 kDa subunit (referred to as AQDQ or the NDUFS4 subunit) was shown to become phosphorylated in elevated levels of intramitochondrial cAMP, leading to a twofold increase of complex I activity. The mouse, bovine, human and *Neurospora crassa* sequences of this protein have consensus phosphorylation sites in both the leader sequence and the mature protein [166], and are phosphorylated by the cAMP-dependent PKA [167,168]. The importance of the PKA recognition site within the NDUFS4 subunit was shown by studies on patients with a fatal neurological Leigh-like syndrome. An inherited autosomal recessive mutation of a 5 bp duplication in the *NDUFS4* gene results in a 14 amino-acid extension of the mature NDUFS4 form which causes the loss of the phosphorylation site required to activate complex I without affecting the assembly nor the activity of other complexes (i.e., cytochrome *c* oxidase) [169,170]. This decreased activity of complex I resulted in the increased production of ROS [171,172]. In addition to regulating complex I activity, PKA also phosphorylates a serine residue in the C-terminal part of NDUFS4 that is necessary for its translocation from the cytosol into the mitochondria [173].

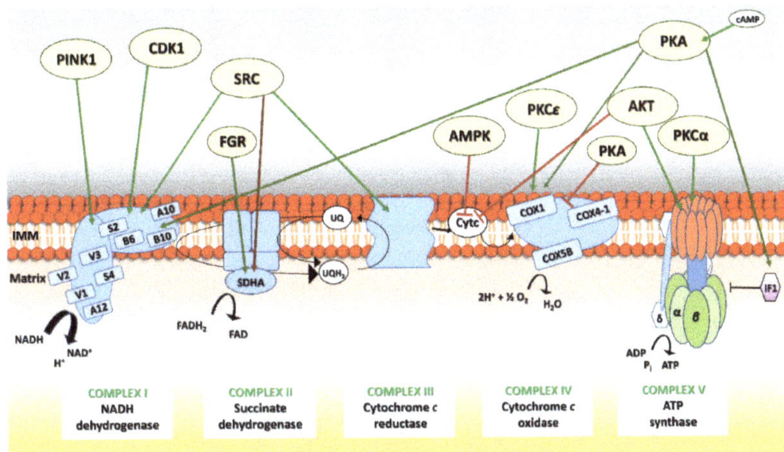

Figure 2. The phosphorylation of mammalian OXPHOS system components. The OXPHOS system consists of five protein complexes (NADH dehydrogenase, succinate dehydrogenase, cytochrome c reductase, cytochrome c oxidase and ATP synthase) embedded in the IMM. The protein kinase-regulated subunits are highlighted as follows: stimulation in green, repression in red and no effect in brown. Phosphorylation by PKA, PINK1, CDK1 and SRC increases the activity of complex I (NADH dehydrogenase). Succinate dehydrogenase (complex II) is regulated by the action of the FGR and SRC kinases. The activity of cytochrome c (Cytc) oxidase (complex III) is upregulated by phosphorylation by the SRC kinase. The activity of cytochrome c is impaired upon phosphorylation by AKT and AMPK. Complex IV (cytochrome c reductase) is negatively regulated by PKA (independently of cAMP levels) and positively by the cAMP-dependent action of PKA and PKC ε. Phosphorylation of ATP synthase (complex V) is upregulated by AKT and PKCα. Phosphorylation of the inhibition factor IF1 by PKA abolishes IF1 binding and its inhibition of ATP synthase activity. IMM, inner mitochondrial membrane.

In humans, the complex I subunit NDUFA10 is phosphorylated at Ser250 by PINK1 [174]. Studies in fly, mouse and human models demonstrated that *PINK1* mutations affect the activity of complex I and overall mitochondrial homeostasis by causing an impaired mitochondrial membrane potential, increased sensitivity to apoptotic stress and decreased complex I activity [174,175]. Mutations in PINK1 that prevent it from phosphorylating its substrates is one of the most significant causes of a recessive form of Parkinson's disease (PD) [174].

Other complex I subunits are phosphorylated by the cyclin B1/CDK1 kinase [176]. In mouse and human cells, subunits NDUFV1, NDUFV3, NDUFS2, NDUFB6 and NDUFA12 have been observed to be phosphorylated; the phosphorylated proteins increase mitochondrial respiration and energy production, which is greatly needed in the later stages of the cell cycle [176]. Furthermore, complex I subunits, NDUFV2 and NDUFB10, were shown to be phosphorylated by the SRC tyrosine kinase [25,26]. Phosphorylation of both subunits increased the levels of respiration, electron transfer and cellular ATP content [25,26]. In humans, inhibiting SRC leads to a decrease in mitochondrial respiration [26], while higher levels of SRC were observed in various types of tumors, indicating that the SRC-mediated maintenance of mitochondrial functions may play a role in tumor cell proliferation [25].

Succinate dehydrogenase (SDH; complex II) is an enzyme with important dual roles in the respiratory chain and TCA cycle. The SDHA flavoprotein is phosphorylated at Tyr215 by c-SRC in vitro and in vivo [26]. Phosphorylation of SDHA has no effect on its respiratory activity but is important for efficient electron transfer through electron transfer chain (ETC) complex II and preventing ROS generation [26]. In male mice with defective SDHA Tyr215 phosphorylation, a low humoral immune response was observed with the suppression of both Ig production and formation of a germinal center against antigens [177]. SDHA is also regulated by the FGR kinase-mediated phosphorylation of Tyr604, which is highly conserved among mammals [28,29]. The absence or inhibition of respiratory complex I (e.g., by rotenone) leads to increased SDHA phosphorylation and a concomitant increase in complex II activity, which may be an efficient compensatory mechanism for electron transfer and $FADH_2$ processing [28].

Not much is known about the phosphorylation of respiratory complex III. Several subunits were among the phosphoproteins identified by proteomic studies in human cells [178], although the functional significance of individual phosphorylations have not yet been elucidated. Here again, the SRC tyrosine kinase seems to be responsible for generally upregulating complex III activity. The same was shown for complexes I and IV, but not for the SRC-mediated decrease in activity of respiratory complex V [179].

Cytochrome *c* (Cytc) is a small protein containing a covalently bound heme group that functions as an electron carrier between mitochondrial ETC complexes III and IV. Mammalian Cytc contains several highly tissue-specific phosphorylation sites. Phosphorylated Tyr97, Tyr48, and Thr28 were found in heart, liver and kidney, respectively [180–182]. In rat renal Cytc, AMPK is responsible for the phosphorylation of Thr28, which lies close to the heme group and reduces its ability to transfer electrons to complex IV [180]. A similar effect was observed after introducing phosphorylation at Tyr48 [183]. Additionally, a phosphorylated Thr58 conserved in mammalian somatic Cytc isoforms (replaced with isoleucine in testis) was found in kidney and a site-specific Ser47 phosphorylated by AKT kinase (protein kinase B) in brain [182,184]. All these Cytc phosphorylations reduce the level of mitochondrial respiration, lower ROS generation, lower the mitochondrial membrane potential and decrease apoptotic activity (see Section 8 below) [180,182–184].

Human respiratory complex IV is composed of 13 subunits, three of which are encoded by mitochondrial DNA [185]. Various studies suggest that the phosphorylation of different subunits may lead to both inhibition and activation of complex IV under various pathophysiological conditions. Tyr304 in cytochrome *c* oxidase catalytic subunit I (COX1) is the target of a cAMP-dependent phosphorylation by an as yet unidentified tyrosine kinase [186]. Tumor necrosis factor-α (TNFα), a pro-inflammatory cytokine participating in the immune response in mammals, seems to be responsible for triggering the signal transduction leading to the activation of an unknown mitochondrial kinase modifying Tyr304 [187]. Like the Cytc phosphorylation mentioned above, phosphorylation of COX1 also leads to decreases in complex IV activity, ATP production and mitochondrial membrane potential [186,187].

Respiratory complex IV is also regulated by PKA, which has a regulatory role in hypoxia and ischemia. Elevated levels of PKA under hypoxic conditions or myocardial ischemia cause the hyperphosphorylation of several COX subunits, including COX1, COX4-1 and COX5b, which is associated with decreased complex IV activity [188]. Hypoxia-activated PKA, independently of mitochondrial cAMP levels, phosphorylates the COX5b subunit. Accordingly, the substitution of phosphorylated COX5b with a phosphorylation-resistant protein retained the activity of complex IV. Thereby, hypoxia-activated PKA responds to increased ROS production, induces PKA substrates phosphorylation and mediates signal for complex IV inhibition [189]. Under hypoxic and ischemic conditions, the levels of COX1, COX4-1 and COX5b are reduced [188]. Recent studies have shown that the mitochondrial protease Lon, which regulates the levels of mitochondrial nucleoid-associated proteins [190], also degrades COX subunits [191]. Lon, induced by hypoxia and ischemia, has a direct role in degrading phosphorylated COX4-1 and COX5b, whereas non-phosphorylated forms of these subunits were relatively resistant to Lon proteolysis [191]. In contrast, Acin-Perez et al. [192] described a PKA-mediated phosphorylation of COX4-1 on Ser58 that increased COX activity, probably by altering its ADP/ATP binding [192]. Other studies found that COX is regulated by protein kinase C, a Ser/Thr kinase PKC-ε [193,194]. Activated PKC is translocated to various subcellular sites depending on cellular conditions.

In mitochondria, hypoxia-induced PKC-ε interacts with and phosphorylates COX4, which may have a protective role against hypoxia by activating complex IV [193,194].

ATP synthase (complex V) is the last enzyme in the OXPHOS system. It uses the energy stored in the proton gradient across the IMM created by the ETC to drive the synthesis of ATP. In potato tuber mitochondria, two phosphorylated complex V proteins have been identified, the δ′-subunit of the F1-ATPase and the β-subunit of the ATPase [195]. In human skeletal muscle, seven potential phosphorylation sites were found in the β-subunit of F1-ATP synthase [196]. Interestingly, patients with type 2 diabetes had downregulated levels of the phosphorylated F1-ATP synthase β-subunit [196]. These results suggested that this phosphorylation may be an important mechanism for regulating electron-coupled ATP synthesis, and that disturbances in it may contribute to the pathogenesis of insulin resistance and type 2 diabetes [197]. Insulin resistance occurs when cells do not respond well to insulin and, at the same time, cannot use glucose from blood for energy. Insulin also stimulates the translocation of AKT to mitochondria, whose activation increases complex V activity [198]. In patients with insulin-resistant type 2 diabetes, a decrease in the mitochondrial translocation of AKT and a related decrease in the activity of ATP synthase were observed [198].

The activity of ATP synthase is also regulated by inhibitory factor 1 (IF1) [199,200]. The binding of IF1 depends on its phosphorylation status. Dephosphorylated IF1, which is abundantly present in human carcinomas, inhibits both the synthase and hydrolase activities of complex V and upregulates aerobic glycolysis. A PKA-mediated phosphorylation at Ser39 abolishes IF1 binding to ATP synthase [199,201]. Further studies revealed that the PKCα-dependent phosphorylation of the α, β and δ subunits of ATP synthase leads to an increase in the activity of ATP synthase, which suggests that PKCα might have a role in the preservation of mitochondrial functions after cell injury [95,202].

ATP synthase subunits were also among those proteins identified by phosphoproteomic studies of human cancers, however the exact function of these phosphorylations is currently not known [118].

Uncoupled respiration is a specific respiratory process that occurs in the mitochondria of brown and beige adipose cells, also known as fat cells [203]. Brown adipose tissue (BAT) is involved in thermogenesis, when energy is dissipated into heat in response to conditions like hibernation, cold stress or food intake [204,205]. Mitochondrial dysfunction in adipocytes can lead to metabolic diseases, including type II diabetes and obesity [204]. Mitochondrial uncoupling protein 1 (UCP1) plays a major role in the uncoupled respiration in brown and activated beige adipocytes. UCP1 transfers protons from the intermembrane space back to the mitochondrial matrix before ATP synthase can use them, thereby producing heat rather than ATP [206,207]. In BAT, UCP1 is inhibited by free purine nucleotides: ATP, ADP, GTP, and GDP [208]. Upon induction of thermogenesis, specific hormones (e.g., norepinephrine) trigger a reaction cascade leading to cAMP production and activation of lipolysis through a PKA-mediated phosphorylation. Subsequently, PKA-phosphorylated enzymes break down lipid droplets into smaller free fatty acids, which in turn activate UCP1 [206,209].

Phosphorylation was also shown to regulate the transcription of the *UCP1* gene in several ways [210]. A phosphorylation reaction cascade involving the cAMP-dependent PKA and the p38 mitogen-activated protein kinase (p38 MAPK) leads to activation of transcription factor 2 (ATF-2), which increases the expression of *UCP1* [211]. PKA also directly upregulates *UCP1* transcription through the cAMP-response element binding protein (CREB) [206,210]. Shinoda et al. [212] also found that the formation of brown and beige adipocytes is suppressed by casein kinase 2 (CK2) and CK2 inhibition stimulates the UCP1-mediated thermogenesis [212]. Yet another study showed that the phosphorylation of UCP1 at Ser51 increases upon cold stress. It is thought that phosphorylation increases the activity of UCP1, though the responsible kinase remains unknown [213].

6. Mitochondrial Quality Control

Mitochondria are highly dynamic cellular structures that have developed their own systems for detecting dysfunction and eliminating the damage. These include quality control mechanisms consisting of a sophisticated network of mitochondrial chaperones and proteases, which are together responsible for stabilizing, folding, refolding or completely removing proteins. When the mitochondrial proteome suffers extreme damage, mitophagy or cell death can occur [214–216]. The main components of the mitochondrial protein quality control (MPQC) system include the heat-shock proteins Hsp60, Hsp70 (mortalin), Hsp90 (TRAP1), the Lon protease [217] and the caseinolytic peptidase ClpXP (absent in *S. cerevisiae* and *S. pombe*). Phosphorylation plays an important role in regulating the functions of all of them, and their inhibition or aberrant activation often leads to pathological conditions.

Several chaperones in plant and mammalian mitochondria are phosphorylated. For example, the small heat-shock protein (sHSP) Hsp22 in *Zea mays* was found to be phosphorylated in vivo [218]. In vegetative tissues, Hsp22 forms large oligomers of 9 or more subunits and is constitutively expressed at very low levels but is strongly induced in response to heat stress and protects mitochondrial respiratory complex I [219–221]. In mammalian systems, the phosphorylation of similar heat shock proteins (e.g., HSP27) is mediated by the activity of the mitogen-activated protein kinase MAPK [222,223], and usually shifts the distribution of sHSPs towards smaller species [224], however, the precise effect of phosphorylation on plant sHSPs remains elusive. Additional proteomic analyses revealed phosphorylations of mitochondrial Hsp70 (*Pisum sativum*), Hsp90 and chaperonin Hsp60 (*Solanum tuberosum*) [149]. Moreover, a tyrosine in Hsp75, a member of the HSP70 family, was found to be phosphorylated in rat hepatoma cells treated with peroxovanadate (a non-specific inhibitor of tyrosine phosphatases) [225].

The human mitochondrial chaperone TRAP1 (TNF receptor-associated protein 1), a member of the highly conserved HSP90 family, is an important heat-shock protein which is involved in regulating mitochondrial energy metabolism in both healthy and damaged cells [226–228]. TRAP1 was found to be constitutively phosphorylated at low levels and phosphorylation significantly increased in response to oxidative stress [229]. Under these conditions, the Ser/Thr kinase PINK1 acts as a major phosphorylation agent. The level of TRAP1 phosphorylation is directly linked to the cytoprotective function of PINK1, which counteracts oxidative-stress induced apoptosis by suppressing the release of cytochrome *c* from the mitochondria [229]. In Parkinson's disease (PD), mutations in PINK1 abolish its ability to phosphorylate TRAP1, thereby increasing Cytc release into the cytosol and triggering the cell death pathway; this might at least partially explain the mitochondrial dysfunction leading to neurodegeneration in PD patients [229].

TRAP1 expression is strongly induced in several types of tumors [230], where it is frequently correlated with progression, metastasis and disease recurrence [226,231]. In carcinogenesis, TRAP1 may function as an effector of the deregulated extracellular signal-regulated protein kinase 1/2 (ERK1/2) signaling pathway [232]. ERK1/2 directly phosphorylates TRAP1 at two site-specific serines, Ser511 and Ser568, which promotes its activation and the concomitant downregulation of mitochondrial respiratory complex II (succinate dehydrogenase) [232]. The resulting accumulation of intracellular succinate causes pseudohypoxia by stabilizing the hypoxia-inducible transcription factor (HIF-1), thereby activating the HIF-1 responsive genes independently of cellular oxygen levels, leading to further metabolic changes and progression to malignancy [227,233].

Interestingly, TRAP1 is also one of the proteins phosphorylated during sperm capacitation in humans, a process which is correlated with an increase in the tyrosine phosphorylation of several proteins by PKA [234]. These include two PKA-membrane anchoring proteins, AKAP3 and AKAP4, and VCP/p97, a member of the AAA family (ATPases associated with diverse cellular activities) that presumably links sperm capacitation and acrosome reaction [234].

Mitochondrial Hsp60, together with its co-chaperone Hsp10, typically serves to maintain mitochondrial homeostasis by assisting in the folding and trafficking of other proteins [235–237]. It also has other roles in mtDNA transactions and apoptosis, and even more roles occur upon its translocation into the cytosol or extracellular space (for rev. see [238]). The phosphorylation of Hsp60 is involved in both physiological and pathological processes: phosphorylation of Tyr227 and Tyr243 were reported in non-small lung cancer cell lines as well as in mouse hyperglycemic myoblasts [239,240]. The SRC-mediated phosphorylation of Tyr227 was also observed during rotavirus infection and resulted in Hsp60 degradation, thereby delaying the import of viral non-structural proteins (NSP4) into the mitochondria. This delay prevents premature host cell apoptosis, thus giving the rotavirus enough time to replicate [241]. A tyrosine-phosphorylated Hsp60 was also found in the surface activation of sperm cell capacitation mentioned above [242].

The mitochondrial ATP-dependent protease Lon is one of the main components of the MPQC system and is involved in the degradation of the damaged, misfolded or unfolded proteins that accumulate in the mitochondrial matrix [216,243]. Lon has been shown to be subjected to at least three different PTMs, including phosphorylation [106,244–246]. Lon expression is induced by oxidative stress, endoplasmic reticulum (ER) stress, proteotoxic stress, starvation or hypoxia [247]. Lon upregulation has been observed in ischemic rabbit hearts and macrophages under hypoxic conditions [248]. In hypoxia associated with cancer, Lon upregulation was linked to the stabilization of HIF-1 and its binding to the *LON* promoter [249,250]. In prostate adenocarcinoma, the upregulation of Lon was connected to the action of the AKT Ser/Thr kinase [245]. AKT phosphorylates Lon at two serines, Ser173 and Ser181, in the N-terminal domain, which increases its proteolytic activity. This enhanced protein degradation in turn restores the activity of ETC complexes II and V, and increases the oxygen consumption rate, ATP production and cell migration [245]. Conversely, plant mitochondrial Lon was found to be phosphorylated at Ser654 in response to a *Xanthomonas citri* infection. This phosphorylation rapidly impairs its proteolytic activity [251], preventing it from degrading HrpG, a transcription regulator of the *hrp/hrc* genes. Since these genes are not expressed, they cannot act against the *X. citri* virulence resulting in necrosis in citrus fruits and green leaves [252].

Yet another mitochondrial protease involved in MPQC is the hetero-oligomeric protein complex, ClpXP, which mediates a proteolytic removal of misfolded proteins [253,254] that are often induced in response to stress [255–257]. A functional Clp protease is formed from two parts, a heptameric peptidase ClpP and a hexameric ATP-dependent chaperone ClpX, which combine to create a higher-order structure that resembles the barrel of the cytoplasmic 26S proteasome [258,259]. ClpP lacks ATPase activity and therefore retains only a low level of peptidase activity when ClpX is missing [260] while ClpX, as a regulatory ATPase, can still perform its chaperone function even in the absence of ClpP [261]. The mutual interaction of both ClpX and ClpP subunits is needed to activate the proteolytic function, in which ClpX unfolds a protein substrate using an ATP-driven translocation through its central pore into the ClpP proteolytic chamber [262].

ClpP has a well-characterized role in the mitochondrial unfolded protein response (mtUPR) of eukaryotes. This is a retrograde signaling pathway that maintains mitochondrial protein homeostasis in response to mitochondrial-specific stress [263]. In mammals, the mtUPR might also be induced by changing the stoichiometry of the proteins encoded by mitochondrial and nuclear genomes. This is correlated with the upregulation of ClpP and Hsp60, which together minimize mitochondrial protein aggregation [255,264]. On the other hand, reduced levels of ClpP alters mitochondrial and cellular morphology, reduces cell proliferation, and results in lower mitochondrial respiration, lower oxygen consumption rate and increased ROS production [263]. MtUPR was shown to be crucial for tumor cell survival during anti-cancer treatment [265], and is dependent on the activation of the cytosolic double-stranded-RNA-activated protein kinase PKR [266]. PKR or protein kinase R is a Ser/Thr kinase constitutively and ubiquitously expressed in vertebrate cells that was first identified as a protector against viral infections [267,268]. In mice lacking a functional PKR, no mtUPR-responsive proteins (e.g., chaperonin 60) were activated upon stress induction [266].

ClpXP was also shown to be dramatically upregulated in primary and disseminated human tumors, which often correlates with reduced patient survival [269]. In these cases, an interaction between ClpXP and TRAP1 were observed with survivin, an inhibitor of apoptosis, which acts as a scaffold bringing ClpP and TRAP1 together [269]. Seo et al. [269] found that succinate dehydrogenase subunit B (SDHB) of respiratory complex II was a key substrate whose loss profoundly affected mitochondrial bioenergetics and lead to decreased oxygen consumption, loss of ATP production and mtUPR activation (no other ETC complex was affected) [265]. Thus, ClpXP, TRAP1 and survivin provide a mechanism for ensuring mitochondrial homeostasis in tumors, making any of these proteins a potential target for anti-cancer therapy.

Proteomic analyses identified phosphorylated residues in both ClpP and ClpX. ClpX was found to be phosphorylated at several serines, threonines and tyrosines, some of which were directly in the ATPase domain. The most frequently phosphorylated residue was Ser617, which was identified by more than 20 phosphoproteomic studies, predominantly in different types of lung cancer tissues [106]. For the proteolytic part of the Clp protease, phosphorylations were found at Ser169, Ser173, Ser181, Ser231, Thr189 and Thr277, all of which, except Thr277, were found in the ClpP protease domain [106]. To date, however, no further experimental data are available on how phosphorylation influences the ClpXP proteolytic or chaperone activities.

7. Nucleoid and Ribosomes (mtDNA Maintenance, Transcription and Protein Synthesis)

Although a large number of mitochondrial proteins are known to be phosphorylated, only a few examples are currently known from those involved in mtDNA maintenance and transactions. Even there several protein kinases could under various, often pathological conditions, translocate into the mitochondria and influence the mtDNA-associated processes.

Mitochondrial DNA is compacted into nucleoids, nucleoprotein complexes that are associated with the IMM [270]. The most abundant group of nucleoid proteins are the mtDNA-packaging proteins which contain one or two high-mobility group (HMG)-box domains that are critically involved in mtDNA binding and compacting (for review see [271]). The most studied and best characterized mitochondrial HMG-box proteins are the *S. cerevisiae* Abf2 (ARS-binding factor 2) and the human mitochondrial transcription factor A (TFAM), both of which are known to be phosphorylated [272–274]. Indeed, nuclear HMG-box proteins are frequently phosphorylated in response to changes in physiological state. In the nucleus, the Ca^{2+}-phospholipid-dependent protein kinase C (PKC) seems to be responsible for many of the phosphorylation events, and has been identified as the modifier of several chromatin-binding HMG-box proteins including mammalian HMG1/2 [275,276], insect HMG1 [277], human HMG14/17 [278] and both human and murine HMGY/I isoforms (these are also phosphorylated by the protein kinases Cdc2 and MAPK [279,280]). In all these proteins, phosphorylation has a negative effect on their DNA-binding properties, significantly weakening even abolishing their affinity for various DNA substrates [277–279].

Yeast mitochondrial Abf2 can be modified by PKA, which rapidly and specifically phosphorylates Thr21 and Ser22 in the N-terminal segment of the Abf2 HMG-box 1 in vitro [272]. Like its nuclear counterparts, phosphorylation in this region inhibits its DNA-binding affinity for various types of DNA substrates and the DNA super-coiling activity of Abf2 itself. As the *ABF2* gene seems to be constitutively expressed, it is tempting to speculate that the inhibition of its activities might be the means by which mtDNA transactions are regulated in yeast. Supporting this hypothesis is the fact that the amount of Abf2 protein in mitochondria changes more as a consequence of fluctuations in mtDNA copy number rather than in changes in *ABF2* gene expression [281]. Despite this, such a challenging hypothesis still requires much more thorough investigation.

Like Abf2, TFAM, the mammalian mtDNA-packaging protein, possesses several potential phosphorylation sites in its two HMG boxes. For example, in vitro PKA-mediated phosphorylation affects several serines, with Ser55, Ser56 and Ser61 in HMG-box 1 being the most notable [273]. In the crystal structure of the TFAM–LSP (light-strand promoter) DNA complex [282], these residues were in direct contact with the DNA and could thus play a decisive role in mtDNA-binding regulation. Indeed, phosphorylation in HMG1 impaired the ability of TFAM to bind DNA and to activate the transcription of mitochondrial DNA [273]. In the mitochondrial transcription complex, TFAM acts as an auxiliary transcription factor in recruiting the mitochondrial RNA polymerase (POLRMT) and mitochondrial transcription factor B (TFB2M) to the mtDNA promoter sites [283]. The mass spectrometry analysis of Wang et al. [274] identified Ser177 in HMG2 as the ERK1/2 phosphorylation site. Elevated ERK kinase activity has been shown in the degenerating substantia nigra neurons of patients with PD and Lewy body disease (LBD), however it is still unclear whether activated ERK localizes to the mitochondria normally or only gains entry following mitochondrial injury [86]. Nonetheless, its presence within the mitochondria is sufficient to induce mitophagy [284]. In cancer, ERK-dependent signalling enhances aerobic glycolysis, a metabolic pathway preferred by highly proliferating cells because it provides intermediates compatible with rapid growth [285]. Mimicking the ERK-mediated TFAM phosphorylation on Ser177 again negatively affects its DNA-binding ability, especially the light-strand promoter sequence, with consequent effects on the levels of mtDNA transcription, mitochondrial respiration and respiratory chain subunit expression. This may reveal an additional mechanism by which ERK1/2 downregulates mitochondrial functions [274].

In *Arabidopsis thaliana* chloroplasts, MFP1 (MAR-binding filament-like protein 1) serves as the major DNA-binding protein and has functions comparable to those of mitochondrial Abf2 or TFAM [286]. MFP1 was shown to associate with the thylakoid membranes and bind plastid DNA in a non-specific manner [286]. Its DNA-binding activity is largely inhibited after the phosphorylation of either of two adjacent serine residues, Ser604 or Ser605, by serine protein kinase CK2, whose α subunit homolog, cpCK2α, is a component of the chloroplast transcription apparatus [287]. Three of the 10 currently known nucleoid-associated proteins in plastids [288] have a DNA-binding activity that can be influenced by phosphorylation. In addition to MFP1, sulfate reductase (SiR), a bifunctional enzyme acting in DNA compaction and sulfur assimilation [289,290] and SWIB-4, a domain of SWI/SNF complex B functioning in cpDNA packaging [291] were both shown to lose their DNA-binding abilities when phosphorylated [291,292].

Several proteins involved in mtDNA transcription and replication have either been found in a phosphorylated form or have been shown to be substrates of a particular protein kinase. In *S. cerevisiae*, a PKA-mediated phosphorylation has been reported for MTF1 (mitochondrial transcription factor 1) [293], a protein that interacts with the mtRNA polymerase and facilitates promoter-specific transcription initiation [294]. MTF1, evolutionarily related to bacterial rRNA methyltransferases, is a functional homolog of human TFB2M [295,296] and binds DNA in a non-sequence-specific manner. Two independent MS analyses [297,298] found a single phosphorylation on Ser93, however direct information on its consequences for mtDNA promoter binding or mitochondrial transcription initiation remain unknown.

Additionally, the activity of mTERF1 (mitochondrial termination factor 1), a factor mediating mtDNA transcription termination in mammalian mitochondria, was shown to be modulated by several phosphorylations and only the phosphorylated protein allowed accurate transcription ending to occur [299]. Interestingly, dephosphorylation of mTERF1 abolishes its termination activity, but not its DNA-binding ability [299], showing that these two activities are separable and the binding of the protein to DNA is necessary but not sufficient to terminate transcription [300].

Mammalian CREB (cAMP-response element binding protein) is also activated by phosphorylation. CREB normally resides in the nucleus, where it acts as a transcription factor in the signaling pathways involved in synaptic transmission and neuronal survival [248]. Despite having no mitochondrial localization signal, phospho-activated CREB was found in the mitochondria of adult rat brain [301]. Following PKA-mediated activation on Ser133, CREB is able to enter mitochondria (through the TOM complex, with the help of mtHSP70) and bind CRE-like sequences in the non-coding region of the mtDNA [302,303]. Its depletion in mitochondria decreases the expression of mitochondrially encoded genes, resulting in the down-regulation of mitochondrial respiration which is one of the characteristics of the mitochondrial dysfunction associated with neurodegenerative disorders [302].

In the maturation of mitochondrial mRNA transcripts prior to translation in *S. cerevisiae*, an as yet unidentified protein kinase phosphorylates the 55-kDa RNA-binding protein DBP (dodecamer-binding protein) [304]. This phosphorylation activates DBP's high-affinity binding to conserved dodecamer sequences in the 3′ UTR region of mature mitochondrial mRNAs [305]. This region also serves as a target for the mitochondrial degradosome (mtEXO), the major regulator of yeast mitochondrial RNA turnover [306–308]. Interestingly, the precise role of DBP in *S. cerevisiae* mtRNA processing has not been described, although its unique DNA-binding activity is known to be regulated by a reversible phosphorylation [305].

In humans, the mitochondrial translational elongation factor (EF-Tu) is phosphorylated in response to ischemia in rabbit cardiomyocytes [309]. EF-Tu is a GTPase that binds aminoacyl-tRNAs and brings them to the ribosome. In bacteria, phosphorylation of Thr382 of the EF-Tu homolog prevents such ternary complex formation and protein translation is largely inhibited [310]. Similarly, mitochondrial EF-Tu from isolated ischemic hearts had multiple phosphorylation sites, including the same highly conserved threonine, and at least one additional serine residue. The cytosolic Ser/Thr kinase JNK, which translocates into mitochondria upon ischemic heart injury [311], was suggested to be a possible candidate. However, follow-up experiments excluded both JNK and the p38 MAP kinase because no change occurred in the levels of EF-Tu phosphorylation upon their immuno-depletion [309]; the kinase responsible for EF-Tu modification in mitochondria still remains to be identified.

The proper maintenance and expression of the mitochondrial genome in all organisms depends on the faithful copying of mtDNA. The mitochondrial single-stranded DNA-binding protein (mtSSB) is a key component of the mitochondrial replisome [312], and is fundamentally involved in keeping the human mtDNA copy number levels constant. It undergoes protein kinase-mediated changes, though these changes are often related to pathological processes. mtSSB was shown to be phosphorylated by SRC, a mitochondrially localized non-receptor tyrosine kinase. SRC is an oncogene which has high expression or activity in several types of solid tumors, including glioblastoma [313], prostate [314], breast [315], pancreas [316], colon [317], and lung cancers [318], and it is associated with increased invasiveness, metastatic potential and lower patient survival [319]. The presence of SRC in the mitochondria was confirmed in breast cancer, prostate carcinoma and bone osteosarcoma cell lines [25,320], and was correlated with altered mitochondrial respiration [321]. In highly metastatic breast cancer cells, mtSSB was shown to be downregulated [322] and similar downregulation was observed when mtSSB was phosphorylated [24]. Phosphorylation of mtSSB on Tyr73 decreases mtDNA replication activity, leading to a reduction in mtDNA levels [24]. This reduced amount of mtDNA may further lower the expression of mitochondrially-encoded proteins (including the respiratory complex subunits), followed by OXPHOS deficiency and more aggressive phenotypes in breast cancer tumors [24].

8. Mitochondrial Kinases in Apoptosis

The role of mitochondria has been univocally established in apoptosis, a major form of regulated cell death that affects the processes of development, differentiation and tissue homeostasis. The central regulatory role is played by the Bcl-2 protein family that are divided into pro-apoptotic (Bax, Bak, Bok, Bad, Bid, Bik, Bim, Bmf, Noxa, Puma) and anti-apoptotic (Bcl-2, Bcl-xL, Bcl-W, Mcl-1) groups. Their dysregulation is often implicated in a wide variety of diseases, including excessively apoptotic atrophy or the insufficiently controlled cell proliferation of cancer. The activities of the Bcl-2 family and other apoptosis-related proteins must therefore be strictly regulated either on the transcriptional or the post-translational levels, which involves the phosphorylation and dephosphorylation activities of many mitochondrial kinases and phosphatases (Figure 3 for mitochondrial kinases).

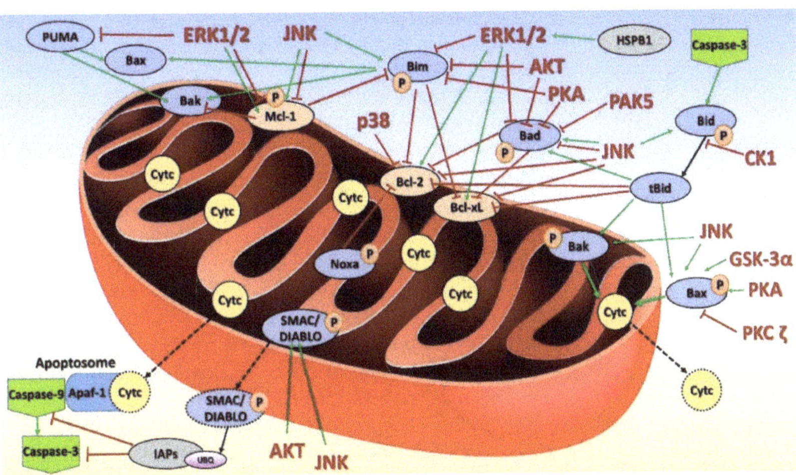

Figure 3. The apoptotic pathways influenced by mitochondrial protein phosphorylation. Pro-apoptotic (circled in blue) and anti-apoptotic proteins (circled in light red) are phosphorylated by the action of various protein kinases (shown in red) as described in Section 8. The dashed black arrows represent a translocation of the indicated protein out of the mitochondrion. Cytc, cytochrome *c*; P, phosphate group; UBQ, ubiquitin.

The mitochondrial pathway of cell death requires mitochondrial outer membrane permeabilization (MOMP) to release soluble proteins from the mitochondrial intermembrane space, especially Cytc, which activates initiator caspase 9, which cleaves and activates the executioner downstream caspases [323]. The other proteins released include Smac/DIABLO and OMI, which block the caspase inhibitor XIAP (X-linked inhibitor-of-apoptosis protein), thereby enabling apoptosis. In healthy cells, Bax localizes to the cytosol and Bak to the mitochondria, and both are inactive and able to shuttle between those compartments [324–326]. Bax and Bak can be directly activated by binding a subclass of BH3-only proteins (Bid, PUMA and Bim) [327]. Upon activation, Bax accumulates in the mitochondria. The phosphorylation of Bid at Ser64 and Ser66 in mouse mitochondria (Ser65 and Ser67 in humans) by CK1 prevented its activation and subsequent binding to Bax and Bak, thereby inhibiting Fas-mediated apoptosis and the release of Cytc from mitochondria [328,329]. Phosphorylation of Bid at Tyr54 is frequently found in phosphoproteomic studies of leukemic, breast and gastric cancer tissues, though its precise role has not yet been described [106].

In early apoptosis, Cytc functions as a peroxidase of cardiolipin, a mitochondria-specific lipid that binds to Cytc, in the presence of H_2O_2 [330,331]. This pro-apoptotic event promotes the dissociation of Cytc from the IMM and its release into the cytosol [332]. Of the five Cytc phosphorylation sites noted above (Section 5), only Tyr48 seems to influence apoptosis. Phosphorylation of Tyr97 showed only a small reduction in caspase-9 activity in vitro [333,334], and a phosphomimetic substitution of Tyr48 reduced caspase 3 activity by ~70% but left the cardiolipin peroxidase activity of Cytc unchanged [183].

As mentioned before, the MAPKs (JNK, ERK1/2, and p38) and PKA exhibit transient mitochondrial localization in response to various cellular stimuli and stresses which alter the regulation of cell death pathways. Their main targets are the members of the Bcl-2 protein family, which are largely phosphorylated [335]. In cerebral ischemia, JNK1 may be an early mitochondrial effector responsible for JNK-mediated apoptosis [4,37]. The numerous kinases and phosphatases localized to the OMM coordinate organellar and cellular physiology by modulating the roles of the Bcl-2 protein family in metabolism, autophagy, and apoptosis. The interplay between ROS and the members of the Bcl-2 family can determine whether a cell undergoes apoptosis [336]. Bcl-2 "senses" the extent of ROS production through the phosphorylation of its Ser70 by JNK, which releases it from mitochondria, and thereby alters ETC function, mitochondrial autophagy and antioxidant capacity [337].

In humans, the pro-apoptotic protein Bad is phosphorylated at four serines, Ser74, Ser75, Ser99 and Ser118. Ser75 seems to be phosphorylated by ERK1/2, which has been correlated with several human cancers [106,115,338–340]. Likewise, Ser99 is a major AKT phosphorylation site [341]. In mice, the JNK kinase phosphorylates Bad at Ser128, which subsequently mediates its oligomerization and promotes its pro-apoptotic effect in primary neurons by antagonizing the ability of growth factors to inhibit Bad-mediated apoptosis [342,343]. On the other hand, JNK reduces apoptosis in interleukin-3-mediated cell survival by phosphorylating Bad at Thr201 and inactivating it [344]. Moreover, in mice, PAK5 can phosphorylate Bad on Ser112 and thereby prevent its migration to mitochondria, which, consequently, increases the levels of the anti-apoptotic protein Bcl-2 and inhibits apoptosis [82]. Mitochondrial PKA and AKT also phosphorylate mouse Bad at Ser112 and Ser155 (Ser75 and Ser118 in humans), as well as Bim, another pro-apoptotic protein, at Ser83 [345] and Ser65 during trophic factor deprivation (TFD) [346,347], both of which reduce their inhibitory effects on the anti-apoptotic proteins Bcl-2 and Bcl-xL, thereby promoting cell survival [348,349]. Bim was also found to be phosphorylated by the action of ERK1/2. Here, the interaction of ERK1/2 with heat shock protein B1 (HSPB1) facilitates the degradation of phosphorylated Bim. This is disturbed in Charcot-Marie-Tooth disease, where HSPB1 mutations lead to high levels of Bim and the cells exhibit increased susceptibility to ER stress-induced cell death [350].

The human pro-apoptotic pore-forming protein Bax is phosphorylated at several threonine (Thr22, Thr86, Thr135, Thr140, Thr167 and Thr174) and serine residues (Ser60, Ser87, Ser163 and Ser184) [106], the physiological effects of most of which remain unclear. The PKA-mediated phosphorylation of Bax at Ser60 is known to promote its mitochondrial translocation and MOMP [351], and the phosphorylation of Ser163 by GSK-3β was observed in the mitochondria of human embryonic kidney cells and cerebellar granule neurons [352]. The phosphorylation of Thr167, located between the last two α-helices, by JNK promotes Bax oligomerization and induces apoptosis [353]. In both small cell and non-small cell lung cancers, however, Bax is phosphorylated at Ser184, presumably by the PKC ζ isoform. Ser184 is inside the last α-helix, which is responsible for mitochondrial membrane localization, and this phosphorylation led to a retention of the protein in the cytoplasm and inhibition of apoptosis [354].

In healthy cells, Bak, the second mitochondrial pore-forming pro-apoptotic protein, is also phosphorylated at several sites, including Tyr108, Tyr110, Ser117 and Thr148 [106]. Dephosphorylation of Tyr108 and Ser117 seems to be an important regulatory step, where dephosphorylated Tyr108 initiates the cell death process by conformational changes in the OMM [355]. Ser117 located in the hydrophobic groove of Bak is involved in binding to other pro-apoptotic partners (BH3-only proteins). Its dephosphorylation mediates further multimerization and, consequently, pore formation, which enables Bak pro-apoptotic function [356].

Bcl-xL is a critical anti-apoptotic protein that plays a role in normal embryogenesis, human erythropoiesis and promotes the survival of differentiating pancreatic cells [357–359]. Apoptosis is promoted when Bcl-xL residues Thr47 and Thr115 are phosphorylated in human myeloid leukemia cells by stress-activated JNK [360]. Outside its function in apoptosis, Bcl-xL was shown to be phosphorylated at Ser49 and Ser62 during normal cell cycle progression and checkpoints [361,362]. The PLK3 (Polo-like kinase 3) phosphorylated Bcl-xL at Ser49 accumulates in the centrosomes during G2 checkpoint induced by DNA damage and during the final stages of mitosis and cytokinesis. Here, phosphorylated Bcl-xL translocates to the microtubule-associated dynein motor protein and prevents chromosome instability [361]. Whereas Bcl-xL phosphorylated at Ser62 by PLK1 and MAPK14/SAPKp38α appears during the early stages of mitosis, and in telophase and cytokinesis is being rapidly dephosphorylated [362]. Phosphorylation of Ser62 was found to have increased after treatment with taxol, an anti-cancer drug used in treating prostate cancer [363].

Bcl-2, a major anti-apoptotic protein, possesses a transmembrane domain (TM), which normally associates the protein with the OMM or IMM. Bcl-2 is predominantly found to be phosphorylated (at Ser70, Ser87, Thr56 and Thr74) by the MAPK kinases (p38 and ERK1/2) leading to a reduction in its anti-apoptotic activities [364–366]. For example, phosphorylation of Ser87 and Thr74 occur in normal human blood cells, but these two residues remain un-phosphorylated in hepatoma tumor cells, suggesting that dephosphorylation promotes Bcl-2′s anti-apoptotic activity in tumor cells [366]. It was also shown that dephosphorylated Ser87 and Thr74 seemed to be a signal for ubiquitin-dependent degradation by the 26S proteasome [367]. Bcl-2 phosphorylation at Ser70, Ser80, and Thr69 by JNK1 shown in human T lymphocytes arrested in G2/M phase in response to microtubule-damaging agents makes cells more susceptible to apoptosis [368].

Similarly, phosphorylation of the pro-survival protein Mcl-1 by JNK on residues Thr163 and Ser121 promotes apoptosis [369,370] while phosphorylation of Ser64 enhances its anti-apoptotic activity [371]; phosphorylation by ERK1/2 exhibited a similar dual effect [372]. The putative phosphorylation of the N-terminal domain of Smac/DIABLO by JNK mediates its release from the mitochondria to ubiquitinylate the inhibitors of apoptosis (IAPs) meant for degradation, thereby maintaining the apoptotic process [373–375]. Smac/DIABLO is also phosphorylated at Ser67 by AKT, which promotes apoptosis in HeLa cells [376]. Overall, the precise mechanism by which JNK and other kinases promote or inhibit cell survival proteins is not entirely clear and often exhibits a dual activating/inhibiting character, where phosphorylation of the same protein by one kinase may lead to opposite effects depending on the various cellular conditions (for rev. see [377].

Although Ser/Thr kinases seem to have the largest role in the Bcl-2 protein family regulated apoptosis, there is a number of other protein kinases which are also involved. The cytosolic form of LYN (cLYN) inhibits the mitochondrial apoptotic pathway, and this effect arises from its kinase activity [378–380]. The deactivation/inhibition of LYN by pharmacological inhibitors or siRNA in apoptosis-resistant cells causes their re-sensitization to chemotherapy-induced apoptosis [381]. Overexpressed PKC δ promotes apoptosis in neoplastic and normal keratinocytes by targeting mitochondria, triggering the release of Cytc and disturbing the membrane potential [13]. Although ABL does not contain a typical mitochondrial localization signal, protein kinase C δ (PKC δ) could bind to ABL in the endoplasmic reticulum. This PKC δ–ABL complex translocates from the ER to the mitochondria and then triggers apoptosis [382]. Cytosolic SRC, normally a proto-oncogene, possesses anti-apoptotic properties and shows increased protein levels and activity in a variety of human cancers [383]. And GSK-3β regulates mitochondria-mediated apoptosis by playing a role in determining the threshold for mitochondrial permeability transition pore opening [384].

9. Conclusions

Kinases and phosphatases act together in mitochondria in a precisely conducted symphony. Any disturbance of their functions by mutations or over- or under-regulation results in severe diseases such as diabetes, cancer, neurodegeneration and, consequently, apoptosis. Here we have listed those kinases that have been found to be associated with the outer mitochondrial membrane, described the effect of phosphorylation on the various components of the mitochondrial import machinery, reviewed the effect of kinases on the enzymes of the tricarboxylic acid cycle and steroid hormone biosynthesis, surveyed what is presently known of the effect of phosphorylation on each component of the oxidative phosphorylation system as well as their influence on the uncoupled respiration that occurs in brown and beige adipose tissue, examined their importance in the various systems responsible for mitochondrial protein quality control, considered their roles in mtDNA maintenance, transcription, and translation, and briefly reviewed what is known of their many roles in apoptosis.

Despite the enormous number of studies published in the last two decades describing the role of kinases in mitochondrial biogenesis, many unanswered or poorly answered questions still remain: What are the interaction partners that support or inhibit the kinases? Are mitochondrial proteins phosphorylated before or after entering the mitochondria or are both alternatives employed? How does phosphorylation or other post-translational modifications of one amino-acid residue influence the phosphorylation of another position? Can two or more kinases phosphorylate proteins simultaneously? Are certain amino acids phosphorylated more rapidly? How does the composition of the mitochondrial membrane influence kinase interactions or membrane insertion? How does phosphorylation influence the dynamics of the mitochondrial nucleoid and thus the replication, repair and transcription of mtDNA? Which kinases interact with the ribosomes and are involved in mtDNA translation? Answering these questions could substantially contribute to better understanding how mitochondria contribute to stress responses and support healthy processes in the cell as well as the background of pathological processes.

Author Contributions: Conceptualization, E.K., N.K.; writing—original draft preparation, V.K., N.K., B.K., G.O., E.K., H.H., V.P.; writing—review and editing, J.A.B., N.K., E.K., V.P.; visualization, E.K., B.K., V.P., N.K.; supervision, E.K.; project administration, V.P.; funding acquisition, E.K. All authors have read and agreed to the published version of the manuscript.

Funding: Research is funded by the Slovak Research and Development Agency (APVV) [APVV-15-0375 and APVV-19-0298], Scientific Grant Agency of the Ministry of Education, Science, Research, and Sport of the Slovak republic (VEGA) [2/0075/18]. NK and BK acknowledge the financial support from the Czech Science Foundation Grant No. 1825144Y. This publication was created with the support of the Interreg V-A Slovakia-Austria program (www.sk-at.eu) for the project StruBioMol, ITMS: 305011X666 and is co-financed by the European Regional Development Fund.

Institutional Review Board Statement: Not applicable.

Informed Consent Statement: Not applicable.

Conflicts of Interest: The authors declare no conflict of interest. The funders had no role in the design of the study; in the collection, analyses, or interpretation of data; in the writing of the manuscript, or in the decision to publish the results.

Abbreviations

AAC	ADP/ATP carrier protein
Abf2	ARS-binding factor 2
ABL	Abelson tyrosine kinase
ACC2	acetyl-CoA carboxylase 2
AD	Alzheimer's disease
AKAP	A-kinase anchoring protein
AKT/PKB	protein kinase B
AMPK	5′-AMP-activated protein kinase
BAT	brown adipose tissue
CAMK	Ca^{2+}/calmodulin-dependent protein kinase
CDK	cyclin-dependent protein kinase
CK	casein kinase
COX	cytochrome *c* oxidase
CREB	cAMP-response element binding protein

Cytc	cytochrome *c*
DBP	dodecamer-binding protein
DRP1	dynamin-related protein 1
EF-Tu	translational elongation factor
ER	endoplasmic reticulum
ERK1/2	extracellular receptor kinase 1/2
ETC	electron transport chain
FDH	formate dehydrogenase
GSK-3β	glycogen synthase kinase 3β
HSP	heat-shock protein
IF	inhibitory factor
IMM	inner mitochondrial membrane
HIF	hypoxia-inducible transcription factor
HMG	high-mobility group
JNK	Jun N-terminal kinase
LRRK2	leucine-rich repeat kinase 2
MAPK	mitogen-activated protein kinase
MFF	mitochondrial fission factor
Mfn	mitofusin
MFP1	MAR-binding filament-like protein 1
MIA	mitochondrial intermembrane space assembly
MIEF1/2	mitochondrial elongation factor 1/2
MIM	mitochondrial import complex
MOMP	mitochondrial outer membrane permeabilization
MPP	mitochondrial processing peptidase
MPQC	mitochondrial protein quality control
mTERF1	mitochondrial termination factor 1
MTF1	mitochondrial transcription factor 1
mTOR	mammalian target of rapamycin
mtSSB	mitochondrial single-stranded DNA-binding protein
mtUPR	mitochondrial unfolded protein response
OMM	outer mitochondrial membrane
OXPHOS	oxidative phosphorylation
PAK5	p21-activated kinase 5
PARL	presenilins-associated rhomboid-like protein
PD	Parkinson's disease
PDC	pyruvate dehydrogenase complex
PDH	pyruvate dehydrogenase
PDK	pyruvate dehydrogenase kinase
PDP	pyruvate dehydrogenase phosphatase
PGK1	phosphoglycerate kinase 1
PKA	protein kinase A
PKC	protein kinase C
PKR	protein kinase R
PINK1	*PTEN*-induced putative kinase 1
PLK3	Polo-like kinase 3
POLRMT	mitochondrial RNA polymerase
PTM	post-translational modification
ROS	reactive-oxygen species
SAM	sorting and assembly machinery
SAPK	stress-induced protein kinase
SDH	succinate dehydrogenase
SDHA	succinate dehydrogenase complex subunit A
sHSP	small heat-shock protein
StAR	steroidogenic acute regulatory protein
TCA	tricarboxylic acid cycle
TFAM	mitochondrial transcription factor A
TFB2M	mitochondrial transcription factor B

TFD	trophic factor deprivation
TIM	translocase of the inner mitochondrial membrane
TNFα	tumor necrosis factor-α
TOM	translocase of the outer mitochondrial membrane
TPKI	Tau protein kinase I
TRAP1	TNF receptor-associated protein 1
UCP1	uncoupling protein 1
VSM	vascular smooth muscle
XIAP	X-linked inhibitor-of-apoptosis protein.

References

1. Kruse, R.; Hojlund, K. Mitochondrial phosphoproteomics of mammalian tissues. *Mitochondrion* **2017**, *33*, 45–57. [CrossRef] [PubMed]
2. Giorgianni, F.; Koirala, D.; Weber, K.T.; Beranova-Giorgianni, S. Proteome analysis of subsarcolemmal cardiomyocyte mitochondria: A comparison of different analytical platforms. *Int. J. Mol. Sci.* **2014**, *15*, 9285–9301. [CrossRef] [PubMed]
3. Padrao, A.I.; Vitorino, R.; Duarte, J.A.; Ferreira, R.; Amado, F. Unraveling the phosphoproteome dynamics in mammal mitochondria from a network perspective. *J. Proteome Res.* **2013**, *12*, 4257–4267. [CrossRef] [PubMed]
4. Lucero, M.; Suarez, A.E.; Chambers, J.W. Phosphoregulation on mitochondria: Integration of cell and organelle responses. *CNS Neurosci. Ther.* **2019**, *25*, 837–858. [CrossRef] [PubMed]
5. Pagliarini, D.J.; Dixon, J.E. Mitochondrial modulation: Reversible phosphorylation takes center stage? *Trends Biochem. Sci.* **2006**, *31*, 26–34. [CrossRef] [PubMed]
6. Salvi, M.; Brunati, A.M.; Toninello, A. Tyrosine phosphorylation in mitochondria: A new frontier in mitochondrial signaling. *Free Radic. Biol. Med.* **2005**, *38*, 1267–1277. [CrossRef] [PubMed]
7. Sorriento, D.; Gambardella, J.; Fiordelisi, A.; Trimarco, B.; Ciccarelli, M.; Iaccarino, G.; Santulli, G. Mechanistic Role of Kinases in the Regulation of Mitochondrial Fitness. *Adv. Exp. Med. Biol.* **2017**, *982*, 521–528. [CrossRef] [PubMed]
8. Corum, D.G.; Tsichlis, P.N.; Muise-Helmericks, R.C. AKT3 controls mitochondrial biogenesis and autophagy via regulation of the major nuclear export protein CRM-1. *FASEB J.* **2014**, *28*, 395–407. [CrossRef] [PubMed]
9. Shaerzadeh, F.; Motamedi, F.; Khodagholi, F. Inhibition of akt phosphorylation diminishes mitochondrial biogenesis regulators, tricarboxylic acid cycle activity and exacerbates recognition memory deficit in rat model of Alzheimer's disease. *Cell Mol. Neurobiol.* **2014**, *34*, 1223–1233. [CrossRef]
10. Gerbeth, C.; Mikropoulou, D.; Meisinger, C. From inventory to functional mechanisms: Regulation of the mitochondrial protein import machinery by phosphorylation. *FEBS J.* **2013**, *280*, 4933–4942. [CrossRef]
11. Opalinska, M.; Meisinger, C. Mitochondrial protein import under kinase surveillance. *Microb. Cell* **2014**, *1*, 51–57. [CrossRef] [PubMed]
12. Yang, K.; Chen, Z.; Gao, J.; Shi, W.; Li, L.; Jiang, S.; Hu, H.; Liu, Z.; Xu, D.; Wu, L. The Key Roles of GSK-3beta in Regulating Mitochondrial Activity. *Cell Physiol. Biochem.* **2017**, *44*, 1445–1459. [CrossRef] [PubMed]
13. Li, L.; Lorenzo, P.S.; Bogi, K.; Blumberg, P.M.; Yuspa, S.H. Protein kinase Cdelta targets mitochondria, alters mitochondrial membrane potential, and induces apoptosis in normal and neoplastic keratinocytes when overexpressed by an adenoviral vector. *Mol. Cell Biol.* **1999**, *19*, 8547–8558. [CrossRef] [PubMed]
14. Nowak, G.; Bakajsova, D.; Clifton, G.L. Protein kinase C-epsilon modulates mitochondrial function and active Na+ transport after oxidant injury in renal cells. *Am. J. Physiol. Renal Physiol.* **2004**, *286*, F307–316. [CrossRef] [PubMed]
15. Matsuda, S.; Kitagishi, Y.; Kobayashi, M. Function and characteristics of PINK1 in mitochondria. *Oxid Med. Cell Longev.* **2013**, *2013*, 601587. [CrossRef] [PubMed]
16. Plun-Favreau, H.; Hardy, J. PINK1 in mitochondrial function. *Proc. Natl. Acad. Sci. USA* **2008**, *105*, 11041–11042. [CrossRef]
17. Kitagishi, Y.; Nakano, N.; Ogino, M.; Ichimura, M.; Minami, A.; Matsuda, S. PINK1 signaling in mitochondrial homeostasis and in aging (Review). *Int. J. Mol. Med.* **2017**, *39*, 3–8. [CrossRef]
18. Debattisti, V.; Gerencser, A.A.; Saotome, M.; Das, S.; Hajnoczky, G. ROS Control Mitochondrial Motility through p38 and the Motor Adaptor Miro/Trak. *Cell Rep.* **2017**, *21*, 1667–1680. [CrossRef]
19. Dhanasekaran, D.N.; Reddy, E.P. JNK signaling in apoptosis. *Oncogene* **2008**, *27*, 6245–6251. [CrossRef]
20. Weindel, C.G.; Bell, S.L.; Vail, K.J.; West, K.O.; Patrick, K.L.; Watson, R.O. LRRK2 maintains mitochondrial homeostasis and regulates innate immune responses to Mycobacterium tuberculosis. *eLife* **2020**, *9*. [CrossRef]
21. Kumar, S.; Bharti, A.; Mishra, N.C.; Raina, D.; Kharbanda, S.; Saxena, S.; Kufe, D. Targeting of the c-Abl tyrosine kinase to mitochondria in the necrotic cell death response to oxidative stress. *J. Biol. Chem.* **2001**, *276*, 17281–17285. [CrossRef] [PubMed]
22. Zhou, L.; Zhang, Q.; Zhang, P.; Sun, L.; Peng, C.; Yuan, Z.; Cheng, J. c-Abl-mediated Drp1 phosphorylation promotes oxidative stress-induced mitochondrial fragmentation and neuronal cell death. *Cell Death Dis.* **2017**, *8*, e3117. [CrossRef] [PubMed]
23. Koc, E.C.; Miller-Lee, J.L.; Koc, H. Fyn kinase regulates translation in mammalian mitochondria. *Biochim. Biophys. Acta Gen. Subj.* **2017**, *1861*, 533–540. [CrossRef]

24. Djeungoue-Petga, M.A.; Lurette, O.; Jean, S.; Hamel-Cote, G.; Martin-Jimenez, R.; Bou, M.; Cannich, A.; Roy, P.; Hebert-Chatelain, E. Intramitochondrial Src kinase links mitochondrial dysfunctions and aggressiveness of breast cancer cells. *Cell Death Dis.* **2019**, *10*, 940. [CrossRef] [PubMed]
25. Hebert-Chatelain, E.; Jose, C.; Gutierrez Cortes, N.; Dupuy, J.W.; Rocher, C.; Dachary-Prigent, J.; Letellier, T. Preservation of NADH ubiquinone-oxidoreductase activity by Src kinase-mediated phosphorylation of NDUFB10. *Biochim. Biophys. Acta* **2012**, *1817*, 718–725. [CrossRef]
26. Ogura, M.; Yamaki, J.; Homma, M.K.; Homma, Y. Mitochondrial c-Src regulates cell survival through phosphorylation of respiratory chain components. *Biochem. J.* **2012**, *447*, 281–289. [CrossRef]
27. Gringeri, E.; Carraro, A.; Tibaldi, E.; D'Amico, F.E.; Mancon, M.; Toninello, A.; Pagano, M.A.; Vio, C.; Cillo, U.; Brunati, A.M. Lyn-mediated mitochondrial tyrosine phosphorylation is required to preserve mitochondrial integrity in early liver regeneration. *Biochem. J.* **2009**, *425*, 401–412. [CrossRef]
28. Acin-Perez, R.; Carrascoso, I.; Baixauli, F.; Roche-Molina, M.; Latorre-Pellicer, A.; Fernandez-Silva, P.; Mittelbrunn, M.; Sanchez-Madrid, F.; Perez-Martos, A.; Lowell, C.A.; et al. ROS-triggered phosphorylation of complex II by Fgr kinase regulates cellular adaptation to fuel use. *Cell Metab.* **2014**, *19*, 1020–1033. [CrossRef]
29. Salvi, M.; Morrice, N.A.; Brunati, A.M.; Toninello, A. Identification of the flavoprotein of succinate dehydrogenase and aconitase as in vitro mitochondrial substrates of Fgr tyrosine kinase. *FEBS Lett.* **2007**, *581*, 5579–5585. [CrossRef]
30. Tibaldi, E.; Brunati, A.M.; Massimino, M.L.; Stringaro, A.; Colone, M.; Agostinelli, E.; Arancia, G.; Toninello, A. Src-Tyrosine kinases are major agents in mitochondrial tyrosine phosphorylation. *J. Cell Biochem.* **2008**, *104*, 840–849. [CrossRef]
31. Che, T.F.; Lin, C.W.; Wu, Y.Y.; Chen, Y.J.; Han, C.L.; Chang, Y.L.; Wu, C.T.; Hsiao, T.H.; Hong, T.M.; Yang, P.C. Mitochondrial translocation of EGFR regulates mitochondria dynamics and promotes metastasis in NSCLC. *Oncotarget* **2015**, *6*, 37349–37366. [CrossRef] [PubMed]
32. Manning, G.; Whyte, D.B.; Martinez, R.; Hunter, T.; Sudarsanam, S. The protein kinase complement of the human genome. *Science* **2002**, *298*, 1912–1934. [CrossRef] [PubMed]
33. Lim, S.; Smith, K.R.; Lim, S.T.; Tian, R.; Lu, J.; Tan, M. Regulation of mitochondrial functions by protein phosphorylation and dephosphorylation. *Cell Biosci.* **2016**, *6*, 25. [CrossRef] [PubMed]
34. Lin, R.Y.; Moss, S.B.; Rubin, C.S. Characterization of S-AKAP84, a novel developmentally regulated A kinase anchor protein of male germ cells. *J. Biol. Chem.* **1995**, *270*, 27804–27811. [CrossRef]
35. Wiltshire, C.; Matsushita, M.; Tsukada, S.; Gillespie, D.A.; May, G.H. A new c-Jun N-terminal kinase (JNK)-interacting protein, Sab (SH3BP5), associates with mitochondria. *Biochem. J.* **2002**, *367*, 577–585. [CrossRef]
36. Chambers, J.W.; Pachori, A.; Howard, S.; Iqbal, S.; LoGrasso, P.V. Inhibition of JNK mitochondrial localization and signaling is protective against ischemia/reperfusion injury in rats. *J. Biol. Chem.* **2013**, *288*, 4000–4011. [CrossRef]
37. Nijboer, C.H.; Bonestroo, H.J.; Zijlstra, J.; Kavelaars, A.; Heijnen, C.J. Mitochondrial JNK phosphorylation as a novel therapeutic target to inhibit neuroinflammation and apoptosis after neonatal ischemic brain damage. *Neurobiol. Dis.* **2013**, *54*, 432–444. [CrossRef]
38. Court, N.W.; Kuo, I.; Quigley, O.; Bogoyevitch, M.A. Phosphorylation of the mitochondrial protein Sab by stress-activated protein kinase 3. *Biochem. Biophys. Res. Commun.* **2004**, *319*, 130–137. [CrossRef]
39. Affaitati, A.; Cardone, L.; de Cristofaro, T.; Carlucci, A.; Ginsberg, M.D.; Varrone, S.; Gottesman, M.E.; Avvedimento, E.V.; Feliciello, A. Essential role of A-kinase anchor protein 121 for cAMP signaling to mitochondria. *J. Biol. Chem.* **2003**, *278*, 4286–4294. [CrossRef]
40. Carnegie, G.K.; Means, C.K.; Scott, J.D. A-kinase anchoring proteins: From protein complexes to physiology and disease. *IUBMB Life* **2009**, *61*, 394–406. [CrossRef]
41. Hoffman, N.J.; Parker, B.L.; Chaudhuri, R.; Fisher-Wellman, K.H.; Kleinert, M.; Humphrey, S.J.; Yang, P.; Holliday, M.; Trefely, S.; Fazakerley, D.J.; et al. Global Phosphoproteomic Analysis of Human Skeletal Muscle Reveals a Network of Exercise-Regulated Kinases and AMPK Substrates. *Cell Metab.* **2015**, *22*, 922–935. [CrossRef] [PubMed]
42. Fullerton, M.D.; Galic, S.; Marcinko, K.; Sikkema, S.; Pulinilkunnil, T.; Chen, Z.P.; O'Neill, H.M.; Ford, R.J.; Palanivel, R.; O'Brien, M.; et al. Single phosphorylation sites in Acc1 and Acc2 regulate lipid homeostasis and the insulin-sensitizing effects of metformin. *Nat. Med.* **2013**, *19*, 1649–1654. [CrossRef] [PubMed]
43. Toyama, E.Q.; Herzig, S.; Courchet, J.; Lewis, T.L., Jr.; Loson, O.C.; Hellberg, K.; Young, N.P.; Chen, H.; Polleux, F.; Chan, D.C.; et al. Metabolism. AMP-activated protein kinase mediates mitochondrial fission in response to energy stress. *Science* **2016**, *351*, 275–281. [CrossRef] [PubMed]
44. Cribbs, J.T.; Strack, S. Reversible phosphorylation of Drp1 by cyclic AMP-dependent protein kinase and calcineurin regulates mitochondrial fission and cell death. *EMBO Rep.* **2007**, *8*, 939–944. [CrossRef] [PubMed]
45. Yu, R.; Liu, T.; Ning, C.; Tan, F.; Jin, S.B.; Lendahl, U.; Zhao, J.; Nister, M. The phosphorylation status of Ser-637 in dynamin-related protein 1 (Drp1) does not determine Drp1 recruitment to mitochondria. *J. Biol. Chem.* **2019**, *294*, 17262–17277. [CrossRef]
46. Tsushima, K.; Bugger, H.; Wende, A.R.; Soto, J.; Jenson, G.A.; Tor, A.R.; McGlauflin, R.; Kenny, H.C.; Zhang, Y.; Souvenir, R.; et al. Mitochondrial Reactive Oxygen Species in Lipotoxic Hearts Induce Post-Translational Modifications of AKAP121, DRP1, and OPA1 That Promote Mitochondrial Fission. *Circ. Res.* **2018**, *122*, 58–73. [CrossRef]

47. Yang, Y.; Tian, Y.; Hu, S.; Bi, S.; Li, S.; Hu, Y.; Kou, J.; Qi, J.; Yu, B. Extract of Sheng-Mai-San Ameliorates Myocardial Ischemia-Induced Heart Failure by Modulating Ca(2+)-Calcineurin-Mediated Drp1 Signaling Pathways. *Int. J. Mol. Sci.* **2017**, *18*, 1825. [CrossRef]
48. Jahani-Asl, A.; Huang, E.; Irrcher, I.; Rashidian, J.; Ishihara, N.; Lagace, D.C.; Slack, R.S.; Park, D.S. CDK5 phosphorylates DRP1 and drives mitochondrial defects in NMDA-induced neuronal death. *Hum. Mol. Genet.* **2015**, *24*, 4573–4583. [CrossRef]
49. Kashatus, J.A.; Nascimento, A.; Myers, L.J.; Sher, A.; Byrne, F.L.; Hoehn, K.L.; Counter, C.M.; Kashatus, D.F. Erk2 phosphorylation of Drp1 promotes mitochondrial fission and MAPK-driven tumor growth. *Mol. Cell* **2015**, *57*, 537–551. [CrossRef]
50. Yan, J.; Liu, X.H.; Han, M.Z.; Wang, Y.M.; Sun, X.L.; Yu, N.; Li, T.; Su, B.; Chen, Z.Y. Blockage of GSK3beta-mediated Drp1 phosphorylation provides neuroprotection in neuronal and mouse models of Alzheimer's disease. *Neurobiol. Aging.* **2015**, *36*, 211–227. [CrossRef]
51. Gui, C.; Ren, Y.; Chen, J.; Wu, X.; Mao, K.; Li, H.; Yu, H.; Zou, F.; Li, W. p38 MAPK-DRP1 signaling is involved in mitochondrial dysfunction and cell death in mutant A53T alpha-synuclein model of Parkinson's disease. *Toxicol. Appl. Pharmacol.* **2020**, *388*, 114874. [CrossRef] [PubMed]
52. Han, H.; Tan, J.; Wang, R.; Wan, H.; He, Y.; Yan, X.; Guo, J.; Gao, Q.; Li, J.; Shang, S.; et al. PINK1 phosphorylates Drp1(S616) to regulate mitophagy-independent mitochondrial dynamics. *EMBO Rep.* **2020**, *21*, e48686. [CrossRef] [PubMed]
53. Pryde, K.R.; Smith, H.L.; Chau, K.Y.; Schapira, A.H. PINK1 disables the anti-fission machinery to segregate damaged mitochondria for mitophagy. *J. Cell Biol.* **2016**, *213*, 163–171. [CrossRef] [PubMed]
54. Schlattner, U.; Tokarska-Schlattner, M.; Ramirez, S.; Bruckner, A.; Kay, L.; Polge, C.; Epand, R.F.; Lee, R.M.; Lacombe, M.L.; Epand, R.M. Mitochondrial kinases and their molecular interaction with cardiolipin. *Biochim. Biophys. Acta* **2009**, *1788*, 2032–2047. [CrossRef] [PubMed]
55. Cotteret, S.; Chernoff, J. Nucleocytoplasmic shuttling of Pak5 regulates its antiapoptotic properties. *Mol. Cell Biol.* **2006**, *26*, 3215–3230. [CrossRef]
56. Greene, A.W.; Grenier, K.; Aguileta, M.A.; Muise, S.; Farazifard, R.; Haque, M.E.; McBride, H.M.; Park, D.S.; Fon, E.A. Mitochondrial processing peptidase regulates PINK1 processing, import and Parkin recruitment. *EMBO Rep.* **2012**, *13*, 378–385. [CrossRef] [PubMed]
57. Jin, Y.; Murata, H.; Sakaguchi, M.; Kataoka, K.; Watanabe, M.; Nasu, Y.; Kumon, H.; Huh, N.H. Partial sensitization of human bladder cancer cells to a gene-therapeutic adenovirus carrying REIC/Dkk-3 by downregulation of BRPK/PINK1. *Oncol. Rep.* **2012**, *27*, 695–699. [CrossRef]
58. Meissner, C.; Lorenz, H.; Weihofen, A.; Selkoe, D.J.; Lemberg, M.K. The mitochondrial intramembrane protease PARL cleaves human Pink1 to regulate Pink1 trafficking. *J. Neurochem.* **2011**, *117*, 856–867. [CrossRef]
59. Voigt, A.; Berlemann, L.A.; Winklhofer, K.F. The mitochondrial kinase PINK1: Functions beyond mitophagy. *J. Neurochem.* **2016**, *139* (Suppl. 1), 232–239. [CrossRef]
60. Lazarou, M.; Jin, S.M.; Kane, L.A.; Youle, R.J. Role of PINK1 binding to the TOM complex and alternate intracellular membranes in recruitment and activation of the E3 ligase Parkin. *Dev. Cell* **2012**, *22*, 320–333. [CrossRef]
61. Okatsu, K.; Oka, T.; Iguchi, M.; Imamura, K.; Kosako, H.; Tani, N.; Kimura, M.; Go, E.; Koyano, F.; Funayama, M.; et al. PINK1 autophosphorylation upon membrane potential dissipation is essential for Parkin recruitment to damaged mitochondria. *Nat. Commun.* **2012**, *3*, 1016. [CrossRef] [PubMed]
62. Kondapalli, C.; Kazlauskaite, A.; Zhang, N.; Woodroof, H.I.; Campbell, D.G.; Gourlay, R.; Burchell, L.; Walden, H.; Macartney, T.J.; Deak, M.; et al. PINK1 is activated by mitochondrial membrane potential depolarization and stimulates Parkin E3 ligase activity by phosphorylating Serine 65. *Open Biol.* **2012**, *2*, 120080. [CrossRef] [PubMed]
63. Dawson, T.M.; Dawson, V.L. The role of parkin in familial and sporadic Parkinson's disease. *Mov. Disord.* **2010**, *25* (Suppl. 1), S32–S39. [CrossRef] [PubMed]
64. Ishihara-Paul, L.; Hulihan, M.M.; Kachergus, J.; Upmanyu, R.; Warren, L.; Amouri, R.; Elango, R.; Prinjha, R.K.; Soto, A.; Kefi, M.; et al. PINK1 mutations and parkinsonism. *Neurology* **2008**, *71*, 896–902. [CrossRef] [PubMed]
65. Kitada, T.; Asakawa, S.; Hattori, N.; Matsumine, H.; Yamamura, Y.; Minoshima, S.; Yokochi, M.; Mizuno, Y.; Shimizu, N. Mutations in the parkin gene cause autosomal recessive juvenile parkinsonism. *Nature* **1998**, *392*, 605–608. [CrossRef] [PubMed]
66. Pilcher, H. Parkin implicated in sporadic Parkinson's disease. *Lancet Neurol.* **2005**, *4*, 798. [CrossRef]
67. Greggio, E.; Jain, S.; Kingsbury, A.; Bandopadhyay, R.; Lewis, P.; Kaganovich, A.; van der Brug, M.P.; Beilina, A.; Blackinton, J.; Thomas, K.J.; et al. Kinase activity is required for the toxic effects of mutant LRRK2/dardarin. *Neurobiol. Dis.* **2006**, *23*, 329–341. [CrossRef]
68. Di Maio, R.; Hoffman, E.K.; Rocha, E.M.; Keeney, M.T.; Sanders, L.H.; De Miranda, B.R.; Zharikov, A.; Van Laar, A.; Stepan, A.F.; Lanz, T.A.; et al. LRRK2 activation in idiopathic Parkinson's disease. *Sci. Transl. Med.* **2018**, *10*. [CrossRef]
69. Angeles, D.C.; Ho, P.; Chua, L.L.; Wang, C.; Yap, Y.W.; Ng, C.; Zhou, Z.; Lim, K.L.; Wszolek, Z.K.; Wang, H.Y.; et al. Thiol peroxidases ameliorate LRRK2 mutant-induced mitochondrial and dopaminergic neuronal degeneration in Drosophila. *Hum. Mol. Genet.* **2014**, *23*, 3157–3165. [CrossRef]
70. Wang, X.; Yan, M.H.; Fujioka, H.; Liu, J.; Wilson-Delfosse, A.; Chen, S.G.; Perry, G.; Casadesus, G.; Zhu, X. LRRK2 regulates mitochondrial dynamics and function through direct interaction with DLP1. *Hum. Mol. Genet.* **2012**, *21*, 1931–1944. [CrossRef]

71. Hsieh, C.H.; Shaltouki, A.; Gonzalez, A.E.; Bettencourt da Cruz, A.; Burbulla, L.F.; St Lawrence, E.; Schule, B.; Krainc, D.; Palmer, T.D.; Wang, X. Functional Impairment in Miro Degradation and Mitophagy Is a Shared Feature in Familial and Sporadic Parkinson's Disease. *Cell Stem Cell* **2016**, *19*, 709–724. [CrossRef] [PubMed]
72. Verma, M.; Callio, J.; Otero, P.A.; Sekler, I.; Wills, Z.P.; Chu, C.T. Mitochondrial Calcium Dysregulation Contributes to Dendrite Degeneration Mediated by PD/LBD-Associated LRRK2 Mutants. *J. Neurosci.* **2017**, *37*, 11151–11165. [CrossRef] [PubMed]
73. Pyakurel, A.; Savoia, C.; Hess, D.; Scorrano, L. Extracellular regulated kinase phosphorylates mitofusin 1 to control mitochondrial morphology and apoptosis. *Mol. Cell* **2015**, *58*, 244–254. [CrossRef] [PubMed]
74. Leboucher, G.P.; Tsai, Y.C.; Yang, M.; Shaw, K.C.; Zhou, M.; Veenstra, T.D.; Glickman, M.H.; Weissman, A.M. Stress-induced phosphorylation and proteasomal degradation of mitofusin 2 facilitates mitochondrial fragmentation and apoptosis. *Mol. Cell* **2012**, *47*, 547–557. [CrossRef]
75. Zhou, W.; Chen, K.H.; Cao, W.; Zeng, J.; Liao, H.; Zhao, L.; Guo, X. Mutation of the protein kinase A phosphorylation site influences the anti-proliferative activity of mitofusin 2. *Atherosclerosis* **2010**, *211*, 216–223. [CrossRef]
76. de la Cruz Lopez, K.G.; Toledo Guzman, M.E.; Sanchez, E.O.; Garcia Carranca, A. mTORC1 as a Regulator of Mitochondrial Functions and a Therapeutic Target in Cancer. *Front. Oncol.* **2019**, *9*, 1373. [CrossRef]
77. Desai, B.N.; Myers, B.R.; Schreiber, S.L. FKBP12-rapamycin-associated protein associates with mitochondria and senses osmotic stress via mitochondrial dysfunction. *Proc. Natl. Acad. Sci. USA* **2002**, *99*, 4319–4324. [CrossRef]
78. Schieke, S.M.; Phillips, D.; McCoy, J.P., Jr.; Aponte, A.M.; Shen, R.F.; Balaban, R.S.; Finkel, T. The mammalian target of rapamycin (mTOR) pathway regulates mitochondrial oxygen consumption and oxidative capacity. *J. Biol. Chem.* **2006**, *281*, 27643–27652. [CrossRef]
79. Ramanathan, A.; Schreiber, S.L. Direct control of mitochondrial function by mTOR. *Proc. Natl. Acad. Sci. USA* **2009**, *106*, 22229–22232. [CrossRef]
80. Lu, C.L.; Qin, L.; Liu, H.C.; Candas, D.; Fan, M.; Li, J.J. Tumor cells switch to mitochondrial oxidative phosphorylation under radiation via mTOR-mediated hexokinase II inhibition–a Warburg-reversing effect. *PLoS ONE* **2015**, *10*, e0121046. [CrossRef]
81. Roberts, D.J.; Tan-Sah, V.P.; Ding, E.Y.; Smith, J.M.; Miyamoto, S. Hexokinase-II positively regulates glucose starvation-induced autophagy through TORC1 inhibition. *Mol. Cell* **2014**, *53*, 521–533. [CrossRef] [PubMed]
82. Cotteret, S.; Jaffer, Z.M.; Beeser, A.; Chernoff, J. p21-Activated kinase 5 (Pak5) localizes to mitochondria and inhibits apoptosis by phosphorylating BAD. *Mol. Cell Biol.* **2003**, *23*, 5526–5539. [CrossRef] [PubMed]
83. Ma, D.; Bai, X.; Zou, H.; Lai, Y.; Jiang, Y. Rheb GTPase controls apoptosis by regulating interaction of FKBP38 with Bcl-2 and Bcl-XL. *J. Biol. Chem.* **2010**, *285*, 8621–8627. [CrossRef] [PubMed]
84. Tasken, K.; Aandahl, E.M. Localized effects of cAMP mediated by distinct routes of protein kinase A. *Physiol. Rev.* **2004**, *84*, 137–167. [CrossRef] [PubMed]
85. Ito, Y.; Mishra, N.C.; Yoshida, K.; Kharbanda, S.; Saxena, S.; Kufe, D. Mitochondrial targeting of JNK/SAPK in the phorbol ester response of myeloid leukemia cells. *Cell Death Differ.* **2001**, *8*, 794–800. [CrossRef] [PubMed]
86. Zhu, J.H.; Guo, F.; Shelburne, J.; Watkins, S.; Chu, C.T. Localization of phosphorylated ERK/MAP kinases to mitochondria and autophagosomes in Lewy body diseases. *Brain Pathol.* **2003**, *13*, 473–481. [CrossRef]
87. Ballard-Croft, C.; Kristo, G.; Yoshimura, Y.; Reid, E.; Keith, B.J.; Mentzer, R.M., Jr.; Lasley, R.D. Acute adenosine preconditioning is mediated by p38 MAPK activation in discrete subcellular compartments. *Am. J. Physiol. Heart Circ. Physiol.* **2005**, *288*, H1359–H1366. [CrossRef]
88. Zhou, C.; Huang, Y.; Shao, Y.; May, J.; Prou, D.; Perier, C.; Dauer, W.; Schon, E.A.; Przedborski, S. The kinase domain of mitochondrial PINK1 faces the cytoplasm. *Proc. Natl. Acad. Sci. USA* **2008**, *105*, 12022–12027. [CrossRef]
89. Biskup, S.; Moore, D.J.; Celsi, F.; Higashi, S.; West, A.B.; Andrabi, S.A.; Kurkinen, K.; Yu, S.W.; Savitt, J.M.; Waldvogel, H.J.; et al. Localization of LRRK2 to membranous and vesicular structures in mammalian brain. *Ann. Neurol.* **2006**, *60*, 557–569. [CrossRef]
90. Liang, J.; Xu, Z.X.; Ding, Z.; Lu, Y.; Yu, Q.; Werle, K.D.; Zhou, G.; Park, Y.Y.; Peng, G.; Gambello, M.J.; et al. Myristoylation confers noncanonical AMPK functions in autophagy selectivity and mitochondrial surveillance. *Nat. Commun.* **2015**, *6*, 7926. [CrossRef]
91. Feng, Y.; Ariza, M.E.; Goulet, A.C.; Shi, J.; Nelson, M.A. Death-signal-induced relocalization of cyclin-dependent kinase 11 to mitochondria. *Biochem. J.* **2005**, *392*, 65–73. [CrossRef] [PubMed]
92. Bordin, L.; Cattapan, F.; Clari, G.; Toninello, A.; Siliprandi, N.; Moret, V. Spermine-mediated casein kinase II-uptake by rat liver mitochondria. *Biochim. Biophys. Acta* **1994**, *1199*, 266–270. [CrossRef]
93. Clari, G.; Toninello, A.; Bordin, L.; Cattapan, F.; Piccinelli-Siliprandi, D.; Moret, V. Spermine effect on the binding of casein kinase I to the rat liver mitochondrial structures. *Biochem. Biophys. Res. Commun.* **1994**, *205*, 389–395. [CrossRef] [PubMed]
94. Gordon, R.; Singh, N.; Lawana, V.; Ghosh, A.; Harischandra, D.S.; Jin, H.; Hogan, C.; Sarkar, S.; Rokad, D.; Panicker, N.; et al. Protein kinase Cdelta upregulation in microglia drives neuroinflammatory responses and dopaminergic neurodegeneration in experimental models of Parkinson's disease. *Neurobiol. Dis.* **2016**, *93*, 96–114. [CrossRef]
95. Nowak, G.; Bakajsova, D. Protein kinase C-alpha interaction with F0F1-ATPase promotes F0F1-ATPase activity and reduces energy deficits in injured renal cells. *J. Biol. Chem.* **2015**, *290*, 7054–7066. [CrossRef]
96. Rathore, R.; Zheng, Y.M.; Li, X.Q.; Wang, Q.S.; Liu, Q.H.; Ginnan, R.; Singer, H.A.; Ho, Y.S.; Wang, Y.X. Mitochondrial ROS-PKCepsilon signaling axis is uniquely involved in hypoxic increase in [Ca2+]i in pulmonary artery smooth muscle cells. *Biochem. Biophys. Res. Commun.* **2006**, *351*, 784–790. [CrossRef]

97. Rubio, M.A.; Hopper, A.K. Transfer RNA travels from the cytoplasm to organelles. *Wiley Interdiscip Rev. RNA* **2011**, *2*, 802–817. [CrossRef]
98. Sieber, F.; Duchene, A.M.; Marechal-Drouard, L. Mitochondrial RNA import: From diversity of natural mechanisms to potential applications. *Int. Rev. Cell Mol. Biol.* **2011**, *287*, 145–190. [CrossRef]
99. Wiedemann, N.; Pfanner, N. Mitochondrial Machineries for Protein Import and Assembly. *Annu. Rev. Biochem.* **2017**, *86*, 685–714. [CrossRef]
100. Law, Y.S.; Ngan, L.; Yan, J.; Kwok, L.Y.; Sun, Y.; Cheng, S.; Schwenkert, S.; Lim, B.L. Multiple Kinases Can Phosphorylate the N-Terminal Sequences of Mitochondrial Proteins in Arabidopsis thaliana. *Front. Plant. Sci.* **2018**, *9*, 982. [CrossRef]
101. Moulin, C.; Caumont-Sarcos, A.; Ieva, R. Mitochondrial presequence import: Multiple regulatory knobs fine-tune mitochondrial biogenesis and homeostasis. *Biochim. Biophys. Acta Mol. Cell Res.* **2019**, *1866*, 930–944. [CrossRef] [PubMed]
102. Becker, T.; Vogtle, F.N.; Stojanovski, D.; Meisinger, C. Sorting and assembly of mitochondrial outer membrane proteins. *Biochim. Biophys. Acta* **2008**, *1777*, 557–563. [CrossRef] [PubMed]
103. Heazlewood, J.L.; Durek, P.; Hummel, J.; Selbig, J.; Weckwerth, W.; Walther, D.; Schulze, W.X. PhosPhAt: A database of phosphorylation sites in Arabidopsis thaliana and a plant-specific phosphorylation site predictor. *Nucleic Acids Res.* **2008**, *36*, D1015–1021. [CrossRef] [PubMed]
104. Rao, S.; Schmidt, O.; Harbauer, A.B.; Schonfisch, B.; Guiard, B.; Pfanner, N.; Meisinger, C. Biogenesis of the preprotein translocase of the outer mitochondrial membrane: Protein kinase A phosphorylates the precursor of Tom40 and impairs its import. *Mol. Biol. Cell* **2012**, *23*, 1618–1627. [CrossRef]
105. Schmidt, O.; Harbauer, A.B.; Rao, S.; Eyrich, B.; Zahedi, R.P.; Stojanovski, D.; Schonfisch, B.; Guiard, B.; Sickmann, A.; Pfanner, N.; et al. Regulation of mitochondrial protein import by cytosolic kinases. *Cell* **2011**, *144*, 227–239. [CrossRef]
106. Hornbeck, P.V.; Zhang, B.; Murray, B.; Kornhauser, J.M.; Latham, V.; Skrzypek, E. PhosphoSitePlus, 2014: Mutations, PTMs and recalibrations. *Nucleic Acids Res.* **2015**, *43*, D512–D520. [CrossRef]
107. Gerbeth, C.; Schmidt, O.; Rao, S.; Harbauer, A.B.; Mikropoulou, D.; Opalinska, M.; Guiard, B.; Pfanner, N.; Meisinger, C. Glucose-induced regulation of protein import receptor Tom22 by cytosolic and mitochondria-bound kinases. *Cell Metab* **2013**, *18*, 578–587. [CrossRef]
108. Chacinska, A.; Koehler, C.M.; Milenkovic, D.; Lithgow, T.; Pfanner, N. Importing mitochondrial proteins: Machineries and mechanisms. *Cell* **2009**, *138*, 628–644. [CrossRef]
109. Neupert, W.; Herrmann, J.M. Translocation of proteins into mitochondria. *Annu. Rev. Biochem.* **2007**, *76*, 723–749. [CrossRef]
110. Becker, T.; Guiard, B.; Thornton, N.; Zufall, N.; Stroud, D.A.; Wiedemann, N.; Pfanner, N. Assembly of the mitochondrial protein import channel: Role of Tom5 in two-stage interaction of Tom40 with the SAM complex. *Mol. Biol. Cell* **2010**, *21*, 3106–3113. [CrossRef]
111. Zaman, S.; Lippman, S.I.; Zhao, X.; Broach, J.R. How Saccharomyces responds to nutrients. *Annu. Rev. Genet.* **2008**, *42*, 27–81. [CrossRef] [PubMed]
112. Becker, T.; Pfannschmidt, S.; Guiard, B.; Stojanovski, D.; Milenkovic, D.; Kutik, S.; Pfanner, N.; Meisinger, C.; Wiedemann, N. Biogenesis of the mitochondrial TOM complex: Mim1 promotes insertion and assembly of signal-anchored receptors. *J. Biol. Chem.* **2008**, *283*, 120–127. [CrossRef] [PubMed]
113. Becker, T.; Wenz, L.S.; Kruger, V.; Lehmann, W.; Muller, J.M.; Goroncy, L.; Zufall, N.; Lithgow, T.; Guiard, B.; Chacinska, A.; et al. The mitochondrial import protein Mim1 promotes biogenesis of multispanning outer membrane proteins. *J. Cell Biol.* **2011**, *194*, 387–395. [CrossRef]
114. Kettenbach, A.N.; Gerber, S.A. Rapid and reproducible single-stage phosphopeptide enrichment of complex peptide mixtures: Application to general and phosphotyrosine-specific phosphoproteomics experiments. *Anal. Chem.* **2011**, *83*, 7635–7644. [CrossRef] [PubMed]
115. Mertins, P.; Mani, D.R.; Ruggles, K.V.; Gillette, M.A.; Clauser, K.R.; Wang, P.; Wang, X.; Qiao, J.W.; Cao, S.; Petralia, F.; et al. Proteogenomics connects somatic mutations to signalling in breast cancer. *Nature* **2016**, *534*, 55–62. [CrossRef]
116. Sharma, K.; D'Souza, R.C.; Tyanova, S.; Schaab, C.; Wisniewski, J.R.; Cox, J.; Mann, M. Ultradeep human phosphoproteome reveals a distinct regulatory nature of Tyr and Ser/Thr-based signaling. *Cell Rep.* **2014**, *8*, 1583–1594. [CrossRef]
117. Tsai, C.F.; Wang, Y.T.; Yen, H.Y.; Tsou, C.C.; Ku, W.C.; Lin, P.Y.; Chen, H.Y.; Nesvizhskii, A.I.; Ishihama, Y.; Chen, Y.J. Large-scale determination of absolute phosphorylation stoichiometries in human cells by motif-targeting quantitative proteomics. *Nat. Commun.* **2015**, *6*, 6622. [CrossRef]
118. Zhou, H.; Di Palma, S.; Preisinger, C.; Peng, M.; Polat, A.N.; Heck, A.J.; Mohammed, S. Toward a comprehensive characterization of a human cancer cell phosphoproteome. *J. Proteome Res.* **2013**, *12*, 260–271. [CrossRef]
119. Zhou, Q.; Lam, P.Y.; Han, D.; Cadenas, E. c-Jun N-terminal kinase regulates mitochondrial bioenergetics by modulating pyruvate dehydrogenase activity in primary cortical neurons. *J. Neurochem.* **2008**, *104*, 325–335. [CrossRef]
120. Seifert, F.; Ciszak, E.; Korotchkina, L.; Golbik, R.; Spinka, M.; Dominiak, P.; Sidhu, S.; Brauer, J.; Patel, M.S.; Tittmann, K. Phosphorylation of serine 264 impedes active site accessibility in the E1 component of the human pyruvate dehydrogenase multienzyme complex. *Biochemistry* **2007**, *46*, 6277–6287. [CrossRef]
121. Sugden, M.C.; Holness, M.J. Recent advances in mechanisms regulating glucose oxidation at the level of the pyruvate dehydrogenase complex by PDKs. *Am. J. Physiol. Endocrinol. Metab* **2003**, *284*, E855–862. [CrossRef] [PubMed]

122. Zhou, Z.H.; McCarthy, D.B.; O'Connor, C.M.; Reed, L.J.; Stoops, J.K. The remarkable structural and functional organization of the eukaryotic pyruvate dehydrogenase complexes. *Proc. Natl. Acad. Sci. USA* **2001**, *98*, 14802–14807. [CrossRef] [PubMed]
123. Linn, T.C.; Pettit, F.H.; Reed, L.J. Alpha-keto acid dehydrogenase complexes. X. Regulation of the activity of the pyruvate dehydrogenase complex from beef kidney mitochondria by phosphorylation and dephosphorylation. *Proc. Natl. Acad. Sci. USA* **1969**, *62*, 234–241. [CrossRef] [PubMed]
124. Bowker-Kinley, M.; Popov, K.M. Evidence that pyruvate dehydrogenase kinase belongs to the ATPase/kinase superfamily. *Biochem. J.* **1999**, *344 Pt. 1*, 47–53. [CrossRef]
125. Bowker-Kinley, M.M.; Davis, W.I.; Wu, P.; Harris, R.A.; Popov, K.M. Evidence for existence of tissue-specific regulation of the mammalian pyruvate dehydrogenase complex. *Biochem. J.* **1998**, *329 Pt. 1*, 191–196. [CrossRef]
126. Steussy, C.N.; Popov, K.M.; Bowker-Kinley, M.M.; Sloan, R.B., Jr.; Harris, R.A.; Hamilton, J.A. Structure of pyruvate dehydrogenase kinase. Novel folding pattern for a serine protein kinase. *J. Biol Chem* **2001**, *276*, 37443–37450. [CrossRef]
127. Teague, W.M.; Pettit, F.H.; Yeaman, S.J.; Reed, L.J. Function of phosphorylation sites on pyruvate dehydrogenase. *Biochem. Biophys. Res. Commun.* **1979**, *87*, 244–252. [CrossRef]
128. Korotchkina, L.G.; Patel, M.S. Site specificity of four pyruvate dehydrogenase kinase isoenzymes toward the three phosphorylation sites of human pyruvate dehydrogenase. *J. Biol. Chem.* **2001**, *276*, 37223–37229. [CrossRef]
129. Korotchkina, L.G.; Patel, M.S. Probing the mechanism of inactivation of human pyruvate dehydrogenase by phosphorylation of three sites. *J. Biol. Chem.* **2001**, *276*, 5731–5738. [CrossRef]
130. Kolobova, E.; Tuganova, A.; Boulatnikov, I.; Popov, K.M. Regulation of pyruvate dehydrogenase activity through phosphorylation at multiple sites. *Biochem. J.* **2001**, *358*, 69–77. [CrossRef]
131. Kerbey, A.L.; Randle, P.J.; Cooper, R.H.; Whitehouse, S.; Pask, H.T.; Denton, R.M. Regulation of pyruvate dehydrogenase in rat heart. Mechanism of regulation of proportions of dephosphorylated and phosphorylated enzyme by oxidation of fatty acids and ketone bodies and of effects of diabetes: Role of coenzyme A, acetyl-coenzyme A and reduced and oxidized nicotinamide-adenine dinucleotide. *Biochem. J.* **1976**, *154*, 327–348. [CrossRef] [PubMed]
132. Nakai, N.; Sato, Y.; Oshida, Y.; Yoshimura, A.; Fujitsuka, N.; Sugiyama, S.; Shimomura, Y. Effects of aging on the activities of pyruvate dehydrogenase complex and its kinase in rat heart. *Life Sci.* **1997**, *60*, 2309–2314. [CrossRef]
133. Zhou, Q.; Lam, P.Y.; Han, D.; Cadenas, E. Activation of c-Jun-N-terminal kinase and decline of mitochondrial pyruvate dehydrogenase activity during brain aging. *FEBS Lett.* **2009**, *583*, 1132–1140. [CrossRef] [PubMed]
134. Peng, J.; Andersen, J.K. The role of c-Jun N-terminal kinase (JNK) in Parkinson's disease. *IUBMB Life* **2003**, *55*, 267–271. [CrossRef] [PubMed]
135. Zhu, X.; Raina, A.K.; Rottkamp, C.A.; Aliev, G.; Perry, G.; Boux, H.; Smith, M.A. Activation and redistribution of c-jun N-terminal kinase/stress activated protein kinase in degenerating neurons in Alzheimer's disease. *J. Neurochem.* **2001**, *76*, 435–441. [CrossRef] [PubMed]
136. Li, X.; Jiang, Y.; Meisenhelder, J.; Yang, W.; Hawke, D.H.; Zheng, Y.; Xia, Y.; Aldape, K.; He, J.; Hunter, T.; et al. Mitochondria-Translocated PGK1 Functions as a Protein Kinase to Coordinate Glycolysis and the TCA Cycle in Tumorigenesis. *Mol. Cell* **2016**, *61*, 705–719. [CrossRef]
137. Warburg, O. On respiratory impairment in cancer cells. *Science* **1956**, *124*, 269–270.
138. Yang, W.; Lu, Z. Regulation and function of pyruvate kinase M2 in cancer. *Cancer Lett.* **2013**, *339*, 153–158. [CrossRef]
139. Yang, W.; Lu, Z. Pyruvate kinase M2 at a glance. *J. Cell Sci.* **2015**, *128*, 1655–1660. [CrossRef]
140. Bernstein, B.E.; Hol, W.G. Crystal structures of substrates and products bound to the phosphoglycerate kinase active site reveal the catalytic mechanism. *Biochemistry* **1998**, *37*, 4429–4436. [CrossRef]
141. Zhang, D.; Tai, L.K.; Wong, L.L.; Chiu, L.L.; Sethi, S.K.; Koay, E.S. Proteomic study reveals that proteins involved in metabolic and detoxification pathways are highly expressed in HER-2/neu-positive breast cancer. *Mol. Cell Proteom.* **2005**, *4*, 1686–1696. [CrossRef] [PubMed]
142. Hwang, T.L.; Liang, Y.; Chien, K.Y.; Yu, J.S. Overexpression and elevated serum levels of phosphoglycerate kinase 1 in pancreatic ductal adenocarcinoma. *Proteomics* **2006**, *6*, 2259–2272. [CrossRef] [PubMed]
143. Duan, Z.; Lamendola, D.E.; Yusuf, R.Z.; Penson, R.T.; Preffer, F.I.; Seiden, M.V. Overexpression of human phosphoglycerate kinase 1 (PGK1) induces a multidrug resistance phenotype. *Anticancer Res.* **2002**, *22*, 1933–1941. [PubMed]
144. Ahmad, S.S.; Glatzle, J.; Bajaeifer, K.; Buhler, S.; Lehmann, T.; Konigsrainer, I.; Vollmer, J.P.; Sipos, B.; Ahmad, S.S.; Northoff, H.; et al. Phosphoglycerate kinase 1 as a promoter of metastasis in colon cancer. *Int. J. Oncol.* **2013**, *43*, 586–590. [CrossRef] [PubMed]
145. Ai, J.; Huang, H.; Lv, X.; Tang, Z.; Chen, M.; Chen, T.; Duan, W.; Sun, H.; Li, Q.; Tan, R.; et al. FLNA and PGK1 are two potential markers for progression in hepatocellular carcinoma. *Cell Physiol. Biochem.* **2011**, *27*, 207–216. [CrossRef]
146. Zieker, D.; Konigsrainer, I.; Tritschler, I.; Loffler, M.; Beckert, S.; Traub, F.; Nieselt, K.; Buhler, S.; Weller, M.; Gaedcke, J.; et al. Phosphoglycerate kinase 1 a promoting enzyme for peritoneal dissemination in gastric cancer. *Int. J. Cancer* **2010**, *126*, 1513–1520. [CrossRef]
147. Hoshi, M.; Takashima, A.; Noguchi, K.; Murayama, M.; Sato, M.; Kondo, S.; Saitoh, Y.; Ishiguro, K.; Hoshino, T.; Imahori, K. Regulation of mitochondrial pyruvate dehydrogenase activity by tau protein kinase I/glycogen synthase kinase 3beta in brain. *Proc. Natl. Acad. Sci. USA* **1996**, *93*, 2719–2723. [CrossRef]
148. Song, J.S.; Yang, S.D. Tau protein kinase I/GSK-3 beta/kinase FA in heparin phosphorylates tau on Ser199, Thr231, Ser235, Ser262, Ser369, and Ser400 sites phosphorylated in Alzheimer disease brain. *J. Protein Chem.* **1995**, *14*, 95–105. [CrossRef]

149. Bykova, N.V.; Stensballe, A.; Egsgaard, H.; Jensen, O.N.; Moller, I.M. Phosphorylation of formate dehydrogenase in potato tuber mitochondria. *J. Biol. Chem.* **2003**, *278*, 26021–26030. [CrossRef]
150. Chen, X.J.; Butow, R.A. The organization and inheritance of the mitochondrial genome. *Nat. Rev. Genet.* **2005**, *6*, 815–825. [CrossRef]
151. Shadel, G.S. Mitochondrial DNA, aconitase 'wraps' it up. *Trends Biochem. Sci.* **2005**, *30*, 294–296. [CrossRef] [PubMed]
152. Lewandrowski, U.; Sickmann, A.; Cesaro, L.; Brunati, A.M.; Toninello, A.; Salvi, M. Identification of new tyrosine phosphorylated proteins in rat brain mitochondria. *FEBS Lett.* **2008**, *582*, 1104–1110. [CrossRef] [PubMed]
153. Guo, X.; Niemi, N.M.; Hutchins, P.D.; Condon, S.G.F.; Jochem, A.; Ulbrich, A.; Higbee, A.J.; Russell, J.D.; Senes, A.; Coon, J.J.; et al. Ptc7p Dephosphorylates Select Mitochondrial Proteins to Enhance Metabolic Function. *Cell Rep.* **2017**, *18*, 307–313. [CrossRef] [PubMed]
154. Hofer, A.; Wenz, T. Post-translational modification of mitochondria as a novel mode of regulation. *Exp. Gerontol.* **2014**, *56*, 202–220. [CrossRef] [PubMed]
155. Duarte, A.; Castillo, A.F.; Podesta, E.J.; Poderoso, C. Mitochondrial fusion and ERK activity regulate steroidogenic acute regulatory protein localization in mitochondria. *PLoS ONE* **2014**, *9*, e100387. [CrossRef] [PubMed]
156. Bose, H.S.; Lingappa, V.R.; Miller, W.L. Rapid regulation of steroidogenesis by mitochondrial protein import. *Nature* **2002**, *417*, 87–91. [CrossRef]
157. Castillo, A.F.; Orlando, U.; Helfenberger, K.E.; Poderoso, C.; Podesta, E.J. The role of mitochondrial fusion and StAR phosphorylation in the regulation of StAR activity and steroidogenesis. *Mol. Cell Endocrinol.* **2015**, *408*, 73–79. [CrossRef]
158. Arakane, F.; King, S.R.; Du, Y.; Kallen, C.B.; Walsh, L.P.; Watari, H.; Stocco, D.M.; Strauss, J.F., 3rd. Phosphorylation of steroidogenic acute regulatory protein (StAR) modulates its steroidogenic activity. *J. Biol. Chem.* **1997**, *272*, 32656–32662. [CrossRef]
159. Fleury, A.; Mathieu, A.P.; Ducharme, L.; Hales, D.B.; LeHoux, J.G. Phosphorylation and function of the hamster adrenal steroidogenic acute regulatory protein (StAR). *J. Steroid Biochem. Mol. Biol.* **2004**, *91*, 259–271. [CrossRef]
160. Granot, Z.; Kobiler, O.; Melamed-Book, N.; Eimerl, S.; Bahat, A.; Lu, B.; Braun, S.; Maurizi, M.R.; Suzuki, C.K.; Oppenheim, A.B.; et al. Turnover of mitochondrial steroidogenic acute regulatory (StAR) protein by Lon protease: The unexpected effect of proteasome inhibitors. *Mol. Endocrinol.* **2007**, *21*, 2164–2177. [CrossRef]
161. van den Heuvel, L.; Smeitink, J. The oxidative phosphorylation (OXPHOS) system: Nuclear genes and human genetic diseases. *Bioessays* **2001**, *23*, 518–525. [CrossRef] [PubMed]
162. Nosek, J.; Fukuhara, H. NADH dehydrogenase subunit genes in the mitochondrial DNA of yeasts. *J. Bacteriol.* **1994**, *176*, 5622–5630. [CrossRef] [PubMed]
163. Chen, R.; Fearnley, I.M.; Peak-Chew, S.Y.; Walker, J.E. The phosphorylation of subunits of complex I from bovine heart mitochondria. *J. Biol. Chem.* **2004**, *279*, 26036–26045. [CrossRef] [PubMed]
164. Papa, S.; De Rasmo, D.; Scacco, S.; Signorile, A.; Technikova-Dobrova, Z.; Palmisano, G.; Sardanelli, A.M.; Papa, F.; Panelli, D.; Scaringi, R.; et al. Mammalian complex I: A regulable and vulnerable pacemaker in mitochondrial respiratory function. *Biochim. Biophys. Acta* **2008**, *1777*, 719–728. [CrossRef] [PubMed]
165. Gowthami, N.; Sunitha, B.; Kumar, M.; Keshava Prasad, T.S.; Gayathri, N.; Padmanabhan, B.; Srinivas Bharath, M.M. Mapping the protein phosphorylation sites in human mitochondrial complex I (NADH: Ubiquinone oxidoreductase): A bioinformatics study with implications for brain aging and neurodegeneration. *J. Chem. Neuroanat* **2019**, *95*, 13–28. [CrossRef] [PubMed]
166. Scacco, S.; Vergari, R.; Scarpulla, R.C.; Technikova-Dobrova, Z.; Sardanelli, A.; Lambo, R.; Lorusso, V.; Papa, S. cAMP-dependent phosphorylation of the nuclear encoded 18-kDa (IP) subunit of respiratory complex I and activation of the complex in serum-starved mouse fibroblast cultures. *J. Biol. Chem.* **2000**, *275*, 17578–17582. [CrossRef] [PubMed]
167. De Rasmo, D.; Palmisano, G.; Scacco, S.; Technikova-Dobrova, Z.; Panelli, D.; Cocco, T.; Sardanelli, A.M.; Gnoni, A.; Micelli, L.; Trani, A.; et al. Phosphorylation pattern of the NDUFS4 subunit of complex I of the mammalian respiratory chain. *Mitochondrion* **2010**, *10*, 464–471. [CrossRef]
168. Papa, S.; Sardanelli, A.M.; Cocco, T.; Speranza, F.; Scacco, S.C.; Technikova-Dobrova, Z. The nuclear-encoded 18 kDa (IP) AQDQ subunit of bovine heart complex I is phosphorylated by the mitochondrial cAMP-dependent protein kinase. *FEBS Lett.* **1996**, *379*, 299–301. [CrossRef]
169. Papa, S.; Scacco, S.; Sardanelli, A.M.; Vergari, R.; Papa, F.; Budde, S.; van den Heuvel, L.; Smeitink, J. Mutation in the NDUFS4 gene of complex I abolishes cAMP-dependent activation of the complex in a child with fatal neurological syndrome. *FEBS Lett.* **2001**, *489*, 259–262. [CrossRef]
170. van den Heuvel, L.; Ruitenbeek, W.; Smeets, R.; Gelman-Kohan, Z.; Elpeleg, O.; Loeffen, J.; Trijbels, F.; Mariman, E.; de Bruijn, D.; Smeitink, J. Demonstration of a new pathogenic mutation in human complex I deficiency: A 5-bp duplication in the nuclear gene encoding the 18-kD (AQDQ) subunit. *Am. J. Hum. Genet.* **1998**, *62*, 262–268. [CrossRef]
171. De Rasmo, D.; Signorile, A.; Larizza, M.; Pacelli, C.; Cocco, T.; Papa, S. Activation of the cAMP cascade in human fibroblast cultures rescues the activity of oxidatively damaged complex I. *Free Radic. Biol. Med.* **2012**, *52*, 757–764. [CrossRef] [PubMed]
172. Piccoli, C.; Scacco, S.; Bellomo, F.; Signorile, A.; Iuso, A.; Boffoli, D.; Scrima, R.; Capitanio, N.; Papa, S. cAMP controls oxygen metabolism in mammalian cells. *FEBS Lett.* **2006**, *580*, 4539–4543. [CrossRef]
173. De Rasmo, D.; Panelli, D.; Sardanelli, A.M.; Papa, S. cAMP-dependent protein kinase regulates the mitochondrial import of the nuclear encoded NDUFS4 subunit of complex I. *Cell Signal.* **2008**, *20*, 989–997. [CrossRef] [PubMed]

174. Morais, V.A.; Haddad, D.; Craessaerts, K.; De Bock, P.J.; Swerts, J.; Vilain, S.; Aerts, L.; Overbergh, L.; Grunewald, A.; Seibler, P.; et al. PINK1 loss-of-function mutations affect mitochondrial complex I activity via NdufA10 ubiquinone uncoupling. *Science* **2014**, *344*, 203–207. [CrossRef] [PubMed]
175. Morais, V.A.; Verstreken, P.; Roethig, A.; Smet, J.; Snellinx, A.; Vanbrabant, M.; Haddad, D.; Frezza, C.; Mandemakers, W.; Vogt-Weisenhorn, D.; et al. Parkinson's disease mutations in PINK1 result in decreased Complex I activity and deficient synaptic function. *EMBO Mol. Med.* **2009**, *1*, 99–111. [CrossRef]
176. Wang, Z.; Fan, M.; Candas, D.; Zhang, T.Q.; Qin, L.; Eldridge, A.; Wachsmann-Hogiu, S.; Ahmed, K.M.; Chromy, B.A.; Nantajit, D.; et al. Cyclin B1/Cdk1 coordinates mitochondrial respiration for cell-cycle G2/M progression. *Dev. Cell* **2014**, *29*, 217–232. [CrossRef] [PubMed]
177. Ogura, M.; Inoue, T.; Yamaki, J.; Homma, M.K.; Kurosaki, T.; Homma, Y. Mitochondrial reactive oxygen species suppress humoral immune response through reduction of CD19 expression in B cells in mice. *Eur. J. Immunol.* **2017**, *47*, 406–418. [CrossRef]
178. Zhao, X.; Leon, I.R.; Bak, S.; Mogensen, M.; Wrzesinski, K.; Hojlund, K.; Jensen, O.N. Phosphoproteome analysis of functional mitochondria isolated from resting human muscle reveals extensive phosphorylation of inner membrane protein complexes and enzymes. *Mol. Cell Proteom.* **2011**, *10*, M110–000299. [CrossRef]
179. Arachiche, A.; Augereau, O.; Decossas, M.; Pertuiset, C.; Gontier, E.; Letellier, T.; Dachary-Prigent, J. Localization of PTP-1B, SHP-2, and Src exclusively in rat brain mitochondria and functional consequences. *J. Biol. Chem.* **2008**, *283*, 24406–24411. [CrossRef]
180. Mahapatra, G.; Varughese, A.; Ji, Q.; Lee, I.; Liu, J.; Vaishnav, A.; Sinkler, C.; Kapralov, A.A.; Moraes, C.T.; Sanderson, T.H.; et al. Phosphorylation of Cytochrome c Threonine 28 Regulates Electron Transport Chain Activity in Kidney: Implications for Amp Kinase. *J. Biol. Chem.* **2017**, *292*, 64–79. [CrossRef]
181. Pecina, P.; Borisenko, G.G.; Belikova, N.A.; Tyurina, Y.Y.; Pecinova, A.; Lee, I.; Samhan-Arias, A.K.; Przyklenk, K.; Kagan, V.E.; Huttemann, M. Phosphomimetic substitution of cytochrome C tyrosine 48 decreases respiration and binding to cardiolipin and abolishes ability to trigger downstream caspase activation. *Biochemistry* **2010**, *49*, 6705–6714. [CrossRef] [PubMed]
182. Wan, J.; Kalpage, H.A.; Vaishnav, A.; Liu, J.; Lee, I.; Mahapatra, G.; Turner, A.A.; Zurek, M.P.; Ji, Q.; Moraes, C.T.; et al. Regulation of Respiration and Apoptosis by Cytochrome c Threonine 58 Phosphorylation. *Sci. Rep.* **2019**, *9*, 15815. [CrossRef] [PubMed]
183. Moreno-Beltran, B.; Guerra-Castellano, A.; Diaz-Quintana, A.; Del Conte, R.; Garcia-Maurino, S.M.; Diaz-Moreno, S.; Gonzalez-Arzola, K.; Santos-Ocana, C.; Velazquez-Campoy, A.; De la Rosa, M.A.; et al. Structural basis of mitochondrial dysfunction in response to cytochrome c phosphorylation at tyrosine 48. *Proc. Natl. Acad. Sci. USA* **2017**, *114*, E3041–E3050. [CrossRef] [PubMed]
184. Kalpage, H.A.; Wan, J.; Morse, P.T.; Lee, I.; Huttemann, M. Brain-Specific Serine-47 Modification of Cytochrome c Regulates Cytochrome c Oxidase Activity Attenuating ROS Production and Cell Death: Implications for Ischemia/Reperfusion Injury and Akt Signaling. *Cells* **2020**, *9*, 1843. [CrossRef] [PubMed]
185. Taanman, J.W. Human cytochrome c oxidase: Structure, function, and deficiency. *J. Bioenerg. Biomembr.* **1997**, *29*, 151–163. [CrossRef] [PubMed]
186. Lee, I.; Salomon, A.R.; Ficarro, S.; Mathes, I.; Lottspeich, F.; Grossman, L.I.; Huttemann, M. cAMP-dependent tyrosine phosphorylation of subunit I inhibits cytochrome c oxidase activity. *J. Biol. Chem.* **2005**, *280*, 6094–6100. [CrossRef]
187. Samavati, L.; Lee, I.; Mathes, I.; Lottspeich, F.; Huttemann, M. Tumor necrosis factor alpha inhibits oxidative phosphorylation through tyrosine phosphorylation at subunit I of cytochrome c oxidase. *J. Biol. Chem.* **2008**, *283*, 21134–21144. [CrossRef]
188. Prabu, S.K.; Anandatheerthavarada, H.K.; Raza, H.; Srinivasan, S.; Spear, J.F.; Avadhani, N.G. Protein kinase A-mediated phosphorylation modulates cytochrome c oxidase function and augments hypoxia and myocardial ischemia-related injury. *J. Biol. Chem.* **2006**, *281*, 2061–2070. [CrossRef]
189. Srinivasan, S.; Spear, J.; Chandran, K.; Joseph, J.; Kalyanaraman, B.; Avadhani, N.G. Oxidative stress induced mitochondrial protein kinase A mediates cytochrome c oxidase dysfunction. *PLoS ONE* **2013**, *8*, e77129. [CrossRef]
190. Kunova, N.; Ondrovicova, G.; Bauer, J.A.; Bellova, J.; Ambro, L.; Martinakova, L.; Kotrasova, V.; Kutejova, E.; Pevala, V. The role of Lon-mediated proteolysis in the dynamics of mitochondrial nucleic acid-protein complexes. *Sci. Rep.* **2017**, *7*, 631. [CrossRef]
191. Sepuri, N.B.V.; Angireddy, R.; Srinivasan, S.; Guha, M.; Spear, J.; Lu, B.; Anandatheerthavarada, H.K.; Suzuki, C.K.; Avadhani, N.G. Mitochondrial LON protease-dependent degradation of cytochrome c oxidase subunits under hypoxia and myocardial ischemia. *Biochim. Biophys. Acta Bioenerg.* **2017**, *1858*, 519–528. [CrossRef] [PubMed]
192. Acin-Perez, R.; Gatti, D.L.; Bai, Y.; Manfredi, G. Protein phosphorylation and prevention of cytochrome oxidase inhibition by ATP: Coupled mechanisms of energy metabolism regulation. *Cell Metab.* **2011**, *13*, 712–719. [CrossRef] [PubMed]
193. Barnett, M.; Lin, D.; Akoyev, V.; Willard, L.; Takemoto, D. Protein kinase C epsilon activates lens mitochondrial cytochrome c oxidase subunit IV during hypoxia. *Exp. Eye Res.* **2008**, *86*, 226–234. [CrossRef] [PubMed]
194. Ogbi, M.; Johnson, J.A. Protein kinase Cepsilon interacts with cytochrome c oxidase subunit IV and enhances cytochrome c oxidase activity in neonatal cardiac myocyte preconditioning. *Biochem. J.* **2006**, *393*, 191–199. [CrossRef] [PubMed]
195. Struglics, A.; Fredlund, K.M.; Moller, I.M.; Allen, J.F. Two subunits of the F0F1-ATPase are phosphorylated in the inner mitochondrial membrane. *Biochem. Biophys. Res. Commun.* **1998**, *243*, 664–668. [CrossRef]
196. Hojlund, K.; Wrzesinski, K.; Larsen, P.M.; Fey, S.J.; Roepstorff, P.; Handberg, A.; Dela, F.; Vinten, J.; McCormack, J.G.; Reynet, C.; et al. Proteome analysis reveals phosphorylation of ATP synthase beta -subunit in human skeletal muscle and proteins with potential roles in type 2 diabetes. *J. Biol. Chem.* **2003**, *278*, 10436–10442. [CrossRef]

197. Hojlund, K.; Yi, Z.; Lefort, N.; Langlais, P.; Bowen, B.; Levin, K.; Beck-Nielsen, H.; Mandarino, L.J. Human ATP synthase beta is phosphorylated at multiple sites and shows abnormal phosphorylation at specific sites in insulin-resistant muscle. *Diabetologia* **2010**, *53*, 541–551. [CrossRef]
198. Yang, J.Y.; Deng, W.; Chen, Y.; Fan, W.; Baldwin, K.M.; Jope, R.S.; Wallace, D.C.; Wang, P.H. Impaired translocation and activation of mitochondrial Akt1 mitigated mitochondrial oxidative phosphorylation Complex V activity in diabetic myocardium. *J. Mol. Cell Cardiol.* **2013**, *59*, 167–175. [CrossRef]
199. Garcia-Bermudez, J.; Sanchez-Arago, M.; Soldevilla, B.; Del Arco, A.; Nuevo-Tapioles, C.; Cuezva, J.M. PKA Phosphorylates the ATPase Inhibitory Factor 1 and Inactivates Its Capacity to Bind and Inhibit the Mitochondrial H(+)-ATP Synthase. *Cell Rep.* **2015**, *12*, 2143–2155. [CrossRef]
200. Pullman, M.E.; Monroy, G.C. A Naturally Occurring Inhibitor of Mitochondrial Adenosine Triphosphatase. *J. Biol. Chem.* **1963**, *238*, 3762–3769. [CrossRef]
201. Garcia-Aguilar, A.; Cuezva, J.M. A Review of the Inhibition of the Mitochondrial ATP Synthase by IF1 in vivo: Reprogramming Energy Metabolism and Inducing Mitohormesis. *Front. Physiol.* **2018**, *9*, 1322. [CrossRef] [PubMed]
202. Castellanos, E.; Lanning, N.J. Phosphorylation of OXPHOS Machinery Subunits: Functional Implications in Cell Biology and Disease. *Yale J. Biol. Med.* **2019**, *92*, 523–531. [PubMed]
203. Rousset, S.; Alves-Guerra, M.C.; Mozo, J.; Miroux, B.; Cassard-Doulcier, A.M.; Bouillaud, F.; Ricquier, D. The biology of mitochondrial uncoupling proteins. *Diabetes* **2004**, *53* (Suppl. 1), S130–S135. [CrossRef] [PubMed]
204. Lee, J.H.; Park, A.; Oh, K.J.; Lee, S.C.; Kim, W.K.; Bae, K.H. The Role of Adipose Tissue Mitochondria: Regulation of Mitochondrial Function for the Treatment of Metabolic Diseases. *Int. J. Mol. Sci.* **2019**, *20*, 4924. [CrossRef] [PubMed]
205. Ricquier, D. Uncoupling protein 1 of brown adipocytes, the only uncoupler: A historical perspective. *Front. Endocrinol* **2011**, *2*, 85. [CrossRef]
206. Jezek, P.; Jaburek, M.; Porter, R.K. Uncoupling mechanism and redox regulation of mitochondrial uncoupling protein 1 (UCP1). *Biochim. Biophys. Acta Bioenerg.* **2019**, *1860*, 259–269. [CrossRef]
207. Nicholls, D.G.; Locke, R.M. Thermogenic mechanisms in brown fat. *Physiol. Rev.* **1984**, *64*, 1–64. [CrossRef]
208. Macher, G.; Koehler, M.; Rupprecht, A.; Kreiter, J.; Hinterdorfer, P.; Pohl, E.E. Inhibition of mitochondrial UCP1 and UCP3 by purine nucleotides and phosphate. *Biochim. Biophys. Acta Biomembr.* **2018**, *1860*, 664–672. [CrossRef]
209. Bast-Habersbrunner, A.; Fromme, T. Purine Nucleotides in the Regulation of Brown Adipose Tissue Activity. *Front. Endocrinol.* **2020**, *11*, 118. [CrossRef]
210. Villarroya, F.; Peyrou, M.; Giralt, M. Transcriptional regulation of the uncoupling protein-1 gene. *Biochimie* **2017**, *134*, 86–92. [CrossRef]
211. Cao, W.; Daniel, K.W.; Robidoux, J.; Puigserver, P.; Medvedev, A.V.; Bai, X.; Floering, L.M.; Spiegelman, B.M.; Collins, S. p38 mitogen-activated protein kinase is the central regulator of cyclic AMP-dependent transcription of the brown fat uncoupling protein 1 gene. *Mol. Cell Biol.* **2004**, *24*, 3057–3067. [CrossRef] [PubMed]
212. Shinoda, K.; Ohyama, K.; Hasegawa, Y.; Chang, H.Y.; Ogura, M.; Sato, A.; Hong, H.; Hosono, T.; Sharp, L.Z.; Scheel, D.W.; et al. Phosphoproteomics Identifies CK2 as a Negative Regulator of Beige Adipocyte Thermogenesis and Energy Expenditure. *Cell Metab.* **2015**, *22*, 997–1008. [CrossRef] [PubMed]
213. Carroll, A.M.; Porter, R.K.; Morrice, N.A. Identification of serine phosphorylation in mitochondrial uncoupling protein 1. *Biochim. Biophys. Acta* **2008**, *1777*, 1060–1065. [CrossRef] [PubMed]
214. Anand, R.; Langer, T.; Baker, M.J. Proteolytic control of mitochondrial function and morphogenesis. *Biochim. Biophys. Acta* **2013**, *1833*, 195–204. [CrossRef] [PubMed]
215. Baker, B.M.; Haynes, C.M. Mitochondrial protein quality control during biogenesis and aging. *Trends Biochem. Sci.* **2011**, *36*, 254–261. [CrossRef]
216. Baker, M.J.; Tatsuta, T.; Langer, T. Quality control of mitochondrial proteostasis. *Cold Spring Harb. Perspect Biol.* **2011**, *3*. [CrossRef]
217. Bender, T.; Lewrenz, I.; Franken, S.; Baitzel, C.; Voos, W. Mitochondrial enzymes are protected from stress-induced aggregation by mitochondrial chaperones and the Pim1/LON protease. *Mol. Biol. Cell* **2011**, *22*, 541–554. [CrossRef]
218. Lund, A.A.; Rhoads, D.M.; Lund, A.L.; Cerny, R.L.; Elthon, T.E. In vivo modifications of the maize mitochondrial small heat stress protein, HSP22. *J Biol. Chem.* **2001**, *276*, 29924–29929. [CrossRef]
219. Downs, C.A.; Heckathorn, S.A. The mitochondrial small heat-shock protein protects NADH:ubiquinone oxidoreductase of the electron transport chain during heat stress in plants. *FEBS Lett.* **1998**, *430*, 246–250. [CrossRef]
220. Lund, A.A.; Blum, P.H.; Bhattramakki, D.; Elthon, T.E. Heat-stress response of maize mitochondria. *Plant. Physiol.* **1998**, *116*, 1097–1110. [CrossRef]
221. Lenne, C.; Douce, R. A Low Molecular Mass Heat-Shock Protein Is Localized to Higher Plant Mitochondria. *Plant. Physiol.* **1994**, *105*, 1255–1261. [CrossRef] [PubMed]
222. Chen, H.F.; Chen, C.Y.; Lin, T.H.; Huang, Z.W.; Chi, T.H.; Ma, Y.S.; Wu, S.B.; Wei, Y.H.; Hsieh, M. The protective roles of phosphorylated heat shock protein 27 in human cells harboring myoclonus epilepsy with ragged-red fibers A8344G mtDNA mutation. *FEBS J.* **2012**, *279*, 2987–3001. [CrossRef] [PubMed]
223. Zhou, M.; Lambert, H.; Landry, J. Transient activation of a distinct serine protein kinase is responsible for 27-kDa heat shock protein phosphorylation in mitogen-stimulated and heat-shocked cells. *J. Biol. Chem.* **1993**, *268*, 35–43. [CrossRef]

224. Arrigo, A.P. Human small heat shock proteins: Protein interactomes of homo- and hetero-oligomeric complexes: An update. *FEBS Lett.* **2013**, *587*, 1959–1969. [CrossRef] [PubMed]
225. Hadari, Y.R.; Haring, H.U.; Zick, Y. p75, a member of the heat shock protein family, undergoes tyrosine phosphorylation in response to oxidative stress. *J. Biol. Chem.* **1997**, *272*, 657–662. [CrossRef] [PubMed]
226. Rasola, A.; Neckers, L.; Picard, D. Mitochondrial oxidative phosphorylation TRAP(1)ped in tumor cells. *Trends Cell Biol.* **2014**, *24*, 455–463. [CrossRef]
227. Sciacovelli, M.; Guzzo, G.; Morello, V.; Frezza, C.; Zheng, L.; Nannini, N.; Calabrese, F.; Laudiero, G.; Esposito, F.; Landriscina, M.; et al. The mitochondrial chaperone TRAP1 promotes neoplastic growth by inhibiting succinate dehydrogenase. *Cell Metab* **2013**, *17*, 988–999. [CrossRef] [PubMed]
228. Yoshida, S.; Tsutsumi, S.; Muhlebach, G.; Sourbier, C.; Lee, M.J.; Lee, S.; Vartholomaiou, E.; Tatokoro, M.; Beebe, K.; Miyajima, N.; et al. Molecular chaperone TRAP1 regulates a metabolic switch between mitochondrial respiration and aerobic glycolysis. *Proc. Natl. Acad. Sci. USA* **2013**, *110*, E1604–E1612. [CrossRef]
229. Pridgeon, J.W.; Olzmann, J.A.; Chin, L.S.; Li, L. PINK1 protects against oxidative stress by phosphorylating mitochondrial chaperone TRAP1. *PLoS Biol.* **2007**, *5*, e172. [CrossRef]
230. Kang, B.H.; Plescia, J.; Dohi, T.; Rosa, J.; Doxsey, S.J.; Altieri, D.C. Regulation of tumor cell mitochondrial homeostasis by an organelle-specific Hsp90 chaperone network. *Cell* **2007**, *131*, 257–270. [CrossRef]
231. Kowalik, M.A.; Guzzo, G.; Morandi, A.; Perra, A.; Menegon, S.; Masgras, I.; Trevisan, E.; Angioni, M.M.; Fornari, F.; Quagliata, L.; et al. Metabolic reprogramming identifies the most aggressive lesions at early phases of hepatic carcinogenesis. *Oncotarget* **2016**, *7*, 32375–32393. [CrossRef] [PubMed]
232. Masgras, I.; Sanchez-Martin, C.; Colombo, G.; Rasola, A. The Chaperone TRAP1 as a Modulator of the Mitochondrial Adaptations in Cancer Cells. *Front. Oncol.* **2017**, *7*, 58. [CrossRef] [PubMed]
233. Selak, M.A.; Armour, S.M.; MacKenzie, E.D.; Boulahbel, H.; Watson, D.G.; Mansfield, K.D.; Pan, Y.; Simon, M.C.; Thompson, C.B.; Gottlieb, E. Succinate links TCA cycle dysfunction to oncogenesis by inhibiting HIF-alpha prolyl hydroxylase. *Cancer Cell* **2005**, *7*, 77–85. [CrossRef] [PubMed]
234. Ficarro, S.; Chertihin, O.; Westbrook, V.A.; White, F.; Jayes, F.; Kalab, P.; Marto, J.A.; Shabanowitz, J.; Herr, J.C.; Hunt, D.F.; et al. Phosphoproteome analysis of capacitated human sperm. Evidence of tyrosine phosphorylation of a kinase-anchoring protein 3 and valosin-containing protein/p97 during capacitation. *J. Biol. Chem.* **2003**, *278*, 11579–11589. [CrossRef]
235. Cappello, F.; Marino Gammazza, A.; Palumbo Piccionello, A.; Campanella, C.; Pace, A.; Conway de Macario, E.; Macario, A.J. Hsp60 chaperonopathies and chaperonotherapy: Targets and agents. *Expert Opin. Ther. Targets* **2014**, *18*, 185–208. [CrossRef] [PubMed]
236. Richardson, A.; Landry, S.J.; Georgopoulos, C. The ins and outs of a molecular chaperone machine. *Trends Biochem. Sci.* **1998**, *23*, 138–143. [CrossRef]
237. Vilasi, S.; Bulone, D.; Caruso Bavisotto, C.; Campanella, C.; Marino Gammazza, A.; San Biagio, P.L.; Cappello, F.; Conway de Macario, E.; Macario, A.J.L. Chaperonin of Group I: Oligomeric Spectrum and Biochemical and Biological Implications. *Front. Mol. Biosci.* **2017**, *4*, 99. [CrossRef] [PubMed]
238. Caruso Bavisotto, C.; Alberti, G.; Vitale, A.M.; Paladino, L.; Campanella, C.; Rappa, F.; Gorska, M.; Conway de Macario, E.; Cappello, F.; Macario, A.J.L.; et al. Hsp60 Post-translational Modifications: Functional and Pathological Consequences. *Front. Mol. Biosci.* **2020**, *7*, 95. [CrossRef]
239. Gu, Y.; Ande, S.R.; Mishra, S. Altered O-GlcNAc modification and phosphorylation of mitochondrial proteins in myoblast cells exposed to high glucose. *Arch. Biochem. Biophys.* **2011**, *505*, 98–104. [CrossRef]
240. Rikova, K.; Guo, A.; Zeng, Q.; Possemato, A.; Yu, J.; Haack, H.; Nardone, J.; Lee, K.; Reeves, C.; Li, Y.; et al. Global survey of phosphotyrosine signaling identifies oncogenic kinases in lung cancer. *Cell* **2007**, *131*, 1190–1203. [CrossRef]
241. Chattopadhyay, S.; Mukherjee, A.; Patra, U.; Bhowmick, R.; Basak, T.; Sengupta, S.; Chawla-Sarkar, M. Tyrosine phosphorylation modulates mitochondrial chaperonin Hsp60 and delays rotavirus NSP4-mediated apoptotic signaling in host cells. *Cell Microbiol.* **2017**, *19*. [CrossRef] [PubMed]
242. Asquith, K.L.; Baleato, R.M.; McLaughlin, E.A.; Nixon, B.; Aitken, R.J. Tyrosine phosphorylation activates surface chaperones facilitating sperm-zona recognition. *J. Cell Sci.* **2004**, *117*, 3645–3657. [CrossRef] [PubMed]
243. Desautels, M.; Goldberg, A.L. Liver mitochondria contain an ATP-dependent, vanadate-sensitive pathway for the degradation of proteins. *Proc. Natl. Acad. Sci. USA* **1982**, *79*, 1869–1873. [CrossRef] [PubMed]
244. Akimov, V.; Barrio-Hernandez, I.; Hansen, S.V.F.; Hallenborg, P.; Pedersen, A.K.; Bekker-Jensen, D.B.; Puglia, M.; Christensen, S.D.K.; Vanselow, J.T.; Nielsen, M.M.; et al. UbiSite approach for comprehensive mapping of lysine and N-terminal ubiquitination sites. *Nat. Struct Mol. Biol.* **2018**, *25*, 631–640. [CrossRef] [PubMed]
245. Ghosh, J.C.; Seo, J.H.; Agarwal, E.; Wang, Y.; Kossenkov, A.V.; Tang, H.Y.; Speicher, D.W.; Altieri, D.C. Akt phosphorylation of mitochondrial Lonp1 protease enables oxidative metabolism and advanced tumor traits. *Oncogene* **2019**, *38*, 6926–6939. [CrossRef]
246. Gibellini, L.; Pinti, M.; Beretti, F.; Pierri, C.L.; Onofrio, A.; Riccio, M.; Carnevale, G.; De Biasi, S.; Nasi, M.; Torelli, F.; et al. Sirtuin 3 interacts with Lon protease and regulates its acetylation status. *Mitochondrion* **2014**, *18*, 76–81. [CrossRef]
247. Gibellini, L.; De Gaetano, A.; Mandrioli, M.; Van Tongeren, E.; Bortolotti, C.A.; Cossarizza, A.; Pinti, M. The biology of Lonp1: More than a mitochondrial protease. *Int. Rev. Cell Mol. Biol.* **2020**, *354*, 1–61. [CrossRef]

248. Sepuri, N.B.V.; Tammineni, P.; Mohammed, F.; Paripati, A. Nuclear Transcription Factors in the Mitochondria: A New Paradigm in Fine-Tuning Mitochondrial Metabolism. *Handb. Exp. Pharmacol.* **2017**, *240*, 3–20. [CrossRef]
249. Fukuda, R.; Zhang, H.; Kim, J.W.; Shimoda, L.; Dang, C.V.; Semenza, G.L. HIF-1 regulates cytochrome oxidase subunits to optimize efficiency of respiration in hypoxic cells. *Cell* **2007**, *129*, 111–122. [CrossRef]
250. Goto, M.; Miwa, H.; Suganuma, K.; Tsunekawa-Imai, N.; Shikami, M.; Mizutani, M.; Mizuno, S.; Hanamura, I.; Nitta, M. Adaptation of leukemia cells to hypoxic condition through switching the energy metabolism or avoiding the oxidative stress. *BMC Cancer* **2014**, *14*, 76. [CrossRef]
251. Zhou, X.; Teper, D.; Andrade, M.O.; Zhang, T.; Chen, S.; Song, W.Y.; Wang, N. A Phosphorylation Switch on Lon Protease Regulates Bacterial Type III Secretion System in Host. *mBio* **2018**, *9*. [CrossRef] [PubMed]
252. Brunings, A.M.; Gabriel, D.W. Xanthomonas citri: Breaking the surface. *Mol. Plant. Pathol.* **2003**, *4*, 141–157. [CrossRef] [PubMed]
253. Baker, T.A.; Sauer, R.T. ClpXP, an ATP-powered unfolding and protein-degradation machine. *Biochim. Biophys. Acta* **2012**, *1823*, 15–28. [CrossRef] [PubMed]
254. Kang, S.G.; Ortega, J.; Singh, S.K.; Wang, N.; Huang, N.N.; Steven, A.C.; Maurizi, M.R. Functional proteolytic complexes of the human mitochondrial ATP-dependent protease, hClpXP. *J. Biol. Chem.* **2002**, *277*, 21095–21102. [CrossRef] [PubMed]
255. Haynes, C.M.; Petrova, K.; Benedetti, C.; Yang, Y.; Ron, D. ClpP mediates activation of a mitochondrial unfolded protein response in C. elegans. *Dev. Cell* **2007**, *13*, 467–480. [CrossRef] [PubMed]
256. Haynes, C.M.; Ron, D. The mitochondrial UPR—Protecting organelle protein homeostasis. *J. Cell Sci.* **2010**, *123*, 3849–3855. [CrossRef] [PubMed]
257. Zhao, Q.; Wang, J.; Levichkin, I.V.; Stasinopoulos, S.; Ryan, M.T.; Hoogenraad, N.J. A mitochondrial specific stress response in mammalian cells. *EMBO J.* **2002**, *21*, 4411–4419. [CrossRef] [PubMed]
258. da Fonseca, P.C.; He, J.; Morris, E.P. Molecular model of the human 26S proteasome. *Mol. Cell* **2012**, *46*, 54–66. [CrossRef]
259. Kang, S.G.; Maurizi, M.R.; Thompson, M.; Mueser, T.; Ahvazi, B. Crystallography and mutagenesis point to an essential role for the N-terminus of human mitochondrial ClpP. *J. Struct. Biol.* **2004**, *148*, 338–352. [CrossRef]
260. Kang, S.G.; Dimitrova, M.N.; Ortega, J.; Ginsburg, A.; Maurizi, M.R. Human mitochondrial ClpP is a stable heptamer that assembles into a tetradecamer in the presence of ClpX. *J. Biol. Chem.* **2005**, *280*, 35424–35432. [CrossRef]
261. Kasashima, K.; Sumitani, M.; Endo, H. Maintenance of mitochondrial genome distribution by mitochondrial AAA+ protein ClpX. *Exp. Cell Res.* **2012**, *318*, 2335–2343. [CrossRef] [PubMed]
262. Sauer, R.T.; Bolon, D.N.; Burton, B.M.; Burton, R.E.; Flynn, J.M.; Grant, R.A.; Hersch, G.L.; Joshi, S.A.; Kenniston, J.A.; Levchenko, I.; et al. Sculpting the proteome with AAA(+) proteases and disassembly machines. *Cell* **2004**, *119*, 9–18. [CrossRef] [PubMed]
263. Deepa, S.S.; Bhaskaran, S.; Ranjit, R.; Qaisar, R.; Nair, B.C.; Liu, Y.; Walsh, M.E.; Fok, W.C.; Van Remmen, H. Down-regulation of the mitochondrial matrix peptidase ClpP in muscle cells causes mitochondrial dysfunction and decreases cell proliferation. *Free Radic Biol. Med.* **2016**, *91*, 281–292. [CrossRef] [PubMed]
264. Houtkooper, R.H.; Mouchiroud, L.; Ryu, D.; Moullan, N.; Katsyuba, E.; Knott, G.; Williams, R.W.; Auwerx, J. Mitonuclear protein imbalance as a conserved longevity mechanism. *Nature* **2013**, *497*, 451–457. [CrossRef] [PubMed]
265. Siegelin, M.D.; Dohi, T.; Raskett, C.M.; Orlowski, G.M.; Powers, C.M.; Gilbert, C.A.; Ross, A.H.; Plescia, J.; Altieri, D.C. Exploiting the mitochondrial unfolded protein response for cancer therapy in mice and human cells. *J. Clin. Invest.* **2011**, *121*, 1349–1360. [CrossRef] [PubMed]
266. Rath, E.; Berger, E.; Messlik, A.; Nunes, T.; Liu, B.; Kim, S.C.; Hoogenraad, N.; Sans, M.; Sartor, R.B.; Haller, D. Induction of dsRNA-activated protein kinase links mitochondrial unfolded protein response to the pathogenesis of intestinal inflammation. *Gut* **2012**, *61*, 1269–1278. [CrossRef]
267. Gal-Ben-Ari, S.; Barrera, I.; Ehrlich, M.; Rosenblum, K. PKR: A Kinase to Remember. *Front. Mol. Neurosci.* **2018**, *11*, 480. [CrossRef]
268. Taniuchi, S.; Miyake, M.; Tsugawa, K.; Oyadomari, M.; Oyadomari, S. Integrated stress response of vertebrates is regulated by four eIF2alpha kinases. *Sci. Rep.* **2016**, *6*, 32886. [CrossRef]
269. Seo, J.H.; Rivadeneira, D.B.; Caino, M.C.; Chae, Y.C.; Speicher, D.W.; Tang, H.Y.; Vaira, V.; Bosari, S.; Palleschi, A.; Rampini, P.; et al. The Mitochondrial Unfoldase-Peptidase Complex ClpXP Controls Bioenergetics Stress and Metastasis. *PLoS Biol.* **2016**, *14*, e1002507. [CrossRef]
270. Brown, T.A.; Tkachuk, A.N.; Shtengel, G.; Kopek, B.G.; Bogenhagen, D.F.; Hess, H.F.; Clayton, D.A. Superresolution fluorescence imaging of mitochondrial nucleoids reveals their spatial range, limits, and membrane interaction. *Mol. Cell Biol.* **2011**, *31*, 4994–5010. [CrossRef]
271. Vozarikova, V.; Kunova, N.; Bauer, J.A.; Frankovsky, J.; Kotrasova, V.; Prochazkova, K.; Dzugasova, V.; Kutejova, E.; Pevala, V.; Nosek, J.; et al. Mitochondrial HMG-Box Containing Proteins: From Biochemical Properties to the Roles in Human Diseases. *Biomolecules* **2020**, *10*, 1193. [CrossRef] [PubMed]
272. Cho, J.H.; Lee, Y.K.; Chae, C.B. The modulation of the biological activities of mitochondrial histone Abf2p by yeast PKA and its possible role in the regulation of mitochondrial DNA content during glucose repression. *Biochim. Biophys. Acta* **2001**, *1522*, 175–186. [CrossRef]
273. Lu, B.; Lee, J.; Nie, X.; Li, M.; Morozov, Y.I.; Venkatesh, S.; Bogenhagen, D.F.; Temiakov, D.; Suzuki, C.K. Phosphorylation of human TFAM in mitochondria impairs DNA binding and promotes degradation by the AAA+ Lon protease. *Mol. Cell* **2013**, *49*, 121–132. [CrossRef] [PubMed]

274. Wang, K.Z.; Zhu, J.; Dagda, R.K.; Uechi, G.; Cherra, S.J., 3rd; Gusdon, A.M.; Balasubramani, M.; Chu, C.T. ERK-mediated phosphorylation of TFAM downregulates mitochondrial transcription: Implications for Parkinson's disease. *Mitochondrion* **2014**, *17*, 132–140. [CrossRef] [PubMed]
275. Alami-Ouahabi, N.; Veilleux, S.; Meistrich, M.L.; Boissonneault, G. The testis-specific high-mobility-group protein, a phosphorylation-dependent DNA-packaging factor of elongating and condensing spermatids. *Mol. Cell Biol.* **1996**, *16*, 3720–3729. [CrossRef] [PubMed]
276. Ramachandran, C.; Yau, P.; Bradbury, E.M.; Shyamala, G.; Yasuda, H.; Walsh, D.A. Phosphorylation of high-mobility-group proteins by the calcium-phospholipid-dependent protein kinase and the cyclic AMP-dependent protein kinase. *J. Biol. Chem.* **1984**, *259*, 13495–13503. [CrossRef]
277. Wisniewski, J.R.; Schulze, E. High affinity interaction of dipteran high mobility group (HMG) proteins 1 with DNA is modulated by COOH-terminal regions flanking the HMG box domain. *J. Biol. Chem.* **1994**, *269*, 10713–10719. [CrossRef]
278. Lund, T.; Berg, K. Metaphase-specific phosphorylations weaken the association between chromosomal proteins HMG 14 and 17, and DNA. *FEBS Lett.* **1991**, *289*, 113–116. [CrossRef]
279. Reeves, R.; Langan, T.A.; Nissen, M.S. Phosphorylation of the DNA-binding domain of nonhistone high-mobility group I protein by cdc2 kinase: Reduction of binding affinity. *Proc. Natl. Acad. Sci. USA* **1991**, *88*, 1671–1675. [CrossRef]
280. Schwanbeck, R.; Wisniewski, J.R. Cdc2 and mitogen-activated protein kinases modulate DNA binding properties of the putative transcriptional regulator Chironomus high mobility group protein I. *J. Biol. Chem.* **1997**, *272*, 27476–27483. [CrossRef]
281. Bogenhagen, D.F.; Rousseau, D.; Burke, S. The layered structure of human mitochondrial DNA nucleoids. *J. Biol. Chem.* **2008**, *283*, 3665–3675. [CrossRef] [PubMed]
282. Rubio-Cosials, A.; Sidow, J.F.; Jimenez-Menendez, N.; Fernandez-Millan, P.; Montoya, J.; Jacobs, H.T.; Coll, M.; Bernado, P.; Sola, M. Human mitochondrial transcription factor A induces a U-turn structure in the light strand promoter. *Nat. Struct. Mol. Biol.* **2011**, *18*, 1281–1289. [CrossRef] [PubMed]
283. Gaspari, M.; Falkenberg, M.; Larsson, N.G.; Gustafsson, C.M. The mitochondrial RNA polymerase contributes critically to promoter specificity in mammalian cells. *EMBO J.* **2004**, *23*, 4606–4614. [CrossRef] [PubMed]
284. Dagda, R.K.; Zhu, J.; Kulich, S.M.; Chu, C.T. Mitochondrially localized ERK2 regulates mitophagy and autophagic cell stress: Implications for Parkinson's disease. *Autophagy* **2008**, *4*, 770–782. [CrossRef] [PubMed]
285. Marko, A.J.; Miller, R.A.; Kelman, A.; Frauwirth, K.A. Induction of glucose metabolism in stimulated T lymphocytes is regulated by mitogen-activated protein kinase signaling. *PLoS ONE* **2010**, *5*, e15425. [CrossRef] [PubMed]
286. Jeong, S.Y.; Rose, A.; Meier, I. MFP1 is a thylakoid-associated, nucleoid-binding protein with a coiled-coil structure. *Nucleic Acids Res.* **2003**, *31*, 5175–5185. [CrossRef] [PubMed]
287. Ogrzewalla, K.; Piotrowski, M.; Reinbothe, S.; Link, G. The plastid transcription kinase from mustard (Sinapis alba L.). A nuclear-encoded CK2-type chloroplast enzyme with redox-sensitive function. *Eur. J. Biochem.* **2002**, *269*, 3329–3337. [CrossRef]
288. Powikrowska, M.; Oetke, S.; Jensen, P.E.; Krupinska, K. Dynamic composition, shaping and organization of plastid nucleoids. *Front. Plant. Sci.* **2014**, *5*, 424. [CrossRef]
289. Sekine, K.; Fujiwara, M.; Nakayama, M.; Takao, T.; Hase, T.; Sato, N. DNA binding and partial nucleoid localization of the chloroplast stromal enzyme ferredoxin:sulfite reductase. *FEBS J.* **2007**, *274*, 2054–2069. [CrossRef]
290. Sekine, K.; Hase, T.; Sato, N. Reversible DNA compaction by sulfite reductase regulates transcriptional activity of chloroplast nucleoids. *J. Biol. Chem.* **2002**, *277*, 24399–24404. [CrossRef]
291. Melonek, J.; Matros, A.; Trosch, M.; Mock, H.P.; Krupinska, K. The core of chloroplast nucleoids contains architectural SWIB domain proteins. *Plant. Cell* **2012**, *24*, 3060–3073. [CrossRef] [PubMed]
292. Chi-Ham, C.L.; Keaton, M.A.; Cannon, G.C.; Heinhorst, S. The DNA-compacting protein DCP68 from soybean chloroplasts is ferredoxin:sulfite reductase and co-localizes with the organellar nucleoid. *Plant. Mol. Biol.* **2002**, *49*, 621–631. [CrossRef] [PubMed]
293. Tomaska, L. Phosphorylation of mitochondrial telomere binding protein of Candida parapsilosis by camp-dependent protein kinase. *Biochem. Biophys. Res. Commun.* **1998**, *242*, 457–460. [CrossRef] [PubMed]
294. Matsunaga, M.; Jaehning, J.A. A mutation in the yeast mitochondrial core RNA polymerase, Rpo41, confers defects in both specificity factor interaction and promoter utilization. *J. Biol. Chem.* **2004**, *279*, 2012–2019. [CrossRef]
295. Cotney, J.; Shadel, G.S. Evidence for an early gene duplication event in the evolution of the mitochondrial transcription factor B family and maintenance of rRNA methyltransferase activity in human mtTFB1 and mtTFB2. *J. Mol. Evol.* **2006**, *63*, 707–717. [CrossRef]
296. Shutt, T.E.; Gray, M.W. Homologs of mitochondrial transcription factor B, sparsely distributed within the eukaryotic radiation, are likely derived from the dimethyladenosine methyltransferase of the mitochondrial endosymbiont. *Mol. Biol. Evol.* **2006**, *23*, 1169–1179. [CrossRef]
297. Gnad, F.; de Godoy, L.M.; Cox, J.; Neuhauser, N.; Ren, S.; Olsen, J.V.; Mann, M. High-accuracy identification and bioinformatic analysis of in vivo protein phosphorylation sites in yeast. *Proteomics* **2009**, *9*, 4642–4652. [CrossRef]
298. Soufi, B.; Kelstrup, C.D.; Stoehr, G.; Frohlich, F.; Walther, T.C.; Olsen, J.V. Global analysis of the yeast osmotic stress response by quantitative proteomics. *Mol. Biosyst.* **2009**, *5*, 1337–1346. [CrossRef]
299. Prieto-Martin, A.; Montoya, J.; Martinez-Azorin, F. Phosphorylation of rat mitochondrial transcription termination factor (mTERF) is required for transcription termination but not for binding to DNA. *Nucleic Acids Res.* **2004**, *32*, 2059–2068. [CrossRef]

300. Fernandez-Silva, P.; Martinez-Azorin, F.; Micol, V.; Attardi, G. The human mitochondrial transcription termination factor (mTERF) is a multizipper protein but binds to DNA as a monomer, with evidence pointing to intramolecular leucine zipper interactions. *EMBO J.* **1997**, *16*, 1066–1079. [CrossRef]
301. Cammarota, M.; Paratcha, G.; Bevilaqua, L.R.; Levi de Stein, M.; Lopez, M.; Pellegrino de Iraldi, A.; Izquierdo, I.; Medina, J.H. Cyclic AMP-responsive element binding protein in brain mitochondria. *J. Neurochem.* **1999**, *72*, 2272–2277. [CrossRef] [PubMed]
302. Lee, J.; Kim, C.H.; Simon, D.K.; Aminova, L.R.; Andreyev, A.Y.; Kushnareva, Y.E.; Murphy, A.N.; Lonze, B.E.; Kim, K.S.; Ginty, D.D.; et al. Mitochondrial cyclic AMP response element-binding protein (CREB) mediates mitochondrial gene expression and neuronal survival. *J. Biol. Chem.* **2005**, *280*, 40398–40401. [CrossRef] [PubMed]
303. Marinov, G.K.; Wang, Y.E.; Chan, D.; Wold, B.J. Evidence for site-specific occupancy of the mitochondrial genome by nuclear transcription factors. *PLoS ONE* **2014**, *9*, e84713. [CrossRef] [PubMed]
304. Li, H.; Zassenhaus, H.P. Purification and characterization of an RNA dodecamer sequence binding protein from mitochondria of Saccharomyces cerevisiae. *Biochem. Biophys. Res. Commun.* **1999**, *261*, 740–745. [CrossRef]
305. Li, H.; Zassenhaus, H.P. Phosphorylation is required for high-affinity binding of DBP, a yeast mitochondrial site-specific RNA binding protein. *Curr. Genet.* **2000**, *37*, 356–363. [CrossRef]
306. Dziembowski, A.; Piwowarski, J.; Hoser, R.; Minczuk, M.; Dmochowska, A.; Siep, M.; van der Spek, H.; Grivell, L.; Stepien, P.P. The yeast mitochondrial degradosome. Its composition, interplay between RNA helicase and RNase activities and the role in mitochondrial RNA metabolism. *J. Biol. Chem.* **2003**, *278*, 1603–1611. [CrossRef]
307. Hofmann, T.J.; Min, J.; Zassenhaus, H.P. Formation of the 3′ end of yeast mitochondrial mRNAs occurs by site-specific cleavage two bases downstream of a conserved dodecamer sequence. *Yeast* **1993**, *9*, 1319–1330. [CrossRef]
308. Osinga, K.A.; De Vries, E.; Van der Horst, G.; Tabak, H.F. Processing of yeast mitochondrial messenger RNAs at a conserved dodecamer sequence. *EMBO J.* **1984**, *3*, 829–834. [CrossRef]
309. He, H.; Chen, M.; Scheffler, N.K.; Gibson, B.W.; Spremulli, L.L.; Gottlieb, R.A. Phosphorylation of mitochondrial elongation factor Tu in ischemic myocardium: Basis for chloramphenicol-mediated cardioprotection. *Circ. Res.* **2001**, *89*, 461–467. [CrossRef]
310. Lippmann, C.; Lindschau, C.; Vijgenboom, E.; Schroder, W.; Bosch, L.; Erdmann, V.A. Prokaryotic elongation factor Tu is phosphorylated in vivo. *J. Biol. Chem.* **1993**, *268*, 601–607. [CrossRef]
311. He, H.; Li, H.L.; Lin, A.; Gottlieb, R.A. Activation of the JNK pathway is important for cardiomyocyte death in response to simulated ischemia. *Cell Death Differ.* **1999**, *6*, 987–991. [CrossRef] [PubMed]
312. Korhonen, J.A.; Gaspari, M.; Falkenberg, M. TWINKLE Has 5′ -> 3′ DNA helicase activity and is specifically stimulated by mitochondrial single-stranded DNA-binding protein. *J. Biol. Chem.* **2003**, *278*, 48627–48632. [CrossRef] [PubMed]
313. Kleber, S.; Sancho-Martinez, I.; Wiestler, B.; Beisel, A.; Gieffers, C.; Hill, O.; Thiemann, M.; Mueller, W.; Sykora, J.; Kuhn, A.; et al. Yes and PI3K bind CD95 to signal invasion of glioblastoma. *Cancer Cell* **2008**, *13*, 235–248. [CrossRef] [PubMed]
314. Tatarov, O.; Mitchell, T.J.; Seywright, M.; Leung, H.Y.; Brunton, V.G.; Edwards, J. SRC family kinase activity is up-regulated in hormone-refractory prostate cancer. *Clin. Cancer Res.* **2009**, *15*, 3540–3549. [CrossRef] [PubMed]
315. Verbeek, B.S.; Vroom, T.M.; Adriaansen-Slot, S.S.; Ottenhoff-Kalff, A.E.; Geertzema, J.G.; Hennipman, A.; Rijksen, G. c-Src protein expression is increased in human breast cancer. An immunohistochemical and biochemical analysis. *J. Pathol.* **1996**, *180*, 383–388. [CrossRef]
316. Fu, Y.; Zagozdzon, R.; Avraham, R.; Avraham, H.K. CHK negatively regulates Lyn kinase and suppresses pancreatic cancer cell invasion. *Int. J. Oncol.* **2006**, *29*, 1453–1458. [CrossRef]
317. Bolen, J.B.; Veillette, A.; Schwartz, A.M.; Deseau, V.; Rosen, N. Analysis of pp60c-src in human colon carcinoma and normal human colon mucosal cells. *Oncogene Res.* **1987**, *1*, 149–168.
318. Masaki, T.; Igarashi, K.; Tokuda, M.; Yukimasa, S.; Han, F.; Jin, Y.J.; Li, J.Q.; Yoneyama, H.; Uchida, N.; Fujita, J.; et al. pp60c-src activation in lung adenocarcinoma. *Eur. J. Cancer* **2003**, *39*, 1447–1455. [CrossRef]
319. Elsberger, B.; Fullerton, R.; Zino, S.; Jordan, F.; Mitchell, T.J.; Brunton, V.G.; Mallon, E.A.; Shiels, P.G.; Edwards, J. Breast cancer patients' clinical outcome measures are associated with Src kinase family member expression. *Br. J. Cancer* **2010**, *103*, 899–909. [CrossRef]
320. Demory, M.L.; Boerner, J.L.; Davidson, R.; Faust, W.; Miyake, T.; Lee, I.; Huttemann, M.; Douglas, R.; Haddad, G.; Parsons, S.J. Epidermal growth factor receptor translocation to the mitochondria: Regulation and effect. *J. Biol. Chem.* **2009**, *284*, 36592–36604. [CrossRef]
321. Hebert-Chatelain, E. Src kinases are important regulators of mitochondrial functions. *Int. J. Biochem. Cell Biol.* **2013**, *45*, 90–98. [CrossRef] [PubMed]
322. Jiang, H.L.; Sun, H.F.; Gao, S.P.; Li, L.D.; Huang, S.; Hu, X.; Liu, S.; Wu, J.; Shao, Z.M.; Jin, W. SSBP1 Suppresses TGFbeta-Driven Epithelial-to-Mesenchymal Transition and Metastasis in Triple-Negative Breast Cancer by Regulating Mitochondrial Retrograde Signaling. *Cancer Res.* **2016**, *76*, 952–964. [CrossRef] [PubMed]
323. Dorstyn, L.; Akey, C.W.; Kumar, S. New insights into apoptosome structure and function. *Cell Death Differ.* **2018**, *25*, 1194–1208. [CrossRef] [PubMed]
324. Schellenberg, B.; Wang, P.; Keeble, J.A.; Rodriguez-Enriquez, R.; Walker, S.; Owens, T.W.; Foster, F.; Tanianis-Hughes, J.; Brennan, K.; Streuli, C.H.; et al. Bax exists in a dynamic equilibrium between the cytosol and mitochondria to control apoptotic priming. *Mol. Cell* **2013**, *49*, 959–971. [CrossRef] [PubMed]

325. Todt, F.; Cakir, Z.; Reichenbach, F.; Emschermann, F.; Lauterwasser, J.; Kaiser, A.; Ichim, G.; Tait, S.W.; Frank, S.; Langer, H.F.; et al. Differential retrotranslocation of mitochondrial Bax and Bak. *EMBO J.* **2015**, *34*, 67–80. [CrossRef]
326. Edlich, F.; Banerjee, S.; Suzuki, M.; Cleland, M.M.; Arnoult, D.; Wang, C.; Neutzner, A.; Tjandra, N.; Youle, R.J. Bcl-x(L) retrotranslocates Bax from the mitochondria into the cytosol. *Cell* **2011**, *145*, 104–116. [CrossRef]
327. Letai, A.; Bassik, M.C.; Walensky, L.D.; Sorcinelli, M.D.; Weiler, S.; Korsmeyer, S.J. Distinct BH3 domains either sensitize or activate mitochondrial apoptosis, serving as prototype cancer therapeutics. *Cancer Cell* **2002**, *2*, 183–192. [CrossRef]
328. Desagher, S.; Osen-Sand, A.; Montessuit, S.; Magnenat, E.; Vilbois, F.; Hochmann, A.; Journot, L.; Antonsson, B.; Martinou, J.C. Phosphorylation of bid by casein kinases I and II regulates its cleavage by caspase 8. *Mol. Cell* **2001**, *8*, 601–611. [CrossRef]
329. Izeradjene, K.; Douglas, L.; Delaney, A.B.; Houghton, J.A. Casein kinase I attenuates tumor necrosis factor-related apoptosis-inducing ligand-induced apoptosis by regulating the recruitment of fas-associated death domain and procaspase-8 to the death-inducing signaling complex. *Cancer Res.* **2004**, *64*, 8036–8044. [CrossRef]
330. Belikova, N.A.; Vladimirov, Y.A.; Osipov, A.N.; Kapralov, A.A.; Tyurin, V.A.; Potapovich, M.V.; Basova, L.V.; Peterson, J.; Kurnikov, I.V.; Kagan, V.E. Peroxidase activity and structural transitions of cytochrome c bound to cardiolipin-containing membranes. *Biochemistry* **2006**, *45*, 4998–5009. [CrossRef]
331. Kagan, V.E.; Bayir, H.A.; Belikova, N.A.; Kapralov, O.; Tyurina, Y.Y.; Tyurin, V.A.; Jiang, J.; Stoyanovsky, D.A.; Wipf, P.; Kochanek, P.M.; et al. Cytochrome c/cardiolipin relations in mitochondria: A kiss of death. *Free Radic. Biol. Med.* **2009**, *46*, 1439–1453. [CrossRef] [PubMed]
332. Rajagopal, B.S.; Edzuma, A.N.; Hough, M.A.; Blundell, K.L.; Kagan, V.E.; Kapralov, A.A.; Fraser, L.A.; Butt, J.N.; Silkstone, G.G.; Wilson, M.T.; et al. The hydrogen-peroxide-induced radical behaviour in human cytochrome c-phospholipid complexes: Implications for the enhanced pro-apoptotic activity of the G41S mutant. *Biochem. J.* **2013**, *456*, 441–452. [CrossRef] [PubMed]
333. Garcia-Heredia, J.M.; Diaz-Quintana, A.; Salzano, M.; Orzaez, M.; Perez-Paya, E.; Teixeira, M.; De la Rosa, M.A.; Diaz-Moreno, I. Tyrosine phosphorylation turns alkaline transition into a biologically relevant process and makes human cytochrome c behave as an anti-apoptotic switch. *J. Biol. Inorg. Chem.* **2011**, *16*, 1155–1168. [CrossRef] [PubMed]
334. Guerra-Castellano, A.; Diaz-Quintana, A.; Perez-Mejias, G.; Elena-Real, C.A.; Gonzalez-Arzola, K.; Garcia-Maurino, S.M.; De la Rosa, M.A.; Diaz-Moreno, I. Oxidative stress is tightly regulated by cytochrome c phosphorylation and respirasome factors in mitochondria. *Proc. Natl. Acad. Sci. USA* **2018**, *115*, 7955–7960. [CrossRef]
335. Hotamisligil, G.S.; Davis, R.J. Cell Signaling and Stress Responses. *Cold Spring Harb Perspect Biol.* **2016**, *8*. [CrossRef] [PubMed]
336. Pohl, S.O.; Agostino, M.; Dharmarajan, A.; Pervaiz, S. Cross Talk Between Cellular Redox State and the Antiapoptotic Protein Bcl-2. *Antioxid Redox Signal.* **2018**, *29*, 1215–1236. [CrossRef]
337. Low, I.C.; Loh, T.; Huang, Y.; Virshup, D.M.; Pervaiz, S. Ser70 phosphorylation of Bcl-2 by selective tyrosine nitration of PP2A-B56delta stabilizes its antiapoptotic activity. *Blood* **2014**, *124*, 2223–2234. [CrossRef]
338. Rayavarapu, R.R.; Heiden, B.; Pagani, N.; Shaw, M.M.; Shuff, S.; Zhang, S.; Schafer, Z.T. The role of multicellular aggregation in the survival of ErbB2-positive breast cancer cells during extracellular matrix detachment. *J. Biol. Chem.* **2015**, *290*, 8722–8733. [CrossRef]
339. Iqbal, A.; Eckerdt, F.; Bell, J.; Nakano, I.; Giles, F.J.; Cheng, S.Y.; Lulla, R.R.; Goldman, S.; Platanias, L.C. Targeting of glioblastoma cell lines and glioma stem cells by combined PIM kinase and PI3K-p110alpha inhibition. *Oncotarget* **2016**, *7*, 33192–33201. [CrossRef]
340. Nalluri, S.; Ghoshal-Gupta, S.; Kutiyanawalla, A.; Gayatri, S.; Lee, B.R.; Jiwani, S.; Rojiani, A.M.; Rojiani, M.V. TIMP-1 Inhibits Apoptosis in Lung Adenocarcinoma Cells via Interaction with Bcl-2. *PLoS ONE* **2015**, *10*, e0137673. [CrossRef]
341. Polzien, L.; Baljuls, A.; Rennefahrt, U.E.; Fischer, A.; Schmitz, W.; Zahedi, R.P.; Sickmann, A.; Metz, R.; Albert, S.; Benz, R.; et al. Identification of novel in vivo phosphorylation sites of the human proapoptotic protein BAD: Pore-forming activity of BAD is regulated by phosphorylation. *J. Biol. Chem.* **2009**, *284*, 28004–28020. [CrossRef] [PubMed]
342. Bhakar, A.L.; Howell, J.L.; Paul, C.E.; Salehi, A.H.; Becker, E.B.; Said, F.; Bonni, A.; Barker, P.A. Apoptosis induced by p75NTR overexpression requires Jun kinase-dependent phosphorylation of Bad. *J. Neurosci.* **2003**, *23*, 11373–11381. [CrossRef] [PubMed]
343. Donovan, N.; Becker, E.B.; Konishi, Y.; Bonni, A. JNK phosphorylation and activation of BAD couples the stress-activated signaling pathway to the cell death machinery. *J. Biol. Chem.* **2002**, *277*, 40944–40949. [CrossRef] [PubMed]
344. Yu, C.; Minemoto, Y.; Zhang, J.; Liu, J.; Tang, F.; Bui, T.N.; Xiang, J.; Lin, A. JNK suppresses apoptosis via phosphorylation of the proapoptotic Bcl-2 family protein BAD. *Mol. Cell* **2004**, *13*, 329–340. [CrossRef]
345. Moujalled, D.; Weston, R.; Anderton, H.; Ninnis, R.; Goel, P.; Coley, A.; Huang, D.C.; Wu, L.; Strasser, A.; Puthalakath, H. Cyclic-AMP-dependent protein kinase A regulates apoptosis by stabilizing the BH3-only protein Bim. *EMBO Rep.* **2011**, *12*, 77–83. [CrossRef]
346. Putcha, G.V.; Le, S.; Frank, S.; Besirli, C.G.; Clark, K.; Chu, B.; Alix, S.; Youle, R.J.; LaMarche, A.; Maroney, A.C.; et al. JNK-mediated BIM phosphorylation potentiates BAX-dependent apoptosis. *Neuron* **2003**, *38*, 899–914. [CrossRef]
347. Putcha, G.V.; Moulder, K.L.; Golden, J.P.; Bouillet, P.; Adams, J.A.; Strasser, A.; Johnson, E.M. Induction of BIM, a proapoptotic BH3-only BCL-2 family member, is critical for neuronal apoptosis. *Neuron* **2001**, *29*, 615–628. [CrossRef]
348. Datta, S.R.; Dudek, H.; Tao, X.; Masters, S.; Fu, H.; Gotoh, Y.; Greenberg, M.E. Akt phosphorylation of BAD couples survival signals to the cell-intrinsic death machinery. *Cell* **1997**, *91*, 231–241. [CrossRef]
349. Majewski, M.; Nieborowska-Skorska, M.; Salomoni, P.; Slupianek, A.; Reiss, K.; Trotta, R.; Calabretta, B.; Skorski, T. Activation of mitochondrial Raf-1 is involved in the antiapoptotic effects of Akt. *Cancer Res.* **1999**, *59*, 2815–2819.

350. Kennedy, D.; Mnich, K.; Oommen, D.; Chakravarthy, R.; Almeida-Souza, L.; Krols, M.; Saveljeva, S.; Doyle, K.; Gupta, S.; Timmerman, V.; et al. HSPB1 facilitates ERK-mediated phosphorylation and degradation of BIM to attenuate endoplasmic reticulum stress-induced apoptosis. *Cell Death Dis.* **2017**, *8*, e3026. [CrossRef]
351. Arokium, H.; Ouerfelli, H.; Velours, G.; Camougrand, N.; Vallette, F.M.; Manon, S. Substitutions of potentially phosphorylatable serine residues of Bax reveal how they may regulate its interaction with mitochondria. *J. Biol. Chem.* **2007**, *282*, 35104–35112. [CrossRef] [PubMed]
352. Linseman, D.A.; Butts, B.D.; Precht, T.A.; Phelps, R.A.; Le, S.S.; Laessig, T.A.; Bouchard, R.J.; Florez-McClure, M.L.; Heidenreich, K.A. Glycogen synthase kinase-3beta phosphorylates Bax and promotes its mitochondrial localization during neuronal apoptosis. *J. Neurosci.* **2004**, *24*, 9993–10002. [CrossRef] [PubMed]
353. Kim, B.J.; Ryu, S.W.; Song, B.J. JNK- and p38 kinase-mediated phosphorylation of Bax leads to its activation and mitochondrial translocation and to apoptosis of human hepatoma HepG2 cells. *J. Biol. Chem.* **2006**, *281*, 21256–21265. [CrossRef] [PubMed]
354. Xin, M.; Gao, F.; May, W.S.; Flagg, T.; Deng, X. Protein kinase Czeta abrogates the proapoptotic function of Bax through phosphorylation. *J. Biol. Chem.* **2007**, *282*, 21268–21277. [CrossRef] [PubMed]
355. Fox, J.L.; Ismail, F.; Azad, A.; Ternette, N.; Leverrier, S.; Edelmann, M.J.; Kessler, B.M.; Leigh, I.M.; Jackson, S.; Storey, A. Tyrosine dephosphorylation is required for Bak activation in apoptosis. *EMBO J.* **2010**, *29*, 3853–3868. [CrossRef] [PubMed]
356. Azad, A.; Fox, J.; Leverrier, S.; Storey, A. Blockade of the BAK hydrophobic groove by inhibitory phosphorylation regulates commitment to apoptosis. *PLoS ONE* **2012**, *7*, e49601. [CrossRef] [PubMed]
357. Afreen, S.; Bohler, S.; Muller, A.; Demmerath, E.M.; Weiss, J.M.; Jutzi, J.S.; Schachtrup, K.; Kunze, M.; Erlacher, M. BCL-XL expression is essential for human erythropoiesis and engraftment of hematopoietic stem cells. *Cell Death Dis.* **2020**, *11*, 8. [CrossRef]
358. Loo, L.S.W.; Soetedjo, A.A.P.; Lau, H.H.; Ng, N.H.J.; Ghosh, S.; Nguyen, L.; Krishnan, V.G.; Choi, H.; Roca, X.; Hoon, S.; et al. BCL-xL/BCL2L1 is a critical anti-apoptotic protein that promotes the survival of differentiating pancreatic cells from human pluripotent stem cells. *Cell Death Dis.* **2020**, *11*, 378. [CrossRef]
359. Motoyama, N.; Wang, F.; Roth, K.A.; Sawa, H.; Nakayama, K.; Nakayama, K.; Negishi, I.; Senju, S.; Zhang, Q.; Fujii, S.; et al. Massive cell death of immature hematopoietic cells and neurons in Bcl-x-deficient mice. *Science* **1995**, *267*, 1506–1510. [CrossRef]
360. Kharbanda, S.; Saxena, S.; Yoshida, K.; Pandey, P.; Kaneki, M.; Wang, Q.; Cheng, K.; Chen, Y.N.; Campbell, A.; Sudha, T.; et al. Translocation of SAPK/JNK to mitochondria and interaction with Bcl-x(L) in response to DNA damage. *J. Biol. Chem.* **2000**, *275*, 322–327. [CrossRef]
361. Wang, J.; Beauchemin, M.; Bertrand, R. Bcl-xL phosphorylation at Ser49 by polo kinase 3 during cell cycle progression and checkpoints. *Cell Signal.* **2011**, *23*, 2030–2038. [CrossRef] [PubMed]
362. Wang, J.; Beauchemin, M.; Bertrand, R. Phospho-Bcl-x(L)(Ser62) plays a key role at DNA damage-induced G(2) checkpoint. *Cell Cycle* **2012**, *11*, 2159–2169. [CrossRef] [PubMed]
363. Basu, A.; Haldar, S. Identification of a novel Bcl-xL phosphorylation site regulating the sensitivity of taxol- or 2-methoxyestradiol-induced apoptosis. *FEBS Lett.* **2003**, *538*, 41–47. [CrossRef]
364. De Chiara, G.; Marcocci, M.E.; Torcia, M.; Lucibello, M.; Rosini, P.; Bonini, P.; Higashimoto, Y.; Damonte, G.; Armirotti, A.; Amodei, S.; et al. Bcl-2 Phosphorylation by p38 MAPK: Identification of target sites and biologic consequences. *J. Biol. Chem.* **2006**, *281*, 21353–21361. [CrossRef] [PubMed]
365. Nencioni, L.; De Chiara, G.; Sgarbanti, R.; Amatore, D.; Aquilano, K.; Marcocci, M.E.; Serafino, A.; Torcia, M.; Cozzolino, F.; Ciriolo, M.R.; et al. Bcl-2 expression and p38MAPK activity in cells infected with influenza A virus: Impact on virally induced apoptosis and viral replication. *J. Biol. Chem.* **2009**, *284*, 16004–16015. [CrossRef] [PubMed]
366. Tamura, Y.; Simizu, S.; Osada, H. The phosphorylation status and anti-apoptotic activity of Bcl-2 are regulated by ERK and protein phosphatase 2A on the mitochondria. *FEBS Lett.* **2004**, *569*, 249–255. [CrossRef]
367. Breitschopf, K.; Haendeler, J.; Malchow, P.; Zeiher, A.M.; Dimmeler, S. Posttranslational modification of Bcl-2 facilitates its proteasome-dependent degradation: Molecular characterization of the involved signaling pathway. *Mol. Cell Biol.* **2000**, *20*, 1886–1896. [CrossRef]
368. Yamamoto, K.; Ichijo, H.; Korsmeyer, S.J. BCL-2 is phosphorylated and inactivated by an ASK1/Jun N-terminal protein kinase pathway normally activated at G(2)/M. *Mol. Cell Biol.* **1999**, *19*, 8469–8478. [CrossRef]
369. Inoshita, S.; Takeda, K.; Hatai, T.; Terada, Y.; Sano, M.; Hata, J.; Umezawa, A.; Ichijo, H. Phosphorylation and inactivation of myeloid cell leukemia 1 by JNK in response to oxidative stress. *J. Biol. Chem.* **2002**, *277*, 43730–43734. [CrossRef]
370. Xu, P.; Das, M.; Reilly, J.; Davis, R.J. JNK regulates FoxO-dependent autophagy in neurons. *Genes Dev.* **2011**, *25*, 310–322. [CrossRef]
371. Kobayashi, S.; Lee, S.H.; Meng, X.W.; Mott, J.L.; Bronk, S.F.; Werneburg, N.W.; Craig, R.W.; Kaufmann, S.H.; Gores, G.J. Serine 64 phosphorylation enhances the antiapoptotic function of Mcl-1. *J. Biol. Chem.* **2007**, *282*, 18407–18417. [CrossRef] [PubMed]
372. Chang, S.H.; Hwang, C.S.; Yin, J.H.; Chen, S.D.; Yang, D.I. Oncostatin M-dependent Mcl-1 induction mediated by JAK1/2-STAT1/3 and CREB contributes to bioenergetic improvements and protective effects against mitochondrial dysfunction in cortical neurons. *Biochim. Biophys. Acta* **2015**, *1853*, 2306–2325. [CrossRef] [PubMed]
373. Deng, Y.; Ren, X.; Yang, L.; Lin, Y.; Wu, X. A JNK-dependent pathway is required for TNFalpha-induced apoptosis. *Cell* **2003**, *115*, 61–70. [CrossRef]

374. Park, B. JNK1mediated phosphorylation of Smac/DIABLO at the serine 6 residue is functionally linked to its mitochondrial release during TNFalpha-induced apoptosis of HeLa cells. *Mol. Med. Rep.* **2014**, *10*, 3205–3210. [CrossRef] [PubMed]
375. Nijboer, C.H.; van der Kooij, M.A.; van Bel, F.; Ohl, F.; Heijnen, C.J.; Kavelaars, A. Inhibition of the JNK/AP-1 pathway reduces neuronal death and improves behavioral outcome after neonatal hypoxic-ischemic brain injury. *Brain Behav. Immun.* **2010**, *24*, 812–821. [CrossRef] [PubMed]
376. Jeong, C.H.; Chun, K.S.; Kundu, J.; Park, B. Phosphorylation of Smac by Akt promotes the caspase-3 activation during etoposide-induced apoptosis in HeLa cells. *Mol. Carcinog.* **2015**, *54*, 83–92. [CrossRef]
377. Cook, S.J.; Stuart, K.; Gilley, R.; Sale, M.J. Control of cell death and mitochondrial fission by ERK1/2 MAP kinase signalling. *FEBS J.* **2017**, *284*, 4177–4195. [CrossRef]
378. Gamas, P.; Marchetti, S.; Puissant, A.; Grosso, S.; Jacquel, A.; Colosetti, P.; Pasquet, J.M.; Mahon, F.X.; Cassuto, J.P.; Auberger, P. Inhibition of imatinib-mediated apoptosis by the caspase-cleaved form of the tyrosine kinase Lyn in chronic myelogenous leukemia cells. *Leukemia* **2009**, *23*, 1500–1506. [CrossRef]
379. Luciano, F.; Herrant, M.; Jacquel, A.; Ricci, J.E.; Auberger, P. The P54-cleaved form of the tyrosine kinase Lyn generated by caspases during BCR-induced cell death in B lymphoma acts as a negative regulator of apoptosis. *Faseb J.* **2003**, *17*, 711. [CrossRef]
380. Contri, A.; Brunati, A.M.; Trentin, L.; Cabrelle, A.; Miorin, M.; Cesaro, L.; Pinna, L.A.; Zambello, R.; Semenzato, G.; Donella-Deana, A. Chronic lymphocytic leukemia B cells contain anomalous Lyn tyrosine kinase, a putative contribution to defective apoptosis. *J. Clin. Invest.* **2005**, *115*, 369–378. [CrossRef]
381. Mahon, F.X.; Hayette, S.; Lagarde, V.; Belloc, F.; Turcq, B.; Nicolini, F.; Belanger, C.; Manley, P.W.; Leroy, C.; Etienne, G.; et al. Evidence that resistance to nilotinib may be due to BCR-ABL, Pgp, or Src kinase overexpression. *Cancer Res.* **2008**, *68*, 9809–9816. [CrossRef] [PubMed]
382. Qi, X.; Mochly-Rosen, D. The PKCdelta -Abl complex communicates ER stress to the mitochondria—An essential step in subsequent apoptosis. *J. Cell Sci.* **2008**, *121*, 804–813. [CrossRef] [PubMed]
383. Ishizawar, R.; Parsons, S.J. c-Src and cooperating partners in human cancer. *Cancer Cell* **2004**, *6*, 209–214. [CrossRef] [PubMed]
384. Juhaszova, M.; Zorov, D.B.; Kim, S.H.; Pepe, S.; Fu, Q.; Fishbein, K.W.; Ziman, B.D.; Wang, S.; Ytrehus, K.; Antos, C.L.; et al. Glycogen synthase kinase-3beta mediates convergence of protection signaling to inhibit the mitochondrial permeability transition pore. *J. Clin. Invest.* **2004**, *113*, 1535–1549. [CrossRef] [PubMed]

 life

Review

PINK1: A Bridge between Mitochondria and Parkinson's Disease

Filipa Barroso Gonçalves and Vanessa Alexandra Morais *

Faculdade de Medicina, Instituto de Medicina Molecular—João Lobo Antunes, Universidade de Lisboa, 1649-028 Lisbon, Portugal; filipagoncalves@medicina.ulisboa.pt
* Correspondence: vmorais@medicina.ulisboa.pt

Abstract: Mitochondria are known as highly dynamic organelles essential for energy production. Intriguingly, in the recent years, mitochondria have revealed the ability to maintain cell homeostasis and ultimately regulate cell fate. This regulation is achieved by evoking mitochondrial quality control pathways that are capable of sensing the overall status of the cellular environment. In a first instance, actions to maintain a robust pool of mitochondria take place; however, if unsuccessful, measures that lead to overall cell death occur. One of the central key players of these mitochondrial quality control pathways is PINK1 (PTEN-induce putative kinase), a mitochondrial targeted kinase. PINK1 is known to interact with several substrates to regulate mitochondrial functions, and not only is responsible for triggering mitochondrial clearance via mitophagy, but also participates in maintenance of mitochondrial functions and homeostasis, under healthy conditions. Moreover, PINK1 has been associated with the familial form of Parkinson's disease (PD). Growing evidence has strongly linked mitochondrial homeostasis to the central nervous system (CNS), a system that is replenished with high energy demanding long-lasting neuronal cells. Moreover, sporadic cases of PD have also revealed mitochondrial impairments. Thus, one could speculate that mitochondrial homeostasis is the common denominator in these two forms of the disease, and PINK1 may play a central role in maintaining mitochondrial homeostasis. In this review, we will discuss the role of PINK1 in the mitochondrial physiology and scrutinize its role in the cascade of PD pathology.

Keywords: PINK1; Parkinson's disease; mitochondria homeostasis

Citation: Gonçalves, F.B.; Morais, V.A. PINK1: A Bridge between Mitochondria and Parkinson's Disease. *Life* **2021**, *11*, 371. https://doi.org/10.3390/life11050371

Academic Editor: Francesco Bruni

Received: 2 April 2021
Accepted: 19 April 2021
Published: 21 April 2021

1. Introduction

Mitochondria are essential for life; they play an intimate role in almost all aspects of cellular function, since they are responsible for calcium intracellular homeostasis, metabolite biosynthesis, cell proliferation, differentiation and apoptosis, and, their most known purpose, energy production. This close interaction between mitochondria and the cell is the main reason why mitochondrial dysregulation is one of the underlying causes of several pathologies, such as cancer, neurological diseases and metabolic disorders. However, mitochondria are highly dynamic organelles that evoke several mechanisms to respond to bioenergetic challenges, such as biogenesis, events of fusion/fission and ultimately mitophagy, a pathway that controls their overall fate.

2. Mitochondria Homeostasis

Mitochondria are highly dynamic organelles that can change their morphological characteristics depending on their host cells environment and needs. Mitochondrial dynamics is a balance dictated by fission (division of one mitochondria by Drp1) and fusion (merging of different mitochondria by Mitofusins 1/2 and Opa1) events [1]. Interestingly, these events may occur to remove damaged portions within mitochondria or even to allow mitochondria movement to a high demanding metabolic area [1]. Additionally, changes in ATP/ADP ratio trigger mitochondria biogenesis, where PGC-1alpha is the major regulator of mitochondrial mass increase [2]. Important to note that all these processes require a coordination of several mechanisms, including expression of mtDNA genes, nuclear–mitochondrial communication and protein expression and import. Disruption of any of these processes can lead to a defective mitochondrial function. However, cells have evolved mechanisms to sequester and remove potentially damaged mitochondria, a form of autophagy called mitophagy [3]. As all these pathways are intertwined, one could argue that mitochondrial homeostasis and mitochondrial quality control are key regulatory features to maintain a healthy and robust pool of mitochondria in our cells. Consequently, when mitochondria go astray due to impairment of these pathways, the overall cell fate is dictated. A clear example of this is neurons, where mitochondrial dysfunction in these high energy demanding cells leads to neuron loss and ultimately to the development of neurodegenerative disorders.

At present, a complete picture of the molecular mechanisms involved in mitochondrial quality control is still to be unraveled. However, a key protein that has revealed to play a vital role in the regulation of several mitochondrial pathways is PINK1 (Phosphatase and Tensin homologue-induced kinase 1). This Serine/Threonine kinase is of special interest since it has different roles depending on the overall mitochondrial steady state, controlling this organelle's fate [4]. Additionally, PINK1 is implicated in the etiology of Parkinson's Disease (PD) as several mutations in this kinase are associated with early onset forms of this disease [5]. Thus, a detailed comprehension of PINK1's biology will not only shed light on etiology of the neurodegenerative disease PD, but will furthermore lead to an enhanced understanding of the molecular pathways that regulate overall mitochondrial homeostasis.

Herein, we will look into the implications of PINK1 in mitochondrial physiology and scrutinize the role this protein plays in PD pathology.

3. PINK1 Helps Maintain Mitochondrial Homeostasis

PINK1 gene was first identified in 2001, through a screen which aimed to analyze the PTEN unregulated targets in endometrial cancer cells [6]. This gene is localized at the human chromosome 1p36.12 and encodes for a Serine/Threonine kinase with 581 amino acid residues, and a molecular weight of approximately 63 kDa 4. *PINK1* was found to be regulated by Foxo3a, a downstream component of the PI3K/AKT pathway, which promotes *PINK1* transcription by binding to its promoter [7]. Additionally, by binding to its promoter, but under stress conditions, the nuclear factor κB up-regulates PINK1 transcription [8] More recently, it was described a downregulation of the *PINK1* gene by two different factors, p53 [9] and ATF3 [10].

PINK1 is composed by a mitochondrial targeting sequence (MTS) located at N-terminal side (Figure 1). Additionally, in the N-terminal, we find what has been named the outer mitochondrial membrane localization signal (OMS), responsible for holding PINK1 to the outer mitochondrial membrane (OMM) upon depolarization [11,12]. The largest region of PINK1 is constituted by its kinase domain, region where the majority of identified mutations associated with PD have been described, a topic that we will elaborate later on this review. This kinase domain is constituted by N-lobe (residues 156–320 in human PINK1; hPINK1) and C-lobe (residues in 321–511 hPINK1). Finally, the C-terminal portion of PINK1 has been associated with the regulation of its kinase activity, since mutations on this terminal downregulate autophosphorylation, but seems to positively affect PINK1 kinase activity towards other substrates, such as histone H1 [13,14].

Despite being a widely studied protein, the resolution of human PINK1 crystal structure is a dilemma due to the poor solubility and rapid degradation of this protein [13–15]. To overcome this the field PINK1 insect orthologs (*Tribolium castaneum*, Tc; and *Pediculus humans corporis*) were used to perform structural studies, as they have high in vitro kinase activity [16] and share a kinase domain sequence identity of approx. 40% to the hPINK1 [17]. However, others have reported different substrate selectively between PINK1 orthologues [18]. Indeed, the experimental conditions used for TcPINK1 are not replicated when using a human form of PINK1, indicating probable regulatory differences between the two orthologues. While accessing these differences, Aerts et al. found that the full-length form of hPINK1 was not able to induce autophosphorylation, although it is catalytic active since it is able to phosphorylate PINK1 substrates, namely Parkin [19]. This suggests that the N-terminal region of PINK1 may have an inhibitory effect. The majority of studies demonstrating PINK1 kinase activity use truncated forms of PINK1, where parts of N-terminal are absent [13–15].

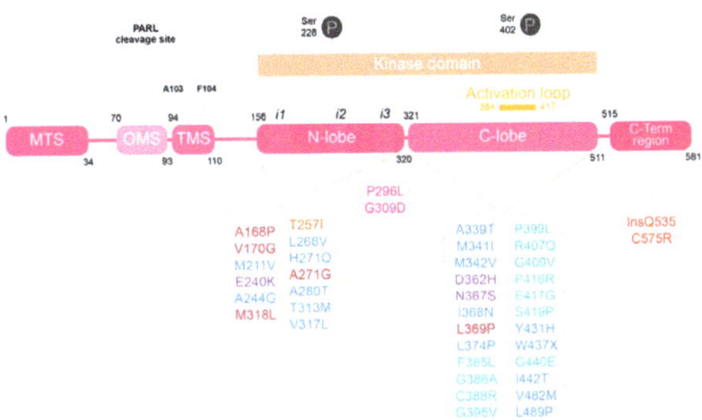

Figure 1. Human PINK1 domains and PD associated mutations. PINK1 is divided into different regions. At the N-terminal are the regions responsible for the processing and delivery of PINK1 to mitochondria: mitochondrial targeting sequence (MTS), the region recently called named the mitochondria membrane localization signal (OMS) and the transmembrane sequence (TMS). Within the TMS resides the PARL cleavage sites. The kinase domain is divided into an N-lobe and C-lobe; it is also the PINK1 domain where the majority of PD-associated mutations are found, and where there are the well described phosphorylation sites [15]. At the N-lobe there are the different inserts (i1, i2 and i3) identified in bioinformatics studies performed using PINK1 insects structure. The activation loop, at the C-lobe, changes the proteins conformation from inactive to active state upon phosphorylation. PINK1-PD mutations can be divided into mutations that affect PINK1's structure, kinase activity or substrate binding, depending on residues and protein regions affected: ATP binding pocket (bordeaux), kinase core (dark blue), catalytic mutations (purple), insert 2 (yellow), insert 3 (pink), activation loop (light blue) and C-terminal region (red).

As reviewed by Rassol and Trempe, important mechanistic information on PINK1's autophosphorylation and substrate binding were achieved, by the structural, biochemical and cellular studies done so far [20]. Bioinformatic and modelling approaches predicted that PINK1 carries additional regions when compared to other kinases, namely three inserts in the N-lobe and two flanking regions on its kinase domain: NT linker and CTE (C-terminal extension). This last one lies on the C-lobe, away from active site and has an important role in PINK1 stabilization [20]. This analysis also reveals that PINK1's autophosphorylation is important to instigate a structural reorganization of insert 3 and consequently ubiquitin (Ub) binding, which needs two different articulations with PINK1, one with the N-lobe and other with the C-lobe, to form the active complex [20]. Although overall structure of PINK1 is likely conserved between species: there are relevant differences between the different animal models, for instance differences in inserts. The N-lobe of PINK1 has three inserts [15], and the length of insert 1 varies between species while insert 3 is the most conserved [16]. Intriguingly, PINK1 is not only a promiscuous kinase that selects its substrate based on the cellular environment it encounters, but it also seems like PINK1 has a different panoply of substrates depending on its host.

PINK1 is responsible for helping in maintaining homeostasis by interacting with several proteins. TRAP1 (TNF receptor associated protein 1/Heat shock protein 75) is an example, it plays a role in reducing reactive oxygen species (ROS) production and assisting protein folding [21,22]. Overexpression of PINK1 leads to an increase in phosphorylation of endogenous TRAP1 [23]. Other reports show an interaction with HtrA2 (high temperature requirement A2), also known as Omi, a protease released in the cytosol during apoptosis [24]; Htr2A is phosphorylated in a PINK1-dependent manner, since PINK1 siRNA leads to a decrease in Htr2A phosphorylation, although it is not clear if it is a direct phosphorylation [24].

Mitochondrial dynamics, a fundamental process to maintain a healthy mitochondrial pool, has shown to be regulated by PINK1. Studies performed using *Drosophila* showed that PINK1 downregulation lead to an increase in mitochondria length, resulting from excessive fusion, while PINK1 overexpression promoted mitochondrial fission [25]. The cytosolic dynamin GTPase Drp1, protein involved in fission events, is regulated by PINK1. It has been proposed that PINK1-mediated fission is dependent on the phosphorylation of Drp1 at residue S616, since in PINK1 null cells and mouse tissue a reduction in phosphorylation at this residue is observed [26]. Interestingly, when in presence of depolarized mitochondria, protein kinase A (PKA), which normally is recruited to OMM by AKAP1 to phosphorylate Drp1 at reside S637 and inhibit fission, is repressed by the presence and accumulation of PINK1, thereby ensuring the fission of damaged mitochondria [27]. Three putative PINK1-phosphorylation sites have been identified in Mitofusin 2 (Mfn2) at residues T111, S378 and S442. Interestingly, the phosphorylation of these residues leads to the suppression of mitochondrial fusion, even though residues T111 and S442 conjugated phosphorylation promotes Mfn2-Parkin binding [28]. Additionally, mitochondrial transport is regulated by PINK1, which can phosphorylate multiple sites of Miro, allowing normal mitochondrial movement in axon terminals [29,30].

The uptake of Ca^{2+} in mitochondria is highly important for several processes, such as oxidative phosphorylation or mitochondrial-induced apoptosis [31,32]. Several studies reveal a role of PINK1 in the regulation of mitochondrial calcium levels, where PINK1 deficiency leads to impaired Ca^{2+} levels in different models: cells [33], cultured neurons [34,35] and zebrafish [36] but affects Ca^{2+} homeostasis evoking different pathways. Heeman et al. reported that PINK1 depletion impaired Ca^{2+} uptake [33], whereas Gandhi et al. reported a regulation of Ca^{2+} levels by PINK1 via Na^+/Ca^{2+} exchanger [34], where PINK1 deficiency causes an accumulation of Ca^{2+}. More recently, dopaminergic neurons from a PINK1-null zebrafish model revealed an inhibition of mitochondrial Ca^{2+} uniporter (MCU) as being directly involved in the neurodegeneration process [36]. Still, new research has been encouraged to unravel this pathway.

Interestingly, PINK1 is enriched in Mitochondria Associated Membranes (MAMs) [37], a widely recognized site of Ca^{2+} exchange with the ER (endoplasmatic reticulum) [38], to maintain cellular bioenergetics and mitochondrial dynamics and transport [39]. Although PINK1's involvement in mitochondria–ER tethering still remains to be explored, it was reported that upon CCCP treatment PINK1 accumulates on MAM's and recruits Beclin1, a pro-autophagic protein, to generate omegasomes [37], which could suggest a new pathway for PINK1 in mitophagy. Additionally, fibroblast derived from patients with PINK1 mutations exhibit an increased ER–mitochondria co-localization, resulting in an abnormal Ca^{2+} signaling [40].

Without a doubt PINK1 is vital for overall mitochondria homeostasis as it regulates several mitochondrial-related pathways within the cell. However, what dictates which pathway will be ameliorated in PINK1 loss-of-function scenarios remains to be clarified.

4. PINK1 in Healthy Mitochondria

PINK1 has been described as a sensor of the overall health status of mitochondria, as the localization and substrate specificity of this kinase is defined when it encounters healthy versus unhealthy mitochondria (Figure 2).

In steady state, PINK1 is targeted to the mitochondria via its MTS. It has been established that MTS-carrying proteins interact with the translocase of outer and the inner membrane (TOM20 and TIM23, respectively) [41]. Along these lines, the conventional theory assumes that PINK1, being a MTS carrying protein, passes through these translocases and enters the matrix, where the MTS is cleaved by a mitochondrial processing protease (MPP) [42]. Then, a second cleavage occur by the mitochondrial rhomboid protease PARL (Presenilins-associated rhomboid like protease), between Ala103 and Phe104 in the transmembrane stretch, reducing the protein size to a 52kDa processed form [43–45], which is released in the cytosol and later degraded by the proteasome [46]. Consequently, PINK1 has a high turnover rate, and it is maintained at a low expression levels within the tissues [47,48].

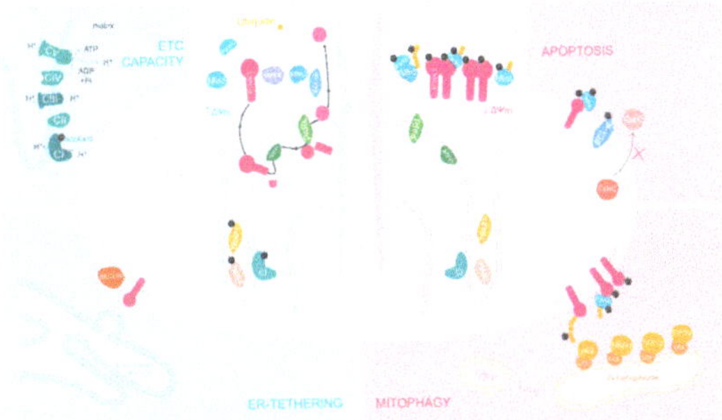

Figure 2. PINK1 has different roles depending on mitochondria's overall state. In the presence of healthy mitochondria, PINK1 is internalized and phosphorylates, among other substrates, the complex I subunit NDUFA10 at the inner mitochondrial membrane (IMM). By regulating the enzymatic activity of complex I, PINK1 modulates the overall electron transport chain (ETC) capacity and, ultimately, the overall output levels of ATP. Afterwards, PINK1 is sequential cleaved by the proteases MPP and PARL and released to the cytosol for degradation. PINK1 also has a protective role, as it phosphorylates Bcl-xl in order to inhibit apoptosis. When mitochondria are depolarized, PINK1 accumulates on the outer mitochondrial membrane (OMM), where it forms homodimers and undergoes autophosphorylation. After this, PINK1 recruits and phosphorylates Parkin, and consequently also phosphorylates ubiquitin. Due to Parkin's E3 ubiquitin ligase activity, Parkin and PINK1 create a positive feedback-loop, recruiting more ubiquitin and Parkin to be phosphorylated, creating poly-ubiquitin chains all around the surface of damaged mitochondria. This targets mitochondria for degradation via mitophagy. Posteriorly, due to the recruitment of autophagic receptors, like LC3, OPTN and NDP52, damaged mitochondria are engulfed and degraded via autophagy. Upon depolarization, ER-tethering to mitochondria is also hampered as PINK1 accumulates on MAM structures recruiting Beclin1 to form omegasomes, which are autophagosome precursors. $\Delta\Psi$m, mitochondrial membrane potential; CI, Complex I; CII, Complex II; CIII, Complex III; CIV, Complex IV; CV, Complex V or ATP synthase; P, phosphorylation.

Despite the increasing number of reports, there are still some conflicting data about the processing of PINK1 and localization of the kinase domain sub-mitochondrially. In fact, in the previously explained processing events, it is not clear how PARL, that resides in the intermembrane mitochondrial space (IMS), can cleave a protein that is associated with the TIM channel. Can PINK1 be partially imported? Some studies have shown that PINK1 or processed forms of PINK1 are located inside mitochondria [13,42,43,49]. While others have shown a partial PINK1 import, only for MPP and PARL cleavage to occur, and the kinase domain remains at the outer membrane facing the cytosol [46,50]. A few reports show that m-AAA proteases can be involved in this second cleavage of PINK1, helping its translocation in the inner mitochondria membrane (IMM) producing a form accessible to PARL [42]. On the other hand, Liu et al. show that PINK1, after being cleaved by PARL, suffers major conformational changes that allow the ubiquitination at Lys137, though an unknown E3 ubiquitin ligase [51]. More studies must be done to clarify PINK1 import and proteolytic processing inside mitochondria. By now we know that PINK1 regulates itself by autophosphorylation, since two phosphorylation sites have been reported, Ser228 and Ser402. Important to stress that Ser402 is located in PINK1's activation loop, a crucial region to kinase activity even towards PINK1 substrates, contrarily to Ser228 that seems to be dispensable. There are other two putative phosphorylation sites, T313 and

T257. Interestingly, the substitution of threonine in residue 313 to a methionine (T313M) is described as a PD mutation, which created a higher curiosity around this residue but, like T257, it does not affect PINK1 autophosphorylation.

Several studies have demonstrated altered mitochondria phenotypes in loss of function PINK1 cells and animal models [52–55]. Drosophila PINK1 mutant displays motor disturbances, abnormal synaptic transmission, structural mitochondrial changes and decreased membrane potential [52,53]. These different phenotypes were later explained as downstream consequences of a reduced Complex I enzymatic activity in Drosophila and mouse models [55]. Moreover, Morais et al. demonstrated with this study the relevance of PINK1 for overall electron transport chain (ETC) function, since wild-type (WT) hPINK1, but not a kinase inactive form of PINK1, was able to fully restore the enzymatic activity of Complex I in PINK1-null models. At present, all evidence points out that when PINK1 encounters a healthy mitochondria, it is internalized, processed and able to phosphorylation mitochondrial-resident substrates, such as the Complex I subunit NDUFA10.

5. PINK1 in Unhealthy Mitochondria

Across the IMS, essential for mitochondrial function, there is a membrane potential, which regulates the ETC and consequently the generation of ATP. In several cases of mitochondria damage, a loss of this potential occurs. When PINK1 encounters depolarized mitochondria its import is halted, and an accumulation of the full-length form occurs on the OMM [56,57]. In this case, PINK1 kinase activity suffers a shift, resulting in autophosphorylation on Ser228 and Ser402 and PINK1 dimerization [19,58,59]. Afterwards, one of the most studied PINK1 interactions occurs, an event first identified in Drosophila as a genetic link to an E3 ubiquitin ligase Parkin [52,53]. Parkin, largely known to cooperate with PINK1 in the removal of damaged mitochondria, interestingly is also known to be mutated in early onset PD cases [60,61]. The Drosophila studies reported that both PINK1 and Parkin mutant flies exhibit indirect flight muscle and dopaminergic neuronal degeneration accompanied by locomotive defects [52,53]. All these defects in PINK1 mutant flies were restored when Parkin was re-introduced. However, the reintroduction of PINK1 did not restore Parkin mutant fly defects, clearly indicating that Parkin works downstream of PINK1.

5.1. *PINK1/Parkin Mediated Mitophagy*

The crystal structure of Parkin revealed that it naturally exists in an autoinhibited conformation [62,63]. However, PINK1 phosphorylated ubiquitin (pUb) molecules interact with Parkin, altering its structure and making Ser65 of the ubiquitin-like domains (UBL) available for PINK1-mediated phosphorylation [64,65]. Only after these two steps is Parkin fully activated. In PINK1-null mammalian cells, Parkin recruitment is completely abolished, and is rescued when a WT but not a kinase-inactive form of PINK1 is re-introduced, confirming that Parkin needs an active PINK1 to be recruited to depolarized mitochondria [56–66].

Parkin's ligase activity will recruit more ubiquitin and Parkin molecules to be phosphorylated by PINK1, creating poly-ubiquitin chains on the OMM [67]. This will further lead to an interaction with autophagy receptors containing an ubiquitin-binding domain and the LC3-interacting region, such as OPTN, NDP52 and NBR1 [68–70], that once connected to the autophagosome membrane will lead to mitochondrial clearance via mitophagy. The p62 or sequestome-1 (SQSTM1) role in this pathway is still controversial, while a number of studies demonstrate that it is recruited to depolarized mitochondria [71,72], others show that it is not essential for mitophagy to occur [73].

Supplementary to mitophagy, some mitochondrial proteins that are ubiquitinated by Parkin are degraded, such as Mitofusin and Miro1 leading to an impaired mitochondrial network [74,75] and mitochondrial movement [29,76,77], respectively.

Important to note that how PINK1 is stabilized on the OMM still remains to be clarified. Previous studies using BN-PAGE analyses revealed that PINK1 accumulates on the OMM in a phosphorylated form and is associated with the TOM40 complex [59,78]. More recently, Sekine et al. suggested that TOM7 may also play a role, as in TOM7 knockout (KO) cells both PINK1 and Parkin recruitment were defective after mitochondrial depolarization treatment [12].

Besides PINK1 stabilization, it is also not clear the mechanism by which Parkin is recruited to mitochondria. A recent study [79] proposes that MITOL (mitochondrial ubiquitin ligase: also known as MARCH5) can ubiquitinate proteins at the OMM, which will facilitate the recruitment of Parkin. Other studies suggest that the phosphorylation of proteins on mitochondria surface by PINK1 will work as "Parkin receptors", like for instance ubiquitin [80], Mfn2 [81] or Miro [77].

Besides the putative phosphorylation of Htr2A in the "healthy mitochondria" pathway, PINK1 has also a protective role against apoptosis upon depolarization, when phosphorylating BCL-xL [82]. This anti-apoptotic protein is localized at OMM, where in a phosphorylated form it prevents the release of cytochrome c and caspases, inhibiting apoptosis [83].

Additionally, in these "unhealthy" mitochondria pathways there was described a link to PD. Zhu and collaborators observed mitochondria in autophagossomes in neurons of PD patients [84]. Later on, several groups reported abnormal mitophagy in different PD models [85–87], facts that we will further explore in the next topic of this review, suggesting a link between a PINK1 regulated pathway and PD.

5.2. PINK1 Independent Mitophagy

Important to note that in the past few years a couple of papers show that PINK1 may not be necessary for mitophagy. In 2017, Koentjoro discovered a mitochondrial autophagy receptor Nip1-like protein X (Nix) responsible for isolating mitochondria into autophagossomes in derived fibroblast from an asymptomatic patient carrying a homozygous Parkin mutation, where all mitochondrial functions and dynamics, including mitophagy, were working [88]. Moreover, Koentjoro showed that Nix compensates for the loss of PINK1/Parkin, as overexpression of Nix restores activation of mitophagy and mitochondrial energy production. Yet, Nix mediated mitophagy, as well as the expression of this protein in PD patient brains or animal models, remains to be determined. McWilliams' team studied the function of PINK1 in vivo using a mito QC reporter mice, which allows the visualization of mitophagy and mitochondrial architecture [89]. Analyzing PINK1 WT or KO mice crossed with mito-QC, they reported that PINK1 was present in all regions of the central nervous system, and that loss of PINK1 did not affect basal levels of mitophagy, when compared to WT mouse brains. Giving the hint that PINK1, mitophagy can occur in a PINK1 independent manner [89]. Nevertheless, it cannot be concluded that PINK1 is dispensable for all forms of mitophagy, indeed the precise role of this protein in vivo still remains to be defined, and PINK1's function may be cell type and content dependent.

Despite this, these recent findings only confirmed the complexity of this process. Therefore, we believe that the study of physiological mitophagy and its alteration when triggered with pathological cascades needs further investigations.

6. PINK1 in PD Pathogenesis

Parkinson's disease is the second most common neurodegenerative disorder, characterized by the loss of dopaminergic neurons (DA) in the substancia nigra and the formation of cytoplasmic inclusion bodies containing alpha-synuclein, called Lewy Bodies (LB). When about 70% of neurons are lost, the communication between brain and muscle cells weakens, resulting in the classic motor symptoms as the resting tremor, stiffness, bradykinesia and postural instability [90]. The etiology of this disease remains to be clarified but is considered to be the result of a combination of genetics and environmental factors, and where aging also has a crucial role. Indeed, PD cases are divided into sporadic and familiar; these last ones carry a heritable disease mutation in genes referred to as PARK genes, and account for 5–10% of all the PD cases [91].

6.1. PINK1 as a Genetic Cause of PD

Mutations in several of these PARK genes also cause mitochondrial dysfunctions, namely PINK1 (PARK6) 5 and Parkin (PARK2) [60,61]. PINK1 mutations cause autosomal recessive PD and certain clinical features are more common in patients with PINK1 mutations. For example, the outstanding difference is the age of onset that in this case is around 30 years of age. Additionally, patients harboring these mutations experience a more benign course with slower progression, more gait and balance difficulties, a better response to L-DOPA and increased dyskinesias [92,93]. Most patients with PINK1-associated PD experience a good response to levodopa and they do not typically develop dementia, although some studies report psychiatric features [94]. It is important to point out the fact that PINK1 was detected in some LBs in sporadic cases as well as in samples carrying only one allele mutant for PINK1, which are clinically and pathologically indistinguishable from the sporadic form [49].

To date it has been reported in the Human Gene Mutation Database (HGMB) at least 130 mutations in the *PINK1* gene, and they can be homozygous or heterozygous; missense mutations, truncating mutations and exon rearrangements [5,16]. A large majority of PINK1 mutations are in kinase domain, demonstrating the importance of PINK1 kinase activity in PD pathogenesis (Figure 1). The firstly identified PINK1 mutations were reported back in 2004 in Spanish (G309D) [5], Italian (W437X) [5], Filipino (L347P) [95], Japanese and Israeli families (R246X/H271Q) [95]. When in 2017 Schubert et al. revealed the structure of PhPINK1 in complex with ubiquitin [96], they were able to pinpoint the location of dozens of PINK1 PD causing mutations (Figure 1). The majority of them are in the kinase core and affect the protein fold or in the activation-loop. Mutations known to affect catalysis were mapped in the ATP binding pocket. These observations are in concordance with previous observations where, for instance, G309D mutations described as in insert 3 are largely described with defective kinase activity [16]. As we already described, PINK1 is involved in several pathways, and some have been proven to be altered in PD.

6.2. PINK1's Link to Dysfunctions in the Respiratory Chain

The most known and described correlation between mitochondria and PD, is the injection of a synthetic heroin, 1-methyl-4-phenyl-1,2,5,6-tetrahydropyridine (MPTP) that rapidly generated Parkinsonism symptoms in drug abusers [97]. This compound has the ability to pass the blood–brain barrier, where it is oxidized to $MPDP^{+}$ by monoamine oxidase B in glia and serotonergic neurons and then is converted to MPP^{+}. This metabolite has a high affinity to dopamine transporters and causes an inhibition of Complex I, and preferentially mediates the degeneration of SN dopaminergic neurons [98,99]. The correlation between mitochondria impairment and PD was later validated when reduced activity in complex I, II and IV was found in brain, skeletal muscle and platelets of patients with sporadic PD [100,101]. Posteriorly, also some pesticides and herbicides, namely rotenone and paraquat that selectively inhibit Complex I, were also shown to cause Parkinsonism in animal models [102].

Several reports that analyzed fibroblasts from patients harboring PINK1 mutations revealed alterations in the ETC and overall oxidative stress levels [103–105]. PINK1 deficient mice reveal impaired dopamine release [106], compromised mitochondrial respiration, increased sensitivity to oxidative stress, progressive weight loss and selective reduction in locomotor activity in older animals [107]. Indeed, Morais et al. demonstrated that the key defect in ETC at the level of Complex I in PINK1 KO mice was due to reduced phosphorylation of the Complex I subunit NDUFA10 [106], confirming that PINK1 phosphorylation is required for Complex I function at the level of ubiquinone reduction. Phosphorylation at this site is a prerequisite for ubiquinone reduction, critical to mitochondrial bioenergetics and ROS production. ROS develops when electrons escape the ETC, especially at complexes I and III [107]. Interesting is the link between dopamine metabolism and ROS, where dopamine degradation causes an increase in ROS production [108], revealing a pathogenic cascade from mitochondrial impairment to neuronal dysfunction, which could explain neuronal death. Additionally, reduced enzymatic activity of Complex I has also been observed in several PD patients. Indeed, an impaired Complex I was one of the first observations made in sporadic PD patients [109], providing a link between mitochondrial dysfunctions and PD. However, whether NDUFA10 phosphorylation is reduced in PD patients harboring PINK1 mutations is still to be shown.

6.3. PINK1 in Mitophagy Context

Mitophagy is a key pathway for maintenance of mitochondrial health and, consequently, to maintain neuronal health. An impaired mitophagy results in an accumulation of damaged mitochondria, leading to neuronal death and consequent neurodegeneration, indeed an abnormal mitophagy has already been linked to neurodegeneration [110]. This pathway was confirmed in PD patient-derived fibroblasts and iPSC-derived neurons [111–113]. Moreover, Lewy bodies contain aggregates of alpha-synuclein, ubiquitin, and other compounds [114], including mitochondria. One study has reported mitochondria within autophagossomes in neurons of PD patients [84]. Abnormal mitophagy was posteriorly observed in several PD models [85–87], but the analysis of mitophagy has to be done with extreme caution, since Rakovic and collaborators demonstrate that the mitophagy process differs between human non-neuronal and neuronal cells, and that the comparison between models with endogenous Parkin expression versus exogenous expression may be misleading [112]. Moreover, the use of different experimental approaches to assess mitophagy have also contributed to contradictory findings in the field [115]. Studies performed in *Drosophila* using mt-Keima, a pH-dependent fluorescent protein fused to a mitochondrial targeting sequence, have shown that mitophagy occurs in muscle and dopaminergic neurons of aged flies [116]. However, in young adult flies no defects in mitophagy are observed, suggesting a role for PINK1 in age-dependent mitophagy. On the other hand, studies using mito-QC, a tandem mCherry-GFP tag fused to a mitochondrial targeting sequence, showed that basal mitophagy is readily detectable and abundant in many tissues of *PINK1* deficient flies [117]. These findings provide evidence that, at least in these experimental conditions, PINK1 and parkin are not essential for basal mitophagy in *Drosophila*, further underpinning that the molecular mechanisms of basal mitophagy remain largely obscure. Nevertheless, fibroblast from PINK1-PD patients confirmed data obtained with other PINK1 models, such as the stabilization and recruitment of Parkin on depolarized mitochondria [105]. Additionally, iPSC-derived neuron cells from an individual with PINK1-PD mutations reported an impaired recruitment of Parkin to mitochondria, and increased mitochondrial copy number [113], indicating that clearance of defective mitochondria is hampered. The relationship between mitochondria autophagy impairment and its contribution to pathogenesis of PD seems more and more evident, but regarding the PINK1/Parkin pathway, although it is critical to regulate mitophagy, the overall mechanism is still unclear. We believe that modulation of this pathway may have a therapeutic potential; nevertheless, investigation is still needed.

7. Conclusions

PINK1 has a pivotal role in several mitochondrial processes. One of these processes is the mitochondrial quality control pathway, known to be related to several diseases, as neurodegenerative diseases, cardiovascular diseases, diabetes or obesity. However, the exact mechanism behind most of PINK1 pathways still remains to be clarified, mostly because of PINK1 low expression levels and fast turn-over. This is something that the field has been working and debating for a while, and at present several working tools are being developed to further strengthen these findings.

Parkinson's disease is the second most common neurodegenerative disorder, thus clarification of its pathogenic mechanism and the development of new diagnostic approaches and effective therapeutics are eagerly awaited. Almost 20 years after PINK1 was linked to this disease a lot of research has characterized this protein and its mutations associated to PD. The discovery of PINK1's link to Complex I activity was one of the major breakthroughs in the field. Important to note is that Complex I was linked to sporadic cases of PD before the PINK1 connection to this disease, as a deficiency in Complex I has already been reported in the *substantia nigra pars compacta* (SNpc) of PD patients. More recently the knowledge that PINK1 is responsible for the phosphorylation of the subunit NDUFA10 of the major ETC complex and that PINK1-PD mutations have Complex I deficiencies, created a new link between sporadic and familiar forms of PD. Understanding deregulated mitochondrial functions through the role of PINK1 mutations in familial forms of PD, may elucidate whether these pathways are eventually also disrupted in sporadic cases. More recently, emerging data focus on PINK1 role in non-neuronal cells, giving rise to the concept that loss of neurons in PD may be due to a brain environmental issue and not neuronal per se. We consider that research, with for instance new culture models, is needed to elucidate the mechanisms underlying PD and to develop potential biomarkers for diagnostic purpose and neuroprotective strategies.

Author Contributions: Writing, F.B.G. and V.A.M. All authors have read and agreed to the published version of the manuscript.

Funding: The authors are supported by Fundação para a Ciência e Tecnologia, Grant references for FBG SFRH/BD/134316/2017, VAM IF/01693/2014, Funding reference: PTDC/BIA-CEL/31230/2017 by Fundação para a Ciência e a Tecnologia (FCT); ERC-StG-679168 by European Research Council; and EMBO-IG/3309 by European Molecular Biology Organization.

Institutional Review Board Statement: Not applicable.

Informed Consent Statement: Not applicable.

Data Availability Statement: Not applicable.

Acknowledgments: We would like to thank Helena Pinheiro (Instituto de Medicina Molecular) for the elaboration of the illustrations.

Conflicts of Interest: The authors declare no conflict of interest.

References

1. Scorrano, L. Keeping mitochondria in shape: A matter of life and death. *Eur. J. Clin. Investig.* **2013**, *43*, 886–893. [CrossRef]
2. Liang, H.; Ward, W.F. PGC-1α: A key regulator of energy metabolism. *Am. J. Physiol.-Adv. Physiol. Educ.* **2006**, *30*, 145–151. [CrossRef] [PubMed]
3. Kim, I.; Rodriguez-Enriquez, S.; Lemasters, J.J. Selective degradation of mitochondria by mitophagy. *Arch. Biochem. Biophys.* **2007**, *462*, 245–253. [CrossRef] [PubMed]
4. Matsuda, S.; Kitagishi, Y.; Kobayashi, M. Function and Characteristics of PINK1 in Mitochondria. *Oxidative Med. Cell. Longev.* **2013**, *2013*, 1–6. [CrossRef] [PubMed]
5. Valente, E.M.; Abou-Sleiman, P.M.; Caputo, V.; Muqit, M.M.K.; Harvey, K.; Gispert, S.; Ali, Z.; Del Turco, D.; Bentivoglio, A.R.; Healy, D.G.; et al. Hereditary Early-Onset Parkinson's Disease Caused by Mutations in PINK1. *Science* **2004**, *304*, 1158–1160. [CrossRef]
6. Unoki, M.; Nakamura, Y. Growth-suppressive effects of BPOZ and EGR2, two genes involved in the PTEN signaling pathway. *Oncogene* **2001**, *20*, 4457–4465. [CrossRef]
7. Mei, Y.; Zhang, Y.; Yamamoto, K.; Xie, W.; Mak, T.W.; You, H. FOXO3a-dependent regulation of Pink1 (Park6) mediates survival signaling in response to cytokine deprivation. *Proc. Natl. Acad. Sci. USA* **2009**, *106*, 5153–5158. [CrossRef] [PubMed]
8. Duan, X.; Tong, J.; Xu, Q.; Wu, Y.; Cai, F.; Li, T.; Song, W. Upregulation of human PINK1 gene expression by NFκB signalling. *Mol. Brain* **2014**, *7*, 57. [CrossRef]
9. Goiran, T.; Duplan, E.; Rouland, L.; El Manaa, W.; Lauritzen, I.; Dunys, J.; You, H.; Checler, F.; Da Costa, C.A. Nuclear p53-mediated repression of autophagy involves PINK1 transcriptional down-regulation. *Cell Death Differ.* **2018**, *25*, 873–884. [CrossRef]
10. Bueno, M.; Brands, J.; Voltz, L.; Fiedler, K.; Mays, B.; Croix, C.S.; Sembrat, J.; Mallampalli, R.K.; Rojas, M.; Mora, A.L. ATF3 represses PINK1 gene transcription in lung epithelial cells to control mitochondrial homeostasis. *Aging Cell* **2018**, *17*, e12720. [CrossRef]
11. Okatsu, K.; Kimura, M.; Oka, T.; Tanaka, K.; Matsuda, N. Unconventional PINK1 localization to the outer membrane of depolarized mitochondria drives Parkin recruitment. *J. Cell Sci.* **2015**, *128*, 964–978. [CrossRef]
12. Sekine, S.; Wang, C.; Sideris, D.P.; Bunker, E.; Zhang, Z.; Youle, R.J. Reciprocal Roles of Tom7 and OMA1 during Mitochondrial Import and Activation of PINK1. *Mol. Cell* **2019**, *73*, 1028–1043.e5. [CrossRef]

13. Silvestri, L.; Caputo, V.; Bellacchio, E.; Atorino, L.; Dallapiccola, B.; Valente, E.M.; Casari, G. Mitochondrial import and enzymatic activity of PINK1 mutants associated to recessive parkinsonism. *Hum. Mol. Genet.* **2005**, *14*, 3477–3492. [CrossRef]
14. Sim, C.H.; Lio, D.S.S.; Mok, S.S.; Masters, C.L.; Hill, A.F.; Culvenor, J.G.; Cheng, H.-C. C-terminal truncation and Parkinson's disease-associated mutations down-regulate the protein serine/threonine kinase activity of PTEN-induced kinase-1. *Hum. Mol. Genet.* **2006**, *15*, 3251–3262. [CrossRef]
15. Beilina, A.; Van Der Brug, M.; Ahmad, R.; Kesavapany, S.; Miller, D.W.; Petsko, G.A.; Cookson, M.R. Mutations in PTEN-induced putative kinase 1 associated with recessive parkinsonism have differential effects on protein stability. *Proc. Natl. Acad. Sci. USA* **2005**, *102*, 5703–5708. [CrossRef] [PubMed]
16. Woodroof, H.I.; Pogson, J.H.; Begley, M.; Cantley, L.C.; Deak, M.; Campbell, D.G.; van Aalten, D.M.F.; Whitworth, A.J.; Alessi, D.R.; Muqit, M.M.K. Discovery of catalytically active orthologues of the Parkinson's disease kinase PINK1: Analysis of substrate specificity and impact of mutations. *Open Biol.* **2011**, *1*, 110012. [CrossRef] [PubMed]
17. Rasool, S.; Soya, N.; Truong, L.; Croteau, N.; Lukacs, G.L.; Trempe, J. PINK 1 autophosphorylation is required for ubiquitin recognition. *EMBO Rep.* **2018**, *19*, e44981. [CrossRef] [PubMed]
18. Aerts, L.; Craessaerts, K.; De Strooper, B.; Morais, V.A. In Vitro Comparison of the Activity Requirements and Substrate Specificity of Human and Triboleum castaneum PINK1 Orthologues. *PLoS ONE* **2016**, *11*, e0146083. [CrossRef]
19. Aerts, L.; Craessaerts, K.; De Strooper, B.; Morais, V.A. PINK1 Kinase Catalytic Activity Is Regulated by Phosphorylation on Serines 228 and 402. *J. Biol. Chem.* **2015**, *290*, 2798–2811. [CrossRef]
20. Rasool, S.; Trempe, J.-F. New insights into the structure of PINK1 and the mechanism of ubiquitin phosphorylation. *Crit. Rev. Biochem. Mol. Biol.* **2018**, *53*, 515–534. [CrossRef]
21. Hua, G.; Zhang, Q.; Fan, Z. Heat Shock Protein 75 (TRAP1) Antagonizes Reactive Oxygen Species Generation and Protects Cells from Granzyme M-mediated Apoptosis. *J. Biol. Chem.* **2007**, *282*, 20553–20560. [CrossRef]
22. Kadye, R.; Kramer, A.H.; Joos-Vandewalle, J.; Parsons, M.; Njengele, Z.; Hoppe, H.; Prinsloo, E. Guardian of the Furnace: Mitochondria, TRAP1, ROS and stem cell maintenance. *IUBMB Life* **2014**, *66*, 42–45. [CrossRef] [PubMed]
23. Pridgeon, J.W.; Olzmann, J.A.; Chin, L.-S.; Li, L. PINK1 Protects against Oxidative Stress by Phosphorylating Mitochondrial Chaperone TRAP1. *PLoS Biol.* **2007**, *5*, e172. [CrossRef] [PubMed]
24. Plun-Favreau, H.; Klupsch, K.; Moisoi, N.; Gandhi, S.; Kjaer, S.; Frith, D.; Harvey, K.; Deas, E.; Harvey, R.J.; McDonald, N.; et al. The mitochondrial protease HtrA2 is regulated by Parkinson's disease-associated kinase PINK1. *Nat. Cell Biol.* **2007**, *9*, 1243–1252. [CrossRef] [PubMed]
25. Yang, Y.; Ouyang, Y.; Yang, L.; Beal, M.F.; McQuibban, A.; Vogel, H.; Lu, B. Pink1 regulates mitochondrial dynamics through interaction with the fission/fusion machinery. *Proc. Natl. Acad. Sci. USA* **2008**, *105*, 7070–7075. [CrossRef] [PubMed]
26. Han, H.; Tan, J.; Wang, R.; Wan, H.; He, Y.; Yan, X.; Guo, J.; Gao, Q.; Li, J.; Shang, S.; et al. PINK 1 phosphorylates Drp1 S616 to regulate mitophagy-independent mitochondrial dynamics. *EMBO Rep.* **2020**, *21*, e48686. [CrossRef] [PubMed]
27. Pryde, K.R.; Smith, H.L.; Chau, K.-Y.; Schapira, A.H.V. PINK1 disables the anti-fission machinery to segregate damaged mitochondria for mitophagy. *J. Cell Biol.* **2016**, *213*, 163–171. [CrossRef] [PubMed]
28. Dorn, G.W. Mitofusins as mitochondrial anchors and tethers. *J. Mol. Cell. Cardiol.* **2020**, *142*, 146–153. [CrossRef] [PubMed]
29. Wang, X.; Winter, D.; Ashrafi, G.; Schlehe, J.; Wong, Y.L.; Selkoe, D.; Rice, S.; Steen, J.; LaVoie, M.J.; Schwarz, T.L. PINK1 and Parkin Target Miro for Phosphorylation and Degradation to Arrest Mitochondrial Motility. *Cell* **2011**, *147*, 893–906. [CrossRef]
30. Tsai, P.-I.; Course, M.M.; Lovas, J.R.; Hsieh, C.-H.; Babic, M.; Zinsmaier, K.E.; Wang, X. PINK1-mediated Phosphorylation of Miro Inhibits Synaptic Growth and Protects Dopaminergic Neurons in Drosophila. *Sci. Rep.* **2014**, *4*, 6962. [CrossRef]
31. Williams, G.S.B.; Boyman, L.; Chikando, A.C.; Khairallah, R.J.; Lederer, W.J. Mitochondrial calcium uptake. *Proc. Natl. Acad. Sci. USA* **2013**, *110*, 10479–10486. [CrossRef]
32. Belosludtsev, K.N.; Dubinin, M.V.; Belosludtseva, N.V.; Mironova, G.D. Mitochondrial Ca2+ Transport: Mechanisms, Molecular Structures, and Role in Cells. *Biochemistry* **2019**, *84*, 593–607. [CrossRef] [PubMed]
33. Heeman, B.; Haute, C.V.D.; Aelvoet, S.-A.; Valsecchi, F.; Rodenburg, R.J.; Reumers, V.; Debyser, Z.; Callewaert, G.; Koopman, W.J.H.; Willems, P.H.G.M.; et al. Depletion of PINK1 affects mitochondrial metabolism, calcium homeostasis and energy maintenance. *J. Cell Sci.* **2011**, *124*, 1115–1125. [CrossRef] [PubMed]
34. Gandhi, S.; Wood-Kaczmar, A.; Yao, Z.; Plun-Favreau, H.; Deas, E.; Klupsch, K.; Downward, J.; Latchman, D.S.; Tabrizi, S.J.; Wood, N.W.; et al. PINK1-Associated Parkinson's Disease Is Caused by Neuronal Vulnerability to Calcium-Induced Cell Death. *Mol. Cell* **2009**, *33*, 627–638. [CrossRef] [PubMed]
35. Kostic, M.; Ludtmann, M.H.; Bading, H.; Hershfinkel, M.; Steer, E.; Chu, C.T.; Abramov, A.Y.; Sekler, I. PKA Phosphorylation of NCLX Reverses Mitochondrial Calcium Overload and Depolarization, Promoting Survival of PINK1-Deficient Dopaminergic Neurons. *Cell Rep.* **2015**, *13*, 376–386. [CrossRef]
36. Soman, S.; Keatinge, M.; Moein, M.; Da Costa, M.; Mortiboys, H.; Skupin, A.; Sugunan, S.; Bazala, M.; Kuznicki, J.; Bandmann, O. Inhibition of the mitochondrial calcium uniporter rescues dopaminergic neurons in pink1−/− zebrafish. *Eur. J. Neurosci.* **2017**, *45*, 528–535. [CrossRef]
37. Gelmetti, V.; De Rosa, P.; Torosantucci, L.; Marini, E.S.; Romagnoli, A.; Di Rienzo, M.; Arena, G.; Vignone, D.; Fimia, G.M.; Valente, E.M. PINK1 and BECN1 relocalize at mitochondria-associated membranes during mitophagy and promote ER-mitochondria tethering and autophagosome formation. *Autophagy* **2017**, *13*, 654–669. [CrossRef]

38. Rizzuto, R.; Pinton, P.; Carrington, W.; Fay, F.S.; Fogarty, K.E.; Lifshitz, L.M.; Tuft, R.A.; Pozzan, T. Close Contacts with the Endoplasmic Reticulum as Determinants of Mitochondrial Ca2+ Responses. *Science* **1998**, *280*, 1763–1766. [CrossRef]
39. Yi, M.; Weaver, D.; Hajnóczky, G. Control of mitochondrial motility and distribution by the calcium signal: A homeostatic circuit. *J. Cell Biol.* **2004**, *167*, 661–672. [CrossRef]
40. Celardo, I.; Costa, A.C.; Lehmann, S.; Jones, C.; Wood, N.; Mencacci, N.E.; Mallucci, G.R.; Loh, S.H.Y.; Martins, L.M. Mitofusin-mediated ER stress triggers neurodegeneration in pink1/parkin models of Parkinson's disease. *Cell Death Dis.* **2016**, *7*, e2271. [CrossRef]
41. Neupert, W.; Herrmann, J.M. Translocation of Proteins into Mitochondria. *Annu. Rev. Biochem.* **2007**, *76*, 723–749. [CrossRef]
42. Greene, A.W.; Grenier, K.; Aguileta, M.A.; Muise, S.; Farazifard, R.; Haque, M.E.; McBride, H.M.; Park, D.S.; Fon, E.A. Mitochondrial processing peptidase regulates PINK1 processing, import and Parkin recruitment. *EMBO Rep.* **2012**, *13*, 378–385. [CrossRef] [PubMed]
43. Jin, S.M.; Lazarou, M.; Wang, C.; Kane, L.A.; Narendra, D.P.; Youle, R.J. Mitochondrial membrane potential regulates PINK1 import and proteolytic destabilization by PARL. *J. Cell Biol.* **2010**, *191*, 933–942. [CrossRef]
44. Deas, E.; Plun-Favreau, H.; Gandhi, S.; Desmond, H.; Kjaer, S.; Loh, S.H.; Renton, A.E.; Harvey, R.J.; Whitworth, A.J.; Martins, L.M.; et al. PINK1 cleavage at position A103 by the mitochondrial protease PARL. *Hum. Mol. Genet.* **2011**, *20*, 867–879. [CrossRef] [PubMed]
45. Meissner, C.; Lorenz, H.; Weihofen, A.; Selkoe, D.J.; Lemberg, M.K. The mitochondrial intramembrane protease PARL cleaves human Pink1 to regulate Pink1 trafficking. *J. Neurochem.* **2011**, *117*, 856–867. [CrossRef]
46. Yamano, K.; Youle, R.J. PINK1 is degraded through the N-end rule pathway. *Autophagy* **2013**, *9*, 1758–1769. [CrossRef]
47. Lin, W.; Kang, U.J. Characterization of PINK1 processing, stability, and subcellular localization. *J. Neurochem.* **2013**, *106*, 464–474. [CrossRef]
48. Takatori, S.; Ito, G.; Iwatsubo, T. Cytoplasmic localization and proteasomal degradation of N-terminally cleaved form of PINK1. *Neurosci. Lett.* **2008**, *430*, 13–17. [CrossRef]
49. Gandhi, S.; Muqit, M.M.K.; Stanyer, L.; Healy, D.G.; Abou-Sleiman, P.M.; Hargreaves, I.; Heales, S.; Ganguly, M.; Parsons, L.; Lees, A.J.; et al. PINK1 protein in normal human brain and Parkinson's disease. *Brain* **2006**, *129*, 1720–1731. [CrossRef] [PubMed]
50. Becker, D.; Richter, J.; Tocilescu, M.A.; Przedborski, S.; Voos, W. Pink1 Kinase and Its Membrane Potential ($\Delta\psi$)-dependent Cleavage Product Both Localize to Outer Mitochondrial Membrane by Unique Targeting Mode. *J. Biol. Chem.* **2012**, *287*, 22969–22987. [CrossRef]
51. Liu, Y.; Guardia-Laguarta, C.; Yin, J.; Erdjument-Bromage, H.; Martin, B.; James, M.; Jiang, X.; Przedborski, S. The Ubiquitination of PINK1 Is Restricted to Its Mature 52-kDa Form. *Cell Rep.* **2017**, *20*, 30–39. [CrossRef]
52. Clark, I.E.; Dodson, M.W.; Jiang, C.; Cao, J.H.; Huh, J.R.; Seol, J.H.; Yoo, S.J.; Hay, B.A.; Guo, M. Drosophila pink1 is required for mitochondrial function and interacts genetically with parkin. *Nature* **2006**, *441*, 1162–1166. [CrossRef] [PubMed]
53. Park, J.; Lee, S.B.; Lee, S.; Kim, Y.; Song, S.; Kim, S.; Bae, E.; Kim, J.; Shong, M.; Kim, J.-M.; et al. Mitochondrial dysfunction in Drosophila PINK1 mutants is complemented by parkin. *Nature* **2006**, *441*, 1157–1161. [CrossRef] [PubMed]
54. Exner, N.; Treske, B.; Paquet, D.; Holmström, K.; Schiesling, C.; Gispert, S.; Carballo-Carbajal, I.; Berg, D.; Hoepken, H.-H.; Gasser, T.; et al. Loss-of-Function of Human PINK1 Results in Mitochondrial Pathology and Can Be Rescued by Parkin. *J. Neurosci.* **2007**, *27*, 12413–12418. [CrossRef] [PubMed]
55. Morais, V.A.; Verstreken, P.; Roethig, A.; Smet, J.; Snellinx, A.; Vanbrabant, M.; Haddad, D.; Frezza, C.; Mandemakers, W.; Vogt-Weisenhorn, D.; et al. Parkinson's disease mutations in PINK1 result in decreased Complex I activity and deficient synaptic function. *EMBO Mol. Med.* **2009**, *1*, 99–111. [CrossRef]
56. Matsuda, N.; Sato, S.; Shiba, K.; Okatsu, K.; Saisho, K.; Gautier, C.A.; Sou, Y.-S.; Saiki, S.; Kawajiri, S.; Sato, F.; et al. PINK1 stabilized by mitochondrial depolarization recruits Parkin to damaged mitochondria and activates latent Parkin for mitophagy. *J. Cell Biol.* **2010**, *189*, 211–221. [CrossRef]
57. Narendra, D.P.; Jin, S.M.; Tanaka, A.; Suen, D.-F.; Gautier, C.A.; Shen, J.; Cookson, M.R.; Youle, R.J. PINK1 Is Selectively Stabilized on Impaired Mitochondria to Activate Parkin. *PLoS Biol.* **2010**, *8*, e1000298. [CrossRef]
58. Okatsu, K.; Oka, T.; Iguchi, M.; Imamura, K.; Kosako, H.; Tani, N.; Kimura, M.; Go, E.; Koyano, F.; Funayama, M.; et al. PINK1 autophosphorylation upon membrane potential dissipation is essential for Parkin recruitment to damaged mitochondria. *Nat. Commun.* **2012**, *3*, 1016. [CrossRef]
59. Okatsu, K.; Uno, M.; Koyano, F.; Go, E.; Kimura, M.; Oka, T.; Tanaka, K.; Matsuda, N. A Dimeric PINK1-containing Complex on Depolarized Mitochondria Stimulates Parkin Recruitment. *J. Biol. Chem.* **2013**, *288*, 36372–36384. [CrossRef]
60. Kitada, T.; Aakawa, S.; Hattori, N.; Matsumine, H.; Yokochi, M.; Mizuno, Y.; Shimizu, N. Mutations in the *parkin* gene cause autosolmal recessive juvenile parkinsonism. *Nat. Lett.* **1998**, *169*, 166–169. [CrossRef]
61. Shimura, H.; Hattori, N.; Kubo, S.-I.; Mizuno, Y.; Asakawa, S.; Minoshima, S.; Shimizu, N.; Iwai, K.; Chiba, T.; Tanaka, K.; et al. Familial Parkinson disease gene product, parkin, is a ubiquitin-protein ligase. *Nat. Genet.* **2000**, *25*, 302–305. [CrossRef]
62. Trempe, J.-F.; Fon, E.A. Structure and Function of Parkin, PINK1, and DJ-1, the Three Musketeers of Neuroprotection. *Front. Neurol.* **2013**, *4*, 38. [CrossRef] [PubMed]
63. Wauer, T.; Komander, D. Structure of the human Parkin ligase domain in an autoinhibited state. *EMBO J.* **2013**, *32*, 2099–2112. [CrossRef]

64. Kane, L.A.; Lazarou, M.; Fogel, A.I.; Li, Y.; Yamano, K.; Sarraf, S.A.; Banerjee, S.; Youle, R.J. PINK1 phosphorylates ubiquitin to activate Parkin E3 ubiquitin ligase activity. *J. Cell Biol.* **2014**, *205*, 143–153. [CrossRef]
65. Koyano, F.; Okatsu, K.; Kosako, H.; Tamura, Y.; Go, E.; Kimura, M.; Kimura, Y.; Tsuchiya, H.; Yoshihara, H.; Hirokawa, T.; et al. Ubiquitin is phosphorylated by PINK1 to activate parkin. *Nature* **2014**, *510*, 162–166. [CrossRef] [PubMed]
66. Narendra, D.; Tanaka, A.; Suen, D.-F.; Youle, R.J. Parkin is recruited selectively to impaired mitochondria and promotes their autophagy. *J. Cell Biol.* **2008**, *183*, 795–803. [CrossRef] [PubMed]
67. Ordureau, A.; Sarraf, S.A.; Duda, D.M.; Heo, J.-M.; Jedrychowski, M.P.; Sviderskiy, V.O.; Olszewski, J.L.; Koerber, J.T.; Xie, T.; Beausoleil, S.A.; et al. Quantitative Proteomics Reveal a Feedforward Mechanism for Mitochondrial PARKIN Translocation and Ubiquitin Chain Synthesis. *Mol. Cell* **2014**, *56*, 360–375. [CrossRef] [PubMed]
68. Wong, Y.C.; Holzbaur, E.L.F. Optineurin is an autophagy receptor for damaged mitochondria in parkin-mediated mitophagy that is disrupted by an ALS-linked mutation. *Proc. Natl. Acad. Sci. USA* **2014**, *111*, E4439–E4448. [CrossRef] [PubMed]
69. Lazarou, M.; Sliter, D.A.; Kane, L.A.; Sarraf, S.A.; Wang, C.; Burman, J.L.; Sideris, D.P.; Fogel, A.I.; Youle, R.J. The ubiquitin kinase PINK1 recruits autophagy receptors to induce mitophagy. *Nature* **2015**, *524*, 309–314. [CrossRef]
70. Heo, J.M.; Ordureau, A.; Paulo, J.A.; Rinehart, J.; Harper, J.W. The PINK1-PARKIN Mitochondrial Ubiquitylation Pathway Drives a Program of OPTN/NDP52 Recruitment and TBK1 Activation to Promote Mitophagy. *Mol. Cell* **2015**, *60*, 7–20. [CrossRef]
71. Geisler, S.; Holmström, K.M.; Skujat, D.; Fiesel, F.C.; Rothfuss, O.C.; Kahle, P.J.; Springer, W. PINK1/Parkin-mediated mitophagy is dependent on VDAC1 and p62/SQSTM1. *Nat. Cell Biol.* **2010**, *12*, 119–131. [CrossRef]
72. Narendra, D.P.; Kane, L.A.; Hauser, D.N.; Fearnley, I.M.; Youle, R.J. p62/SQSTM1 is required for Parkin-induced mitochondrial clustering but not mitophagy; VDAC1 is dispensable for both. *Autophagy* **2010**, *6*, 1090–1106. [CrossRef]
73. Okatsu, K.; Saisho, K.; Shimanuki, M.; Nakada, K.; Shitara, H.; Sou, Y.-S.; Kimura, M.; Sato, S.; Hattori, N.; Komatsu, M.; et al. P62/SQSTM1 cooperates with Parkin for perinuclear clustering of depolarized mitochondria. *Genes Cells* **2010**, *15*, 887–900. [CrossRef] [PubMed]
74. Gegg, M.E.; Cooper, J.M.; Chau, K.-Y.; Rojo, M.; Schapira, A.H.; Taanman, J.-W. Mitofusin 1 and mitofusin 2 are ubiquitinated in a PINK1/parkin-dependent manner upon induction of mitophagy. *Hum. Mol. Genet.* **2010**, *19*, 4861–4870. [CrossRef]
75. Ziviani, E.; Tao, R.N.; Whitworth, A.J. Drosophila Parkin requires PINK1 for mitochondrial translocation and ubiquitinates Mitofusin. *Proc. Natl. Acad. Sci. USA* **2010**, *107*, 5018–5023. [CrossRef] [PubMed]
76. Birsa, N.; Norkett, R.; Wauer, T.; Mevissen, T.E.T.; Wu, H.-C.; Foltynie, T.; Bhatia, K.; Hirst, W.D.; Komander, D.; Plun-Favreau, H.; et al. Lysine 27 Ubiquitination of the Mitochondrial Transport Protein Miro Is Dependent on Serine 65 of the Parkin Ubiquitin Ligase. *J. Biol. Chem.* **2014**, *289*, 14569–14582. [CrossRef] [PubMed]
77. Safiulina, D.; Kuum, M.; Choubey, V.; Gogichaishvili, N.; Liiv, J.; Hickey, M.A.; Cagalinec, M.; Mandel, M.; Zeb, A.; Liiv, M.; et al. Miro proteins prime mitochondria for Parkin translocation and mitophagy. *EMBO J.* **2019**, *38*, 1–18. [CrossRef] [PubMed]
78. Lazarou, M.; Jin, S.M.; Kane, L.A.; Youle, R.J. Role of PINK1 Binding to the TOM Complex and Alternate Intracellular Membranes in Recruitment and Activation of the E3 Ligase Parkin. *Dev. Cell* **2012**, *22*, 320–333. [CrossRef]
79. Koyano, F.; Yamano, K.; Kosako, H.; Kimura, Y.; Kimura, M.; Fujiki, Y.; Tanaka, K.; Matsuda, N. Parkin-mediated ubiquitylation redistributes MITOL/March5 from mitochondria to peroxisomes. *EMBO Rep.* **2019**, *20*, e47728. [CrossRef] [PubMed]
80. Kazlauskaite, A.; Martínez-Torres, R.J.; Wilkie, S.; Kumar, A.; Peltier, J.; Gonzalez, A.; Johnson, C.; Zhang, J.; Hope, A.G.; Peggie, M.; et al. Binding to serine 65-phosphorylated ubiquitin primes Parkin for optimal PINK 1-dependent phosphorylation and activation. *EMBO Rep.* **2015**, *16*, 939–954. [CrossRef]
81. Chen, Y.; Dorn, G.W. PINK1-phosphorylated mitofusin 2 is a Parkin receptor for culling damaged mitochondria. *Science* **2013**, *340*, 471–475. [CrossRef]
82. Arena, G.; Gelmetti, V.; Torosantucci, L.; Vignone, D.; Lamorte, G.; De Rosa, P.; Cilia, E.; Jonas, E.A.; Valente, E.M. PINK1 protects against cell death induced by mitochondrial depolarization, by phosphorylating Bcl-xL and impairing its pro-apoptotic cleavage. *Cell Death Differ.* **2013**, *20*, 920–930. [CrossRef]
83. Adams, J.M.; Cory, S. Bcl-2-regulated apoptosis: Mechanism and therapeutic potential. *Curr. Opin. Immunol.* **2007**, *19*, 488–496. [CrossRef]
84. Zhu, J.-H.; Guo, F.; Shelburne, J.; Watkins, S.; Chu, C.T. Localization of Phosphorylated ERK/MAP Kinases to Mitochondria and Autophagosomes in Lewy Body Diseases. *Brain Pathol.* **2003**, *13*, 473–481. [CrossRef]
85. Dagda, R.K.; Cherra, S.J.; Kulich, S.M.; Tandon, A.; Park, D.; Chu, C.T. Loss of PINK1 Function Promotes Mitophagy through Effects on Oxidative Stress and Mitochondrial Fission. *J. Biol. Chem.* **2009**, *284*, 13843–13855. [CrossRef]
86. Chinta, S.J.; Mallajosyula, J.K.; Rane, A.; Andersen, J.K. Mitochondrial alpha-synuclein accumulation impairs complex I function in dopaminergic neurons and results in increased mitophagy in vivo. *Neurosci. Lett.* **2010**, *486*, 235–239. [CrossRef]
87. Osellame, L.D.; Rahim, A.A.; Hargreaves, I.P.; Gegg, M.E.; Richard-Londt, A.; Brandner, S.; Waddington, S.N.; Schapira, A.H.; Duchen, M.R. Mitochondria and Quality Control Defects in a Mouse Model of Gaucher Disease—Links to Parkinson's Disease. *Cell Metab.* **2013**, *17*, 941–953. [CrossRef] [PubMed]
88. Koentjoro, B.; Park, J.-S.; Sue, C.M. Nix restores mitophagy and mitochondrial function to protect against PINK1/Parkin-related Parkinson's disease. *Sci. Rep.* **2017**, *7*, srep44373. [CrossRef] [PubMed]
89. McWilliams, T.G.; Prescott, A.R.; Montava-Garriga, L.; Ball, G.; Singh, F.; Barini, E.; Muqit, M.M.; Brooks, S.P.; Ganley, I.G. Basal Mitophagy Occurs Independently of PINK1 in Mouse Tissues of High Metabolic Demand. *Cell Metab.* **2018**, *27*, 439–449.e5. [CrossRef] [PubMed]

 life

Article

Exploring the Ability of LARS2 Carboxy-Terminal Domain in Rescuing the MELAS Phenotype

Francesco Capriglia [1,†], Francesca Rizzo [1,†], Giuseppe Petrosillo [2], Veronica Morea [3], Giulia d'Amati [4], Palmiro Cantatore [1], Marina Roberti [1,5], Paola Loguercio Polosa [1,5,*,‡] and Francesco Bruni [1,*,‡]

1 Department of Biosciences, Biotechnologies and Biopharmaceutics, University of Bari Aldo Moro, 70125 Bari, Italy; francesco.capriglia@uniba.it (F.C.); francesca.rizzo1@uniba.it (F.R.); palmiro.cantatore@uniba.it (P.C.); marina.roberti@uniba.it (M.R.)
2 Institute of Biomembranes, Bioenergetics and Molecular Biotechnologies (IBIOM), National Research Council (CNR), 70125 Bari, Italy; g.petrosillo@ibbe.cnr.it
3 Institute of Molecular Biology and Pathology (IMBP), National Research Council (CNR), 00185 Rome, Italy; veronica.morea@cnr.it
4 Department of Radiological, Oncological and Pathological Sciences, Sapienza University of Rome, 00185 Rome, Italy; giulia.damati@uniroma1.it
5 Consorzio Interuniversitario Biotecnologie (CIB), 34149 Trieste, Italy
* Correspondence: paolaannamaria.loguerciopolosa@uniba.it (P.L.P.); francesco.bruni@uniba.it (F.B.)
† These authors contributed equally to this work.
‡ These authors contributed equally to this work.

Abstract: The m.3243A>G mutation within the mitochondrial mt-tRNALeu$^{(UUR)}$ gene is the most prevalent variant linked to mitochondrial encephalopathy with lactic acidosis and stroke-like episodes (MELAS) syndrome. This pathogenic mutation causes severe impairment of mitochondrial protein synthesis due to alterations of the mutated tRNA, such as reduced aminoacylation and a lack of post-transcriptional modification. In transmitochondrial cybrids, overexpression of human mitochondrial leucyl-tRNA synthetase (LARS2) has proven effective in rescuing the phenotype associated with m.3243A>G substitution. The rescuing activity resides in the carboxy-terminal domain (Cterm) of the enzyme; however, the precise molecular mechanisms underlying this process have not been fully elucidated. To deepen our knowledge on the rescuing mechanisms, we demonstrated the interactions of the Cterm with mutated mt-tRNALeu$^{(UUR)}$ and its precursor in MELAS cybrids. Further, the effect of Cterm expression on mitochondrial functions was evaluated. We found that Cterm ameliorates de novo mitochondrial protein synthesis, whilst it has no effect on mt-tRNALeu$^{(UUR)}$ steady-state levels and aminoacylation. Despite the complete recovery of cell viability and the increase in mitochondrial translation, Cterm-overexpressing cybrids were not able to recover bioenergetic competence. These data suggest that, in our MELAS cell model, the beneficial effect of Cterm may be mediated by factors that are independent of the mitochondrial bioenergetics.

Keywords: Cterm; MELAS; transmitochondrial cybrids; aminoacyl-tRNA synthetases; LARS2; mitochondrial disease; therapeutic peptides

Citation: Capriglia, F.; Rizzo, F.; Petrosillo, G.; Morea, V.; d'Amati, G.; Cantatore, P.; Roberti, M.; Loguercio Polosa, P.; Bruni, F. Exploring the Ability of LARS2 Carboxy-Terminal Domain in Rescuing the MELAS Phenotype. *Life* **2021**, *11*, 674. https://doi.org/10.3390/life11070674

Academic Editor: Pascale Bélenguer

Received: 22 June 2021
Accepted: 8 July 2021
Published: 10 July 2021

1. Introduction

Mitochondrial encephalomyopathies are a group of complex and clinically heterogeneous metabolic disorders caused by a wide spectrum of mutations either in mitochondrial DNA (mtDNA) or in nuclear genes encoding proteins that control mitochondrial functions [1]. Such genetic defects predominantly affect the mitochondrial oxidative phosphorylation system (OXPHOS), which is responsible for most of the ATP supply in cells. OXPHOS is located in the mitochondrial inner membrane and consists of five complexes, denoted as complex I–V. Complex I, III, and IV generate a proton-motive force across the inner membrane, which is used by F1Fo-ATP synthase (complex V) to synthesize ATP. Only 13 subunits of the OXPHOS system are encoded by mtDNA, together with 2 rRNAs and 22 tRNAs necessary for mitochondrial translation machinery. The rest of the OXPHOS subunits, as well as the other proteins that constitute the vast majority of the mitochondrial proteome, are nucleus-encoded and transported into mitochondria [2,3].

Mutations in the mt-tRNA genes are associated with several mitochondrial pathologies. An example is the well-documented mitochondrial encephalopathy with lactic acidosis and stroke-like episodes (MELAS) syndrome. MELAS patients typically develop encephalopathy, stroke-like episodes before the age of 40, and lactic acidosis. Additional clinical features involve other neurological symptoms, exercise intolerance, cardiomyopathy, deafness, and diabetes. About 80% of MELAS syndromes are caused by the m.3243A>G mutation in the mt-tRNALeu$^{(UUR)}$ gene [4–6]. The base substitution causes destabilization of the mutated mt-tRNALeu$^{(UUR)}$, thus adversely affecting stability, aminoacylation, and the addition of a taurinomethyl group to the wobble uridine (τm^5U34) [7,8]. These defects contribute to the overall reduction in mitochondrial protein synthesis observed in MELAS disease [9–11].

Despite the increasing knowledge on MELAS pathogenesis, an effective therapy for this and other mitochondrial diseases is still far from being available. As a possible strategy to overcome defects of mt-tRNAs, several molecular approaches have been proposed. One of these consists of the overexpression of human mitochondrial cognate or non-cognate aminoacyl-tRNA synthetases (aaRSs), which proved to rescue the defective viability and energetic competence as well as mitochondrial protein synthesis in transmitochondrial cybrid cell lines, a well-established cellular model of mt-tRNA mutations [12,13]. It was shown in MELAS cybrids that the rescuing activity of human mitochondrial leucyl-tRNA synthetase (LARS2) resided in the non-catalytic carboxy-terminal domain (Cterm, 67 residues long). A more detailed analysis ascribed the rescue capacity of the Cterm fragment to two short β-strand regions, denoted as peptide β30_31 (15 residues in length) and peptide β32_33 (16 residues in length). The correction activity was observed following both overexpression and exogenous administration to cells carrying the m.3243A>G MELAS mutation in the mt-tRNALeu$^{(UUR)}$, as well as the m.8344A>G MERRF mutation in mt-tRNALys. To explain the rescuing effect of the Cterm and its peptides, a chaperonic activity towards mutated tRNAs were proposed on the basis of in vitro evidences, which could result in the stabilization of a wild-type-like conformation of the tRNA [14–17].

In view of the development of a therapeutic strategy based on the use of Cterm-related small molecules, we decided to deepen our knowledge on the mechanisms that underlie the reported rescuing activity of the Cterm by using the cybrid system. Here, we provide evidence that the overexpressed Cterm domain is able to contact the cognate mutated tRNA in cultured cells, thus corroborating the "chaperonic" hypothesis previously formulated mainly on the basis of in vitro findings. We also evaluated the effect of Cterm overexpression on a range of mitochondrial processes such as mitochondrial translation, steady-state level and aminoacylation efficiency of mt-tRNALeu$^{(UUR)}$, bioenergetics, and mitophagy. In agreement with previous findings [14–17], a complete recovery of cell viability was observed. Unexpectedly, despite an improvement of protein synthesis, the other analysed mitochondrial processes were not affected by Cterm expression.

2. Materials and Methods

2.1. Tissue Culture, Transfections and Viability Assay

For RNA immunoprecipitation (RIP) experiments, we used previously established osteosarcoma-derived (143B.TK$^-$) cybrid cell lines from MELAS patients and controls [16]. For all the other experiments, we used MELAS cells provided by Prof. A. Filipovska (University of Western Australia). Cybrids were cultured (37 °C, humidified 5% CO_2) in Dulbecco's Modified Eagle's Medium (DMEM, Lonza, Basel, Switzerland) supplemented with 10% (v/v) foetal bovine serum (FBS, ThermoFisher Scientific, Carlsbad, CA, USA), 1 mM sodium pyruvate, 1× non-essential amino acids, 50 μg/mL uridine, and 1× Antibiotic-Antimycotic (Euroclone, Milan, Italy).

The recombinant expression vector pcDNA6.2 bearing FLAG-tagged Cterm for transient transfections was prepared as follows. The insert was obtained by PCR using the plasmid MTS-Cterm-FLAG [16] as a template and the following primers: For 5′-CACCATGTCCGTCCTGACGCCG-3′ and Rev 5′-CTACTTATCGTCGTCATCCTTGTAATC-3′. The amplicon was cloned into the pcDNA6.2/V5/GW/D-TOPO® vector (Invitrogen, Waltham, MA, USA) according to manufacturer's instructions. Both Cterm and mock (empty vector) constructs were confirmed by sequencing. Transfections were carried out using the DNA constructs and Lipofectamine™ 3000 Transfection Reagent (ThermoFisher Scientific, Carlsbad, CA, USA) following manufacturer's recommendations.

For the cell viability assay, cybrids were transfected in 6-well plates and maintained in glucose standard medium. After 24 h of transfection, cells were harvested, seeded at a lower confluency (1.5×10^5 cells/well), and grown for an additional 24 h in either standard medium or galactose medium [glucose-free DMEM (Sigma Aldrich, St. Louis, MO, USA) supplemented with 5 mM galactose, 1% sodium pyruvate, and 10% FBS]. Finally, cells were harvested and counted in triplicate by a TC20TM automated cell counter (Bio-Rad, Hercules, CA, USA) in the presence of trypan blue solution (Sigma Aldrich, St. Louis, MO, USA).

2.2. Heteroplasmy Determination by PCR/RFLP Analysis

Total DNA was extracted from the cybrids using the Wizard® Genomic DNA Purification Kit (Promega, Madison, WI, USA) following manufacturer's recommendations. For RFLP analysis, the mtDNA 3243 locus was amplified using primers flanking the mutation, as follows: For 5′-CCTCGGAGCAGAACCCAACCT-3′ and Rev 5′-CGAAGGGTTGTAGTAGCCCGT-3′. PCR products were digested by Apa I, separated by a 1.5% agarose gel, and stained with GelRed® (Biotium, Fremont, CA, USA). Heteroplasmy levels were determined as a proportion of mutant (digested) to wild type (undigested) mtDNA and measured by ImageQuant TL software (GE Healthcare Life Sciences, Marlborough, MA, USA).

2.3. RNA Extraction and RT-qPCR

Total and immunoprecipitated RNA was extracted with TRIsure™ reagent (Meridian Bioscience, Cincinnati, OH, USA) and concentration was measured by NanoDrop™ 1000 (ThermoFisher Scientific, Carlsbad, CA, USA). For RT-qPCR, RNA samples were reverse-transcribed using the High-Capacity cDNA Reverse Transcription Kit (ThermoFisher Scientific, Carlsbad, CA, USA), according to the manufacturer's instructions. Primers, as reported in Table 1, were designed using the NCBI Primer-BLAST tool; amplification reactions were performed using the SsoAdvanced Universal SYBR® Green Supermix (Bio-Rad, Hercules, CA, USA) and analysed with the Applied Biosystem 7500 Fast Real-Time PCR System (ThermoFisher Scientific, Carlsbad, CA, USA). Relative quantification of the qPCR products was achieved by comparative Ct method; statistical analysis was performed using the unpaired two-tailed Student's *t*-test.

2.4. Whole Cell Extracts Preparation and Mitochondria Isolation

Whole cell extracts were obtained by solubilization in RIPA buffer (25 mM Tris-HCl, pH 7.6; 150 mM NaCl; 1% NP-40; 1% sodium deoxycholate; 1% SDS) with protease inhibitor cocktail (Sigma Aldrich, St. Louis, MO, USA). Extracts were cleared by 5 min centrifugation at 10,000× *g*, snap-frozen, and stored at −80 °C.

The mitochondria were isolated using differential centrifugation. Briefly, confluent cells ($6–10 \times 10^7$) were washed twice in cold PBS, harvested by scraping, and pelleted at 800× *g* for 7 min at 4 °C. Cells were broken by adding 10 mL of cold mitochondria isolation buffer (MIB: 20 mM HEPES, pH 7.6; 220 mM mannitol; 70 mM sucrose; 1 mM EDTA; 2 mg/mL BSA; 1 mM AEBSF), left on ice for 20 min, and homogenized by 15 strokes using a Thomas homogenizer with a motor-driven Teflon pestle. Homogenate was centrifuged at 800× *g* for 5 min at 4 °C and the supernatant was subsequently spun at 10,000× *g* for 10 min at 4 °C. Crude mitochondrial pellets were washed by suspending in 20 mL of MIB without BSA and centrifuged at 10,000× *g* for 10 min at 4 °C. The mitochondria were suspended in a small volume of MIB without BSA, and protein concentration was determined by the Bradford protein assay. The mitochondria were snap-frozen in liquid nitrogen and stored at −80 °C.

Table 1. Primer sequences used for RT-qPCR analyses.

Amplicon	Primer Sequence
mt-tRNALeu$^{(UUR)}$	For: GTTAAGATGGCAGAGCCC Rev: GAAGAGGAATTGAACCTCTGAC
RNA19	For: TATACCCACACCCACCCAAG Rev: GCGATTAGAATGGGTACAAT
mt-tRNALys	For: ATAGGGCCCGTATTTACCCTA Rev: ATACGGTAGTATTTAGTTGG
mt-tRNATyr	For: TGGTAAAAAGAGGCCTAACCC Rev: ATGGCTGAGTGAAGCATTGG
ND4 mRNA	For: CCATTCTCCTCCTATCCCTCAAC Rev: CACAATCTGATGTTTTGGTTAAAC
12S rRNA	For: ACACTACGAGCCACAGCT Rev: GCTACACCTTGACCTAACGTC
Cterm	For: AAATTCCTGTGCCCCAACAA Rev: CTACTTATCGTCGTCATCCT
18S rRNA	For: GTAACCCGTTGAACCCCATT Rev: CCATCCAATCGGTAGTAGCG

2.5. *RNA Immunoprecipitation*

The MELAS cybrids were stably transfected with either MTS-Cterm-FLAG construct or an empty vector (mock) [16] (90% confluency), grown in 2 × 500 cm^2 square dishes (Corning, Corning, NY, USA), washed with PBS, crosslinked in 1% formaldehyde-PBS solution for 10 min at room temperature, and then treated with 0.125 M glycine, pH 7.0, for 5 min. The mitochondria were isolated as described above, using a different mitochondria isolation buffer (10 mM Tris-HCl, pH 7.4; 600 mM mannitol; 1 mM EGTA; 0.1% BSA; 1 mM PMSF).

Mitochondrial proteins (600 µg) were immunoprecipitated by anti-FLAG M2 affinity gel (Sigma Aldrich, St. Louis, MO, USA), as reported elsewhere [18]. After crosslink reversion, immunoprecipitated RNA was extracted as described above; isopropanol precipitation was carried out overnight in the presence of 5 µg of glycogen (Roche, Basel, Switzerland) as a carrier. Mitochondrial RNAs in the immunoprecipitate were identified and quantified by RT-qPCR.

2.6. *SDS-PAGE, BN-PAGE and Immunoblotting*

For SDS-PAGE, whole cell proteins (40–100 µg, depending on the proteins being probed) were solubilized in 1× Laemmli buffer and separated by 12% Tris-Glycine -SDS minigels.

Blue-native PAGE (BN-PAGE) analysis was performed using NativePAGE Novex Bis-Tris Gel System (Invitrogen, Waltham, MA, USA) according to the manufacturer's recommendations. The mitochondrial pellets (40 µg) were solubilized in 50 µL of the NativePAGE Sample Buffer containing dodecylmaltoside (detergent/protein ratio 6:1) and 1 mM AEBSF. After 20 min on ice, samples were centrifuged at 20,000× *g* for 30 min at 4 °C. Supernatants were supplemented with NativePAGE G250 Sample Additive (concentration in the sample was 1/4th to 1/2nd of the detergent concentration) and 15 µg/lane were fractionated through 3–12% NativePAGE Novex Bis-Tris gel at 4 °C.

For western blotting, gels were electro-transferred at 4 °C for 2 h (OXPHOS subunits) or over-night (all the other proteins) onto polyvinylidene difluoride (PVDF) membranes

(Millipore, Burlington, MA, USA). Immunoblotting was performed according to standard techniques. Primary antibodies (all from Abcam, Cambridge, UK; diluted 1:1000 except where differently indicated) were as follows: total OXPHOS antibody cocktail (1:250), anti-BNIP3, anti-BNIP3L/NIX, anti-citrate synthase, anti-SDHA, anti-COX I, anti-ATP5A, anti-UQCRC2, and anti-NDUFA9. Detection was performed with the HRP-conjugated secondary antibody (Bio-Rad, Hercules, CA, USA). Chemiluminescent detection was achieved using Amersham ECL™ Prime Western blotting detection reagent (GE Healthcare Life Sciences, Marlborough, MA, USA) or Clarity Western ECL substrate (Bio-Rad, Hercules, CA, USA); signals were revealed by ChemiDoc MP Imaging System (Bio-Rad, Hercules, CA, USA).

2.7. [^{35}S]-Labelling of Mitochondrial Translation Products

Exponentially growing cells in a 12-well plate were washed twice with methionine-free DMEM and incubated for 1 h at 37 °C with 300 μCi/mL [^{35}S]-methionine (Perkin Elmer, Waltham, MA, USA) in 300 μL of methionine-free DMEM supplemented with 10% dialysed FBS, emetine (100 μg/mL), and cycloheximide (100 μg/mL). After the radioactive pulse, cells were washed twice with phosphate-buffered saline (PBS) and finally dissolved in denaturing buffer (3% SDS; 60 mM Tris-HCl, pH 8; 10 mM sucrose, 2 mM EDTA, 5% β-mercaptoethanol). Aliquots (20 μg) of total cell proteins were separated by 15% SDS-polyacrylamide gels and signals were detected by Typhoon FLA 9500 phosphorimager (GE Healthcare Life Sciences, Marlborough, MA, USA). Assessment of protein loading was achieved by Coomassie blue staining.

2.8. High Resolution Northern Blot Analysis

High-resolution northerns were carried out as described in [19] with modifications. Total RNA (5 μg) was separated by 15% polyacrylamide-8 M urea gels and electroblotted to Hybond®-N+ nylon membranes (GE Healthcare Life Sciences, Marlborough, MA, USA). After transfer, RNA was crosslinked to the membranes at 37 °C for 1 h with freshly prepared EDAC [1-ethyl-3-(3-dimethylaminopropyl) carbodiimide, Sigma Aldrich, St. Louis, MO, USA] reagent. Probes (Table 2) were labelled with non-radioactive digoxigenin-dUTP using the DIG Oligonucleotide Tailing Kit, 2nd Generation (Roche, Basel, Switzerland), following the manufacturer's recommendations. Pre-hybridization, hybridization, washing, and anti-DIG-AP (Roche, Basel, Switzerland; 1:10,000) incubation were carried out, as detailed elsewhere [20]. Chemiluminescence signals were detected by ChemiDoc MP Imaging System (Bio-Rad, Hercules, CA, USA).

Table 2. Oligonucleotides used for DIG-northern analyses.

Target	Probe Sequence
mt-tRNALeu$^{(UUR)}$	TATGCGATTACCGGGCTCTGCCATCTTAAC
mt-tRNAGlu	TATTCTCGCACGGACTACAACCACGAC
5S rRNA	GGGTGGTATGGCCGTAGAC

For mt-tRNALeu$^{(UUR)}$ aminoacylation analysis, the same procedures as above were followed, except that total RNA was dissolved in acidic buffer (10 mM sodium acetate, 1 mM EDTA, pH 5.2) and electrophoresed at 4 °C through acid (pH 5.2) 12% polyacrylamide-8 M urea gels to separate charged and uncharged tRNA pools.

2.9. Complex IV Activity Determination

Mitochondrial cytochrome c oxidase (complex IV) activity was measured spectrophotometrically, essentially as described in [21] with some modifications. Cells were suspended in hypotonic 20 mM potassium phosphate buffer (pH 7.4) and subjected to three cycles of freezing and thawing. Enzymatic activity was measured at 37 °C in an assay mixture composed of 25 mM phosphate buffer (pH 7.0), 0.75 μM n-dodecyl-β-D-maltoside, and

50 μM reduced cytochrome c. The reaction started following the addition of 50 μg of cell protein in a final volume of 1 mL. The specific activity of the enzyme was expressed as nmol of cytochrome c oxidized/min per mg of cell protein. The cyanide-insensitive rate of cytochrome c oxidation was measured and subtracted.

2.10. Mitochondrial Oxygen Consumption Measurement

The mitochondrial oxygen consumption rate (OCR) was calculated by subtracting non-mitochondrial OCR from cellular OCR. This was measured with a Clark-type oxygen electrode (Hansatech Instruments Ltd., King's Lynn, UK). Oxygen consumption in intact cybrids was measured at 37 °C in 0.5 mL of fresh culture medium lacking glucose, supplemented with 10% foetal bovine serum and 1 mM sodium pyruvate, using a concentration of 2×10^6 cells/mL. Non-mitochondrial oxygen consumption was measured following the addition of 2 μM rotenone and 1.6 μM antimycin A. Results were expressed as nmol O_2/min per 10^6 cells.

2.11. Lactate Assay

L-lactate production in cell culture medium was measured by spectrophotometrically monitoring NADH formation following its oxidation to pyruvate in the presence of lactate dehydrogenase (LDH). Briefly, non-transfected and transfected cybrids were grown in 6-well plates, and media were collected and centrifuged at 10,000× *g* for 5 min. Aliquots of the supernatant (4–12 μL) were added to a reaction mixture containing 2.5 mM NAD^+, 320 mM glycine, pH 9.5, 320 mM hydrazine, and 5.8 U of bovine heart LDH (Sigma Aldrich, St. Louis, MO, USA). Reactions were carried out at 25 °C and followed for approximately 15 min.

3. Results

3.1. Cterm Domain Interacts with the Mutated mt-tRNALeu$^{(UUR)}$ in MELAS Cybrids

The previously reported rescuing activity of the LARS2 carboxy-terminal domain (Cterm) overexpressed in MELAS mutant cybrids was explained by assuming that Cterm is able to interact with mutated mt-tRNALeu$^{(UUR)}$ [14]. However, such interaction had only been characterised in vitro. Therefore, we decided to investigate the capacity of Cterm to bind mt-tRNALeu$^{(UUR)}$ in cultured cells. In addition, we explored whether this molecule is able to interact with other mitochondrial RNA species, since it is well documented that aaRSs also have the ability to bind non-tRNA substrates [22]. We employed a MELAS cybrid line with a >95% mutation load that stably overexpressed the FLAG-tagged Cterm construct. Firstly, we assessed the Cterm expression by RT-qPCR (Figure S1). Next, we performed RNA immunoprecipitation (RIP) analysis in mitochondrial lysates from MELAS cybrids to identify the RNA species specifically contacted by the Cterm domain. RNA was extracted from complexes immunoprecipitated via anti-FLAG M2 Affinity Gel, and the relative levels of mature mt-tRNALeu$^{(UUR)}$ and its precursor RNA19 (16S rRNA + mt-tRNALeu$^{(UUR)}$ + ND1 mRNA) as well as two mitochondrial tRNAs (mt-tRNALys and mt-tRNATyr), ND4 mRNA and 12S rRNA, were measured by RT-qPCR (Figure 1). In the case of mature mt-RNAs, primers were internal to the mature RNA sequences; for RNA19, primers were designed outside the mt-tRNALeu$^{(UUR)}$, with forward and reverse (see Table 1) placed on the 16S rRNA and ND1, respectively.

We found that the mature and precursor forms of mt-tRNALeu$^{(UUR)}$ exhibited a 6- and 8-fold enrichment, respectively, in immunoprecipitated RNA from Cterm-overexpressing cells with respect to cells transfected with an empty vector (mock) (Figure 1 and Figure S2). An enrichment was also measured for the other examined mitochondrial RNA species, although to a lesser extent. These results clearly demonstrate that: (i) a specific interaction occurs between the Cterm domain and its cognate target tRNALeu$^{(UUR)}$ in MELAS cybrids, in agreement with the direct interaction previously demonstrated to occur in vitro [14]; and (ii) the Cterm also interacts with other mt-tRNAs as well as different species of mt-RNAs.

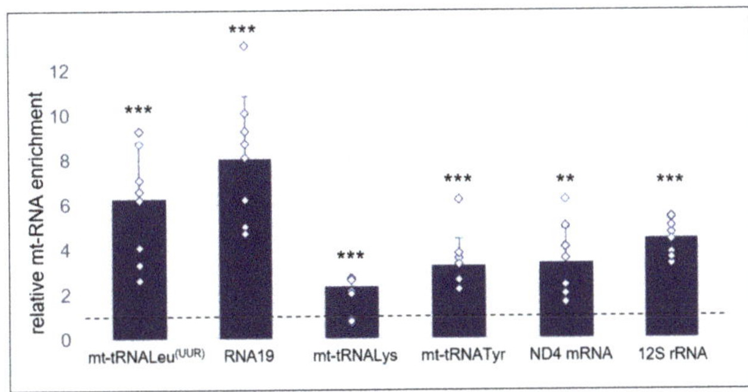

	mt-tRNALeu(UUR)	RNA19	mt-tRNALys	mt-tRNATyr	ND4 mRNA	12S rRNA
mt-tRNALeu(UUR)	-	0.32483	0.00378	0.00482	0.01739	0.28966
RNA19	0.32483	-	0.00002	0.00001	0.00005	0.00215

Figure 1. Cterm domain interacts with mutated mt-tRNALeu$^{(UUR)}$ in MELAS cybrids. Relative quantification of mt-RNA species by RT-qPCR was achieved on the Cterm complexes immunoprecipitated with anti-FLAG antibodies from MELAS cybrids stably expressing FLAG-tagged Cterm. Values are referred to immunoprecipitated RNAs obtained from MELAS cybrids stably transfected with empty vector (mock), fixed as 1-value (dashed line). The amount of analysed mt-RNA species in the input was used as endogenous control. Results are presented as the mean ± S.D. Statistical analysis was performed on three independent biological replicates using two-tailed Student's t test; asterisks (**, $p < 0.01$; ***, $p < 0.001$) indicate the significance of the enrichment of each analysed mt-RNA in Cterm-overexpressing cells with respect to mock. Individual data points are depicted by white diamond-shaped dots. Normal distribution of the six examined mt-RNAs was confirmed by the Shapiro–Wilk test; means were simultaneously compared by one-way ANOVA (α: 0.05). Lower table reports the post-hoc Tukey's test p-values relative to compared pairs of samples (red: statistically significant; green: not statistically significant).

3.2. *Cterm Rescues Viability and Mitochondrial Translation in MELAS Cybrids*

It is known that, in the vast majority of the cases, the MELAS 3243A>G mutation results in a molecular phenotype characterized by a defective mitochondrial protein synthesis [9–11]. Therefore, we decided to assess the impact of Cterm peptide on mitochondrial translation efficiency. Since the cybrid lines used for RIP experiments and in previous work [14–17] do not show defects in mitochondrial protein synthesis, from here on we employed a MELAS cybrid cell line (kindly provided by Prof. A. Filipovska, University of Western Australia) that displays a depressed mitochondrial translation phenotype (see below). In these cells, mtDNA heteroplasmy analysis indicated a 3243A>G mutation load of ~94% (Figure S3). As a first step, we tested how this system of MELAS cybrids responded to Cterm overexpression in terms of cell viability. The experiment was designed as follows. Cells were routinely propagated on glucose; non-transfected (wild-type or control and MELAS) and transiently transfected MELAS cells (empty vector or mock, and Cterm-FLAG overexpressing vector or Cterm) were transferred in either glucose or galactose medium. The latter forces cells to rely on the mitochondrial respiration for ATP synthesis through pyruvate oxidation because conversion of galactose to pyruvate yields no net ATP. On the contrary, the production of pyruvate via glycolytic metabolism of glucose yields two net ATP molecules [23]. Viable cells were counted after 24 h. As expected, a significant growth decline in galactose was observed for MELAS cells as compared to wild-type cybrids (Figure 2a). Most importantly, MELAS cybrids transfected with the Cterm-expressing vector showed a significant increase in viability with respect to mock and non-transfected cells. The restored viability was comparable to that of wild-type cells. For each transfection experiment, Cterm expression was assessed by RT-qPCR (Figure S4). These results indicate that Cterm is able to rescue the cell viability in line with what was previously reported on different MELAS cybrid cell lines [14,16].

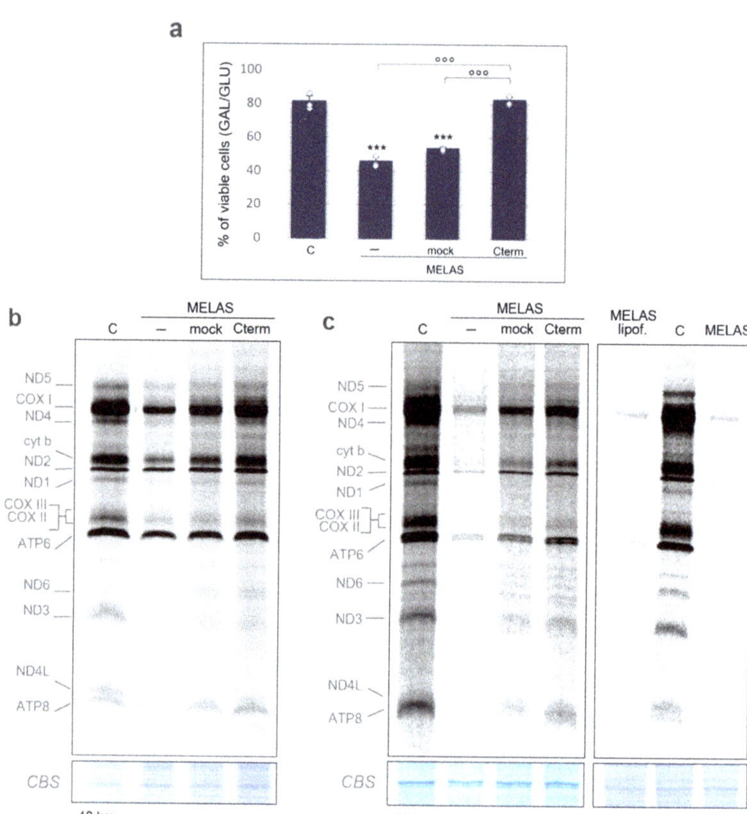

Figure 2. Cterm rescues viability and mitochondrial translation in MELAS cybrids. (**a**) Viability of wild-type control cybrids (C) and MELAS cybrids either non-transfected (-) or transiently transfected with an empty pcDNA6.2 vector (mock) or a Cterm-overexpressing vector (Cterm) evaluated after 24 h-incubation in galactose medium. The number of viable cells in galactose was normalized to the number of viable cells in glucose. Results are presented as the mean ± S.D. Statistical analysis was performed on three independent biological replicates using two-tailed Student's t test; asterisks (***, $p < 0.001$) indicate the significance of the decreased viability in MELAS cells with respect to the control. Normal distribution of MELAS samples was confirmed by the Shapiro–Wilk test and means were simultaneously compared by one-way ANOVA (α: 0.05) followed by post-hoc Tukey's test; degree symbols (°°°, $p < 0.001$) refer to the significance of the increase in cybrids overexpressing Cterm with respect to the non-transfected and mock. Individual data points are depicted by white diamond-shaped dots. (**b**) Metabolic [^{35}S]-methionine labelling of mitochondrial translation products performed in wild-type cybrids (C) and in MELAS cybrids detailed in (**a**). Pulse-labelling (1 h) was carried out 48 h after transfection; total cell protein (20 µg) were separated by 15% SDS-PAA gels. Mitochondrially encoded polypeptides were assigned, as in [24]. Coomassie blue staining (CBS) of the gel was used as loading control. (**c**) Metabolic [^{35}S]-methionine labelling as in (**b**), except that labelling was performed 72 h after transfection (left panel). Labelling of MELAS cells treated with the transfection reagent only (lipof.) was also carried out (right panel).

To assess the Cterm effect on de novo synthesis of mtDNA-encoded polypeptides, protein labelling was performed on 48- and 72-h transfected cells in the presence of [^{35}S]-

methionine and inhibitors of cytosolic translation cycloheximide and emetine. For the 72-h transfection, cells were subjected to additional stress by not replacing the growth medium; this caused medium acidification and accumulation of cellular metabolism by-products.

In all of these experiments, the protein labelling pattern of wild-type cells showed the typical profile of mitochondrial products (Figure 2b,c, Figures S5 and S6). A general consistent decrease in protein synthesis, which was much more pronounced after 72-h transfection, was observed in MELAS cybrids (Figure 2b,c and Figure S7). Interestingly, Cterm transient overexpression was able to significantly rescue most of the protein synthesis defect observed in MELAS cells. Unexpectedly, mock cells exhibited a partial recovery of the newly synthesized polypeptide level. The mock stimulating effect does not seem to depend on the pcDNA6.2 plasmid since two different wild-type plasmids, namely pBAD and pQE60, caused a similar increase (Figure S8). However, transfection with the vector-bearing Cterm construct clearly induced a better recovery of the deficit observed in MELAS cybrids. Finally, to rule out any stimulating effect of the transfection reagent, we performed a protein labelling experiment in the presence of lipofectamine only, showing no effect (Figure 2c, right panel).

3.3. *Cterm Has No Effect on mt-tRNALeu$^{(UUR)}$ Steady-State Level and Aminoacylation*

Additional distinctive features of cells carrying the MELAS mutation are the decreased steady-state level and aminoacylation efficiency of mt-tRNALeu$^{(UUR)}$, due to 3243A>G substitution that would cause structural destabilization of the mutated tRNA, making it prone to degradation. Thus, the interaction of Cterm with mt-tRNALeu$^{(UUR)}$ might serve to protect the tRNA against misfolding, preventing its degradation. In view of this, Cterm overexpression in MELAS cybrids would result in a recovery of the mt-tRNALeu$^{(UUR)}$ steady-state level. To test this hypothesis, we measured the mt-tRNALeu$^{(UUR)}$ content by high-resolution northern blot analysis. As shown in Figure 3a, the steady-state level of mt-tRNALeu$^{(UUR)}$ remarkably decreased in MELAS cells, with no recovery effect in the cells transfected with Cterm and mock as well as with pBAD or pQE60 plasmids (Figure S9). As a control, we showed that the level of mt-tRNAGlu was unchanged in all analysed conditions.

Since the observed increase in mitochondrial protein synthesis apparently did not seem to be related to the mt-tRNALeu$^{(UUR)}$ steady-state level, we investigated whether Cterm overexpression could somehow improve the aminoacylation efficiency of mutated mt-tRNALeu$^{(UUR)}$ by rendering the bound tRNA more accessible to the aminoacyl-tRNA synthetase. Our results, however, seem to rule out such a possibility because we did not find any increase in the level of mt-tRNALeu$^{(UUR)}$ aminoacylation in Cterm-overexpressing cybrids (Figure 3b).

3.4. *Cterm Does Not Affect Mitochondrial Bioenergetic Competence, Mitophagy and Mitochondrial Mass*

We then tested whether the observed rescuing effect of the Cterm domain on mitochondrial protein synthesis reflects into any improvement of the oxidative phosphorylation capacity. First, the steady-state level of the five subunits representative of the OXPHOS system was analysed by SDS-PAGE and immunoblotting (Figure 4a and Figure S10, upper panel).

MELAS cells showed a severe decrease in NDUFB8, a nuclear-encoded subunit of complex I and COX I, a mitochondrial-encoded subunit of complex IV, whilst the nuclear-encoded SDHB (complex II), UQCRC2 (complex III) and ATP5A (complex V) were unchanged. Cterm overexpression was unable to rescue the levels of both NDUFB8 and COX I subunits. Then, we measured the levels of the assembled respiratory chain complexes by immunoblotting of crude mitochondrial fractions solubilized with dodecylmaltoside and resolved by BN-PAGE (Figure 4b and Figure S10, lower panel). This analysis showed that complex IV was markedly diminished in MELAS cells, as was complex I alone and in association with complex III (I+III). This is mostly in line with the data reported in literature, showing that combined complexes I-IV deficiency is a general feature of MELAS patients carrying the m.3243A>G mutation [26]. Upon Cterm overexpression, no increase in the content of the affected respiratory complexes was detected, which is in accordance with the immunoblot data relative to the OXPHOS subunits after SDS-PAGE (see Figure 4a). Immunodetection of complex V in wild-type control cells using antibodies against subunit alpha (ATP5A) revealed the presence of complex V holoenzyme along with a very small amount of F1-containing sub-complexes, most likely V* (F1-ATPase with several c-subunits) and F1-ATPase domain alone (Figure 4b). We observed a similar complex V profile in the MELAS mitochondrial fraction, except that a marked increase in the sub-complex V* was found. Furthermore, Cterm overexpression did not cause any change in the complex V profile in MELAS cells.

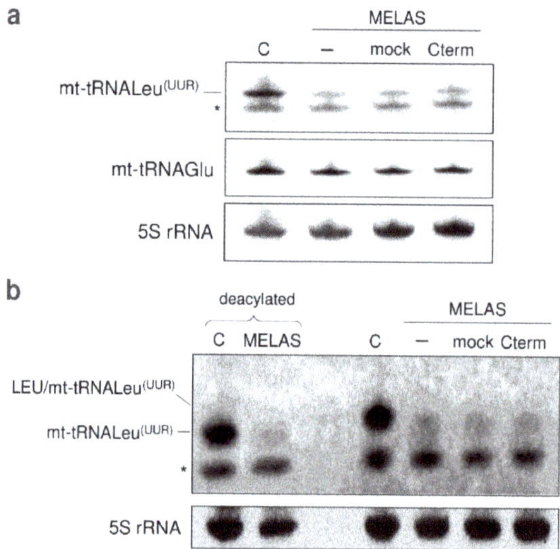

Figure 3. Cterm has no effect on mt-tRNALeu$^{(UUR)}$ steady-state level and aminoacylation. (**a**) Steady-state levels of mt-tRNALeu$^{(UUR)}$ and mt-tRNAGlu in control (C) and MELAS cybrids, either non-transfected (-) or transfected with empty pcDNA6.2 vector (mock) or Cterm-overexpressing vector (Cterm), were determined by high-resolution northern blot analysis with the indicated specific oligonucleotide probes. Each of the three DIG-labelled oligonucleotides detected a band corresponding to the size expected for mt-tRNALeu$^{(UUR)}$ (78 nt), mt-tRNAGlu (72 nt), and 5S rRNA (120 nt, used as loading control), respectively. The asterisk (*) indicates a non-specific band detected by mt-tRNALeu$^{(UUR)}$ probe. (**b**) The aminoacylation levels of mt-tRNALeu$^{(UUR)}$ in wild-type cybrids (C) and MELAS cybrids [same as described in (a)] were estimated by acidic high resolution northern blot analysis. DIG-labelled oligos, as in (a), were probed to mt-tRNALeu$^{(UUR)}$ and 5S rRNA. Positions of aminoacylated and deacylated tRNA (80 °C for 15 min at pH 8) are indicated on the left and are in agreement with Chomyn et al. [25].

We assessed COX activity in control and MELAS cells by spectrophotometric assay (Figure 4c), detecting a decrease in MELAS cybrids as expected, given the strong decline in assembled CIV (see Figure 4a). Following overexpression of Cterm, there was no recovery of COX activity in MELAS cells. In line with these data, the mitochondrial oxygen consumption rate of Cterm-transfected cybrids did not increase (Figure 4d); furthermore, lactate production, which was higher in MELAS cells given their prevalent glycolytic metabolism, was not affected by Cterm overexpression (Figure 4e).

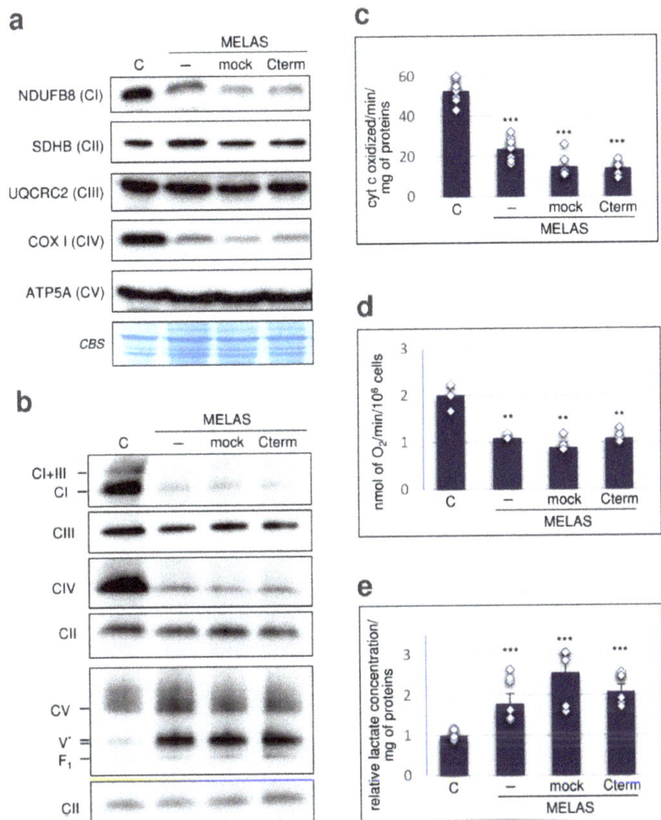

Figure 4. Cterm does not improve mitochondrial bioenergetic competence. (**a**) SDS-PAGE western blot analysis of representative subunits of complex I (NDUFB8), II (SDHB), III (UQCRC2), IV (COX I, mtDNA-encoded) and V (ATP5A or alpha subunit). Immunoblot was carried out using total OXPHOS antibody cocktail (Abcam). Coomassie blue staining (CBS) of the membrane was used as transfer control. (**b**) Blue-native PAGE western blot analysis of the respiratory chain assembled complexes in cybrid cells. Holoenzyme complexes and sub-complexes were visualized by single primary antibodies, as follows: NDUFA9 for complex I, UQCRC2 for complex III, COX I for complex IV, ATP5A for complex V, and SDHA for complex II (used as loading reference). F1-ATPase domain alone (F1) and with several c-subunits (V*) are also indicated. (**c**) Measurement of cytochrome c oxidase (complex IV) activity in whole cybrid cells. Enzymatic activity is expressed as nmol of cytochrome c oxidized/min per mg of proteins. (**d**) Mitochondrial oxygen consumption rate in intact cybrids. Each value is expressed as nmol of O_2/min per 10^6 cells. (**e**) Extracellular lactate levels were measured in cell growth media, normalized to total cellular proteins and expressed as fold change relative to control cells (fixed as 1-value). In panels (**c–e**), results are presented as the mean ± S.D. and statistical analyses were performed on at least three independent biological replicates using two-tailed Student's t test; asterisks (**, $p < 0.01$; ***, $p < 0.001$) indicate the significance of variations in MELAS samples with respect to the control. Individual data points are depicted by white diamond-shaped dots. For all the panels, cells were as previously specified.

To further asses the biochemical effect of Cterm on transfected cybrids, we evaluated mitochondrial degradation in control and MELAS cells by examining the levels of mitophagy receptors BNIP3 and BNIP3-like (BNIP3L)/NIX (Figure 5). The results showed a marked NIX increase in MELAS non-transfected cells with respect to controls, in agreement with previous results, showing that mitophagy activation occurs in MELAS cells to eliminate dysfunctional mitochondria [27,28]. The expression of BNIP3, instead, remained almost unvaried in MELAS cells as compared to controls, suggesting a little impact of this receptor on mitophagy in our MELAS cybrids. However, Cterm overexpression did not substantially alter the expression level of either protein, which remained unchanged with respect to non-transfected MELAS cells. We also examined mitochondrial mass by immunoblot analysis of citrate synthase amount, a mitochondrial matrix enzyme whose level correlates with mitochondrial mass. We found that citrate synthase levels were similar in control and MELAS cells, and did not change in Cterm-overexpressing cells (Figure 5). Collectively, these findings indicate that Cterm overexpression had no impact on mitophagy and mitochondrial content.

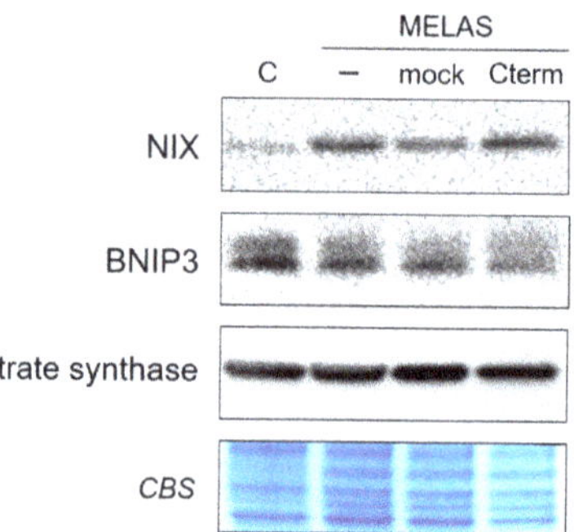

Figure 5. Cterm does not affect mitophagy and mitochondrial mass. Western blot analysis of NIX, BNIP3 and citrate synthase was performed on whole cell lysate. Coomassie blue staining (CBS) of the membrane was used as a loading and transfer control.

4. Discussion

Mitochondrial encephalopathy with lactic acidosis and stroke-like episodes (MELAS) syndrome is among the most studied neurodegenerative mitochondriopathies. In 80% of cases, this pathology is associated with the point mutation A3243G in the mt-tRNALeu$^{(UUR)}$ gene [29,30]. The MELAS mutation disrupts the tertiary structure of the mutant tRNA, resulting in various defects spanning from improper RNA processing with accumulation of a large polycistronic precursor transcript to decreased tRNA stability and both aminoacylation and τm^5U34 wobble posttranscriptional modification deficiency [31–33]. All of these defects are causative factors of impaired mitochondrial translation typical of MELAS syndrome.

Numerous studies performed in yeast [34] and human cell lines [13–16] highlighted the ability of the LARS2 carboxy-terminal domain (Cterm) to interact with several mt-tRNAs, including the cognate mt-tRNALeu$^{(UUR)}$, and to relieve cell growth and respiration defects due to mt-tRNA mutations. Furthermore, a similar rescuing effect by short Cterm-derived peptides was also observed in human MELAS cybrids [17].

The exact molecular mechanism of this rescue activity remains unclear, although a chaperonic effect by Cterm on the mutated tRNA has been proposed, whereby the Cterm would stabilize a functional tRNA conformation and attenuate the detrimental effects of mt-tRNA mutation. In light of this, we decided to undertake a study aimed at elucidating the molecular effect of Cterm on the mitochondrial function in MELAS cybrid-transfected cells.

Here, as a starting point of our investigation, we showed that overexpressed Cterm was capable to interact with the mutated cognate mt-tRNALeu$^{(UUR)}$ in cultured cybrids; this finding is in line with the obtained in vitro results and proposed chaperonic effect of Cterm domain [14–17]. Then, we assessed whether Cterm was able to relieve mitochondrial molecular defects in a MELAS cell line characterised by depressed mitochondrial translation. In parallel with a significant increase in cell growth, we observed a mild but consistent enhancement of de novo mitochondrial protein synthesis. The latter result is likely mediated by the observed interaction between Cterm and the mutated mt-tRNALeu$^{(UUR)}$. However, our results indicate that the amelioration of both cell viability and de novo protein synthesis was not paralleled by a detectable increase in either steady-state level or aminoacylation efficiency of mt-tRNALeu$^{(UUR)}$. Failure to increase the level of mt-tRNALeu$^{(UUR)}$ is actually not surprising because previous reports suggested that the chaperonic activity of Cterm would improve the function of mutated tRNALeu$^{(UUR)}$ by stabilizing its shape rather than preventing the degradation [16]. Mitochondrial tRNAs are known to have great structural instability compared to the bacterial counterparts [35], which explains why stabilizing proteins in general can improve mt-tRNA aminoacylation, especially in the presence of destabilizing mutations (as discussed in [13]). In this framework, it is puzzling that we could not detect a clear enhancement of tRNALeu$^{(UUR)}$ aminoacylation by the stabilizing molecule Cterm. One possible explanation is that, since mutated mt-tRNALeu$^{(UUR)}$ levels in MELAS cells are very low (as we show here) and only a fraction of the tRNA would reasonably undergo aminoacylation by the intervention of Cterm, the signal of aminoacylation enhancement would be below the detection limit of the technique we used in this experiment. The improvement of mitochondrial translation driven by the Cterm domain could also be ascribed to the capacity of the peptide to restore wobble taurine modification on the mutated tRNA or to improve interactions with the ribosome or with other proteins required to exert its function.

We showed here that Cterm is able to bind not only the mature mt-tRNALeu$^{(UUR)}$ but also its precursor, named RNA19. This is a polycistronic transcript including 16S rRNA, mt-tRNALeu$^{(UUR)}$, and ND1 mRNA, which accumulates because of processing defects caused by the mutated tRNA. Increased levels of RNA19 have been described in MELAS cells carrying the A3243G mutation [9,36] and also in cybrids bearing the mt-tRNALeu$^{(UUR)}$ A3302G mutation [37]. Attardi and colleagues [25] provided evidence for defective polysome formation in MELAS cells, which could lead to mRNAs degradation resulting in a decreased translation rate. A possible cause of functionally deficient polysomes could be that RNA19 is incorporated into ribosomes rendering them rate-limiting for translation [31]. In light of this, the observed increase in mitochondrial protein synthesis in Cterm-overexpressing cells might be due to the ability of the Cterm to sequester RNA19 and/or promote its correct processing, minimizing the accumulation of such an abnormal RNA processing transcript. This hypothesis is also in agreement with our RIP data, which show a preferential enrichment of RNA19, notwithstanding the higher physiological amount of mature mt-tRNALeu$^{(UUR)}$ with respect to its precursor.

A puzzling finding of our study is that the observed rescuing did not correspond to an improvement in the analysed bioenergetic aspects. Steady-state levels of representative OXPHOS subunits of complex I and IV remained low in the MELAS cybrids overexpressing Cterm, as did the level of assembled complex I and IV. Accordingly, complex IV activity and mitochondrial oxygen consumption rate did not recover in the presence of the Cterm domain. Moreover, Cterm expression did not influence the higher glycolytic flux peculiar of MELAS cells, as shown by lactate measurement. Interestingly, our MELAS cell line displayed a strong accumulation of ATP synthase F1-c subcomplexes, named V*. It was shown in cultured cells [38] that intermediate subcomplexes V* increase when mitochondrial translation is inhibited by doxycycline, indicating that, in order for the subcomplex to evolve to the fully assembled holoenzyme, the mitochondrially encoded subunits ATP6 and ATP8 are needed. The strong increment of subcomplex V* in MELAS cells is in accordance with the observed decreased level of subunits ATP6 and ATP8 (see Figure 2). Moreover, Cterm did not influence the complex V profile either on a quantitative or qualitative basis. The fact that overexpressed Cterm did not improve these bioenergetic defects could be explained by assuming that, although increased, the observed translation efficiency is still below a threshold for respiratory chain improvement. Even a longer timescale of Cterm expression was not able to produce any increment in the OXPHOS subunit level (data not shown). In light of this, the slight increase in mitochondrial protein synthesis observed in mock cells, which is anyhow lower than that caused by Cterm, has a negligible functional significance, especially considering that no recovery of cell viability occurs upon mock transfection. Finally, Cterm peptide had no impact on mitophagy and mitochondrial mass.

It is important to point out that, unlike what has been here described, some of us reported an increase in oxygen consumption using another MELAS cell line transfected with Cterm [14,16]. This discrepancy suggests that distinct MELAS cybrid lines may have peculiar molecular pathways and compensatory mechanisms, in parallel with the various phenotypes observed in MELAS patients with similar levels of heteroplasmy.

Overall, the reported data tend to exclude that Cterm capacity to restore cell viability to the wild type level derives from a direct involvement of the mitochondrial functions analysed herein, suggesting that other mitochondrial pathways, as well as those extra-mitochondrial, that are possibly triggered by the mitochondrial status are involved. Mitochondrial dysfunction can influence nuclear gene expression in mammalian systems by means of retrograde signalling mediated by ROS, Ca^{2+}, ADP/ATP, NAD/NADH, and non-coding RNAs [39,40]. A study of transcriptional reprogramming conducted in the MELAS cybrids highlighted various metabolic pathways that are differentially triggered by variations in mtDNA 3243 mutation heteroplasmy [41]. These pathways, which represent both adaptive and maladaptive cellular responses, refer to a series of genes related to glycolysis for energy production, antioxidant and redox regulatory systems, autophagy/mitophagy, apoptosis, chromatin remodelling, and compartment-specific unfolded protein stress response. In this context, nuclear mRNAs or non-coding RNAs of either nuclear or mitochondrial origin, such as processed mt-tRNA fragments exported to the cytoplasm or circular RNAs [40,42–44], may represent possible targets of cytosolic RNA-binding proteins for regulation of nuclear gene expression. In addition to mitochondrial retrograde signalling to the cytoplasm, the full recovery of cell viability might be the result of Cterm capacity to bind cytosolic RNA targets. This hypothesis is supported by the observation that: (i) a certain amount of overexpressed Cterm is not imported into mitochondria and remains in the cytosol [13]; and (ii) Cterm interacts also with non-tRNA targets (see Figure 1). This circumstance is in accordance with the hypothesis that the Cterm mode of RNA recognition relies more on structural rather than on specific base–base interactions, as observed in the available 3D structures of whole LeuRS-tRNALeu complexes from bacterial species [45]. This feature is reminiscent of the promiscuous RNA binding mode typical of aaRSs [22,46].

In conclusion, our results on the rescuing ability of Cterm in MELAS cybrids open up to pathways that, in our cellular model, seem to be independent of the mitochondrial bioenergetics. Importantly, the demonstration that Cterm has rescuing activity in MELAS cybrids, in spite of differences in the apparent mechanism of action, strongly supports the therapeutic potential of the Cterm and of molecules derived thereof [17], opening up new perspectives for the treatment of mitochondrial diseases. Further work on additional cellular models, preferably tissue-specific, will be required to fully unveil the molecular basis of Cterm rescuing activity.

Supplementary Materials: The following are available online at https://www.mdpi.com/article/10.3390/life11070674/s1, Figure S1: Stable expression of Cterm-FLAG transcript, Figure S2: Raw data from RIP experiments, Figure S3: RFLP analysis and quantification of m.3243A>G heteroplasmy in MELAS cybrids, Figure S4: Transient expression of Cterm-FLAG transcript, Figure S5: Gel scans and densitometry data of metabolic [^{35}S]-methionine labelling replicates after 48-h transfection, Figure S6: Gel scans and densitometry data of metabolic [^{35}S]-methionine labelling replicates after 72-h transfection; Figure S7: Relative quantification of [^{35}S]-labelling signals, Figure S8: Effect of pBAD and pQE60 transfection on metabolic [^{35}S]-methionine labelling of mitochondrial translation products, Figure S9: Effect of pBAD and pQE60 transfection on mt-tRNALeu$^{(UUR)}$ and mt-tRNAGlu steady-state levels, Figure S10: Relative quantification of SDS-PAGE and BN-PAGE signals.

Author Contributions: Conceptualization, G.d., P.C., M.R., P.L.P. and F.B.; methodology, M.R., P.L.P. and F.B.; formal analysis, F.C., F.R., G.P. and F.B.; investigation, F.C., F.R., G.P. and F.B.; resources, G.d. and P.L.P.; writing—original draft preparation, M.R., P.L.P. and F.B.; writing—review and editing, G.d., V.M., P.C., M.R., P.L.P. and F.B.; supervision, M.R., P.L.P. and F.B.; project administration, M.R., P.L.P. and F.B.; funding acquisition, G.d., M.R., P.L.P. and F.B. All authors have read and agreed to the published version of the manuscript.

Funding: This research was funded by Progetto competitivo "Sviluppo Catalisi dell'Innovazione nelle Biotecnologie" (CIB); Fondi di Ateneo "Contributo Ordinario di Supporto alla Ricerca" 2015/16 and 2017/18 (University of Bari Aldo Moro); AFM Télethon and Istituto Pasteur Italia-Fondazione Cenci Bolognetti.

Institutional Review Board Statement: Not applicable.

Informed Consent Statement: Not applicable.

Data Availability Statement: Not applicable.

Acknowledgments: We are grateful to Aleksandra Filipovska (University of Western Australia) for supplying MELAS transmitochondrial cybrids and to Claudio Lo Giudice (University of Bari Aldo Moro) for his kind help with statistical analyses.

Conflicts of Interest: The authors declare no conflict of interest. The funders had no role in the design of the study; in the collection, analyses, or interpretation of data; in the writing of the manuscript, or in the decision to publish the results.

References

1. Gorman, G.S.; Chinnery, P.F.; DiMauro, S.; Hirano, M.; Koga, Y.; McFarland, R.; Suomalainen, A.; Thorburn, D.R.; Zeviani, M.; Turnbull, D.M. Mitochondrial Diseases. *Nat. Rev. Dis. Primers* **2016**, *2*, 16080. [CrossRef]
2. Shokolenko, I.N.; Alexeyev, M.F. Mitochondrial Transcription in Mammalian Cells. *Front. Biosci.* **2017**, *22*, 835–853. [CrossRef]
3. Pfanner, N.; Warscheid, B.; Wiedemann, N. Mitochondrial Proteins: From Biogenesis to Functional Networks. *Nat. Rev. Mol. Cell Biol.* **2019**, *20*, 267–284. [CrossRef] [PubMed]
4. Majamaa-Voltti, K.; Peuhkurinen, K.; Kortelainen, M.-L.; Hassinen, I.E.; Majamaa, K. Cardiac Abnormalities in Patients with Mitochondrial DNA Mutation 3243A>G. *BMC Cardiovasc. Disord.* **2002**, *2*, 12. [CrossRef]
5. Bates, M.G.D.; Newman, J.H.; Jakovljevic, D.G.; Hollingsworth, K.G.; Alston, C.L.; Zalewski, P.; Klawe, J.J.; Blamire, A.M.; MacGowan, G.A.; Keavney, B.D.; et al. Defining Cardiac Adaptations and Safety of Endurance Training in Patients with m.3243A>G-Related Mitochondrial Disease. *Int. J. Cardiol.* **2013**, *168*, 3599–3608. [CrossRef]
6. El-Hattab, A.W.; Adesina, A.M.; Jones, J.; Scaglia, F. MELAS Syndrome: Clinical Manifestations, Pathogenesis, and Treatment Options. *Mol. Genet. Metab.* **2015**, *116*, 4–12. [CrossRef] [PubMed]
7. Yasukawa, T.; Suzuki, T.; Suzuki, T.; Ueda, T.; Ohta, S.; Watanabe, K. Modification Defect at Anticodon Wobble Nucleotide of Mitochondrial TRNAs Leu (UUR) with Pathogenic Mutations of Mitochondrial Myopathy, Encephalopathy, Lactic Acidosis, and Stroke-like Episodes. *J. Biol. Chem.* **2000**, *275*, 4251–4257. [CrossRef] [PubMed]
8. Kirino, Y.; Goto, Y.-I.; Campos, Y.; Arenas, J.; Suzuki, T. Specific Correlation between the Wobble Modification Deficiency in Mutant TRNAs and the Clinical Features of a Human Mitochondrial Disease. *Proc. Natl. Acad. Sci. USA* **2005**, *102*, 7127–7132. [CrossRef]
9. King, M.P.; Koga, Y.; Davidson, M.; Schon, E.A. Defects in Mitochondrial Protein Synthesis and Respiratory Chain Activity Segregate with the TRNA(Leu(UUR)) Mutation Associated with Mitochondrial Myopathy, Encephalopathy, Lactic Acidosis, and Strokelike Episodes. *Mol. Cell. Biol.* **1992**, *12*, 480–490. [CrossRef] [PubMed]
10. Sasarman, F.; Antonicka, H.; Shoubridge, E.A. The A3243G TRNALeu(UUR) MELAS Mutation Causes Amino Acid Misincorporation and a Combined Respiratory Chain Assembly Defect Partially Suppressed by Overexpression of EFTu and EFG2. *Hum. Mol. Genet.* **2008**, *17*, 3697–3707. [CrossRef] [PubMed]
11. Suzuki, T.; Miyauchi, K.; Suzuki, T.; Yokobori, S.; Shigi, N.; Kondow, A.; Takeuchi, N.; Yamagishi, A.; Watanabe, K. Taurine-Containing Uridine Modifications in TRNA Anticodons Are Required to Decipher Non-Universal Genetic Codes in Ascidian Mitochondria. *J. Biol. Chem.* **2011**, *286*, 35494–35498. [CrossRef]
12. Rorbach, J.; Yusoff, A.A.; Tuppen, H.; Abg-Kamaludin, D.P.; Chrzanowska-Lightowlers, Z.M.A.; Taylor, R.W.; Turnbull, D.M.; McFarland, R.; Lightowlers, R.N. Overexpression of Human Mitochondrial Valyl TRNA Synthetase Can Partially Restore Levels of Cognate Mt-TRNAVal Carrying the Pathogenic C25U Mutation. *Nucleic Acids Res.* **2008**, *36*, 3065–3074. [CrossRef]
13. Hornig-Do, H.T.; Montanari, A.; Rozanska, A.; Tuppen, H.A.; Almalki, A.A.; Abg-Kamaludin, D.P.; Frontali, L.; Francisci, S.; Lightowlers, R.N.; Chrzanowska-Lightowlers, Z.M. Human Mitochondrial Leucyl TRNA Synthetase Can Suppress Non Cognate Pathogenic Mt-TRNA Mutations. *EMBO Mol. Med.* **2014**, *6*, 183–193. [CrossRef]

14. Perli, E.; Giordano, C.; Pisano, A.; Montanari, A.; Campese, A.F.; Reyes, A.; Ghezzi, D.; Nasca, A.; Tuppen, H.A.; Orlandi, M.; et al. The Isolated Carboxy-Terminal Domain of Human Mitochondrial Leucyl-TRNA Synthetase Rescues the Pathological Phenotype of Mitochondrial TRNA Mutations in Human Cells. *EMBO Mol. Med.* **2014**, *6*, 169–182. [CrossRef]
15. Giordano, C.; Morea, V.; Perli, E.; d'Amati, G. The Phenotypic Expression of Mitochondrial TRNA-Mutations Can Be Modulated by Either Mitochondrial Leucyl-TRNA Synthetase or the C-Terminal Domain Thereof. *Front. Genet.* **2015**, *6*, 113. [CrossRef]
16. Perli, E.; Fiorillo, A.; Giordano, C.; Pisano, A.; Montanari, A.; Grazioli, P.; Campese, A.F.; Di Micco, P.; Tuppen, H.A.; Genovese, I.; et al. Short Peptides from Leucyl-TRNA Synthetase Rescue Disease-Causing Mitochondrial TRNA Point Mutations. *Hum. Mol. Genet.* **2016**, *25*, 903–915. [CrossRef]
17. Perli, E.; Pisano, A.; Pignataro, M.G.; Campese, A.F.; Pelullo, M.; Genovese, I.; de Turris, V.; Ghelli, A.M.; Cerbelli, B.; Giordano, C.; et al. Exogenous Peptides Are Able to Penetrate Human Cell and Mitochondrial Membranes, Stabilize Mitochondrial TRNA Structures, and Rescue Severe Mitochondrial Defects. *FASEB J.* **2020**, *34*, 7675–7686. [CrossRef]
18. Cámara, Y.; Asin-Cayuela, J.; Park, C.B.; Metodiev, M.D.; Shi, Y.; Ruzzenente, B.; Kukat, C.; Habermann, B.; Wibom, R.; Hultenby, K.; et al. MTERF4 Regulates Translation by Targeting the Methyltransferase NSUN4 to the Mammalian Mitochondrial Ribosome. *Cell Metab.* **2011**, *13*, 527–539. [CrossRef]
19. Bruni, F.; Proctor-Kent, Y.; Lightowlers, R.N.; Chrzanowska-Lightowlers, Z.M. Messenger RNA Delivery to Mitoribosomes—Hints from a Bacterial Toxin. *FEBS J.* **2021**, *288*, 437–451. [CrossRef]
20. Kim, S.W.; Li, Z.; Moore, P.S.; Monaghan, A.P.; Chang, Y.; Nichols, M.; John, B. A Sensitive Non-Radioactive Northern Blot Method to Detect Small RNAs. *Nucleic Acids Res.* **2010**, *38*, e98. [CrossRef]
21. Spinazzi, M.; Casarin, A.; Pertegato, V.; Salviati, L.; Angelini, C. Assessment of Mitochondrial Respiratory Chain Enzymatic Activities on Tissues and Cultured Cells. *Nat. Protoc.* **2012**, *7*, 1235–1246. [CrossRef] [PubMed]
22. Garin, S.; Levi, O.; Cohen, B.; Golani-Armon, A.; Arava, Y.S. Localization and RNA Binding of Mitochondrial Aminoacyl TRNA Synthetases. *Genes* **2020**, *11*, 1185. [CrossRef] [PubMed]
23. Robinson, B.H.; Petrova-Benedict, R.; Buncic, J.R.; Wallace, D.C. Nonviability of Cells with Oxidative Defects in Galactose Medium: A Screening Test for Affected Patient Fibroblasts. *Biochem. Med. Metab. Biol.* **1992**, *48*, 122–126. [CrossRef]
24. Chomyn, A.; Meola, G.; Bresolin, N.; Lai, S.T.; Scarlato, G.; Attardi, G. In Vitro Genetic Transfer of Protein Synthesis and Respiration Defects to Mitochondrial DNA-Less Cells with Myopathy-Patient Mitochondria. *Mol. Cell Biol.* **1991**, *11*, 2236–2244. [CrossRef]
25. Chomyn, A.; Enriquez, J.A.; Micol, V.; Fernandez-Silva, P.; Attardi, G. The Mitochondrial Myopathy, Encephalopathy, Lactic Acidosis, and Stroke-like Episode Syndrome-Associated Human Mitochondrial TRNA Leu(UUR) Mutation Causes Aminoacylation Deficiency and Concomitant Reduced Association of MRNA with Ribosomes. *J. Biol. Chem.* **2000**, *275*, 19198–19209. [CrossRef] [PubMed]
26. Jaksch, M.; Kleinle, S.; Scharfe, C.; Klopstock, T.; Pongratz, D.; Müller-Höcker, J.; Gerbitz, K.-D.; Liechti-Gallati, S.; Lochmuller, H.; Horvath, R. Frequency of Mitochondrial Transfer RNA Mutations and Deletions in 225 Patients Presenting with Respiratory Chain Deficiencies. *J. Med. Genet.* **2001**, *38*, 665–673. [CrossRef]
27. Cotán, D.; Cordero, M.D.; Garrido-Maraver, J.; Oropesa-Ávila, M.; Rodríguez-Hernández, A.; Gómez Izquierdo, L.; De la Mata, M.; De Miguel, M.; Lorite, J.B.; Infante, E.R.; et al. Secondary Coenzyme Q10 Deficiency Triggers Mitochondria Degradation by Mitophagy in MELAS Fibroblasts. *FASEB J.* **2011**, *25*, 2669–2687. [CrossRef] [PubMed]
28. Garrido-Maraver, J.; Paz, M.V.; Cordero, M.D.; Bautista-Lorite, J.; Oropesa-Ávila, M.; de la Mata, M.; Pavón, A.D.; de Lavera, I.; Alcocer-Gómez, E.; Galán, F.; et al. Critical Role of AMP-Activated Protein Kinase in the Balance between Mitophagy and Mitochondrial Biogenesis in MELAS Disease. *Biochim. Biophys. Acta* **2015**, *1852*, 2535–2553. [CrossRef]
29. Goto, Y.; Nonaka, I.; Horai, S. A Mutation in the TRNA Leu(UUR) Gene Associated with the MELAS Subgroup of Mitochondrial Encephalomyopathies. *Nature* **1990**, *348*, 651–653. [CrossRef] [PubMed]
30. Kobayashi, Y.; Momoi, M.Y.; Tominaga, K.; Momoi, T.; Nihei, K.; Yanagisawa, M.; Kagawa, Y.; Ohta, S. A Point Mutation in the Mitochondrial TRNALeu(UUR) Gene in Melas (Mitochondrial Myopathy, Encephalopathy, Lactic Acidosis and Stroke-like Episodes). *Biochem. Biophys. Res. Commun.* **1990**, *173*, 816–822. [CrossRef]
31. Schon, E.A.; Koga, Y.; Davidson, M.; Moraes, C.T.; King, M.P. The Mitochondrial TRNALeu(UUR) Mutation in MELAS: A Model for Pathogenesis. *Biochim. Et Biophys. Acta BBA Bioenergy* **1992**, *1101*, 206–209. [CrossRef]
32. Park, H.; Davidson, E.; King, M.P. The Pathogenic A3243G Mutation in Human Mitochondrial TRNALeu(UUR) Decreases the Efficiency of Aminoacylation. *Biochemistry* **2003**, *42*, 958–964. [CrossRef]
33. Kazuhito, T.; Wei, F.-Y. Posttranscriptional Modifications in Mitochondrial TRNA and Its Implication in Mitochondrial Translation and Disease. *J. Biochem.* **2020**, *168*, 435–444. [CrossRef]
34. Francisci, S.; Montanari, A.; De Luca, C.; Frontali, L. Peptides from Aminoacyl-TRNA Synthetases Can Cure the Defects Due to Mutations in Mt TRNA Genes. *Mitochondrion* **2011**, *11*, 919–923. [CrossRef]
35. Fender, A.; Gaudry, A.; Jühling, F.; Sissler, M.; Florentz, C. Adaptation of Aminoacylation Identity Rules to Mammalian Mitochondria. *Biochimie* **2012**, *94*, 1090–1097. [CrossRef]
36. Kaufmann, P.; Shanske, S.; Hirano, M.; DiMauro, S.; King, M.P.; Koga, Y.; Schon, E.A. Mitochondrial DNA and RNA Processing in MELAS. *Ann Neurol.* **1996**, *40*, 172–180. [CrossRef]

37. Maniura-Weber, K.; Helm, M.; Engemann, K.; Eckertz, S.; Möllers, M.; Schauen, M.; Hayrapetyan, A.; von Kleist-Retzow, J.-C.; Lightowlers, R.N.; Bindoff, L.A.; et al. Molecular Dysfunction Associated with the Human Mitochondrial 3302A>G Mutation in the MTTL1 (Mt-TRNALeu(UUR)) Gene. *Nucleic Acids Res.* **2006**, *34*, 6404–6415. [CrossRef]
38. Nijtmans, L.G.; Klement, P.; Houstěk, J.; van den Bogert, C. Assembly of Mitochondrial ATP Synthase in Cultured Human Cells: Implications for Mitochondrial Diseases. *Biochim. Biophys. Acta* **1995**, *1272*, 190–198. [CrossRef]
39. Wallace, D.C.; Fan, W. Energetics, Epigenetics, Mitochondrial Genetics. *Mitochondrion* **2010**, *10*, 12–31. [CrossRef] [PubMed]
40. Meseguer, S.; Navarro-González, C.; Panadero, J.; Villarroya, M.; Boutoual, R.; Sánchez-Alcázar, J.A.; Armengod, M.-E. The MELAS Mutation m.3243A>G Alters the Expression of Mitochondrial TRNA Fragments. *Biochim. Biophys. Acta BBA Mol. Cell Res.* **2019**, *1866*, 1433–1449. [CrossRef]
41. Picard, M.; Zhang, J.; Hancock, S.; Derbeneva, O.; Golhar, R.; Golik, P.; O'Hearn, S.; Levy, S.; Potluri, P.; Lvova, M.; et al. Progressive Increase in MtDNA 3243A>G Heteroplasmy Causes Abrupt Transcriptional Reprogramming. *Proc. Natl. Acad. Sci. USA* **2014**, *111*, E4033–E4042. [CrossRef]
42. Lu, M. Circular RNA: Functions, Applications and Prospects. *ExRNA* **2020**, *2*, 1. [CrossRef]
43. Bravo, J.I.; Nozownik, S.; Danthi, P.S.; Benayoun, B.A. Transposable Elements, Circular RNAs and Mitochondrial Transcription in Age-Related Genomic Regulation. *Development* **2020**, *147*, dev175786. [CrossRef] [PubMed]
44. Meseguer, S. MicroRNAs and TRNA-Derived Small Fragments: Key Messengers in Nuclear–Mitochondrial Communication. *Front. Mol. Biosci.* **2021**, *8*, 643575. [CrossRef]
45. Tukalo, M.; Yaremchuk, A.; Fukunaga, R.; Yokoyama, S.; Cusack, S. The Crystal Structure of Leucyl-TRNA Synthetase Complexed with TRNALeu in the Post-Transfer-Editing Conformation. *Nat. Struct. Mol. Biol.* **2005**, *12*, 923–930. [CrossRef] [PubMed]
46. Sarkar, J.; Poruri, K.; Boniecki, M.T.; McTavish, K.K.; Martinis, S.A. Yeast Mitochondrial Leucyl-TRNA Synthetase CP1 Domain Has Functionally Diverged to Accommodate RNA Splicing at Expense of Hydrolytic Editing. *J. Biol. Chem.* **2012**, *287*, 14772–14781. [CrossRef]

 life

MDPI

Article

Subcellular Localization of Fad1p in *Saccharomyces cerevisiae*: A Choice at Post-Transcriptional Level?

Francesco Bruni [1], Teresa Anna Giancaspero [1], Mislav Oreb [2], Maria Tolomeo [1], Piero Leone [1], Eckhard Boles [2], Marina Roberti [1], Michele Caselle [3] and Maria Barile [1,*]

1 Department of Biosciences, Biotechnologies and Biopharmaceutics, University of Bari Aldo Moro, Via Orabona 4, 70125 Bari, Italy; francesco.bruni@uniba.it (F.B.); teresagiancaspero@virgilio.it (T.A.G.); maria.tolomeo@uniba.it (M.T.); pieroleone87@gmail.com (P.L.); marina.roberti@uniba.it (M.R.)
2 Institute of Molecular Biosciences, Goethe-University Frankfurt, Max-von-Laue-Str. 9, 60438 Frankfurt am Main, Germany; m.oreb@bio.uni-frankfurt.de (M.O.); e.boles@bio.uni-frankfurt.de (E.B.)
3 Physics Department, University of Turin and INFN, Via P. Giuria 1, 10125 Turin, Italy; michele.caselle@unito.it
* Correspondence: maria.barile@uniba.it; Tel.: +39-080-544-3604

Abstract: FAD synthase is the last enzyme in the pathway that converts riboflavin into FAD. In *Saccharomyces cerevisiae*, the gene encoding for FAD synthase is *FAD1*, from which a sole protein product (Fad1p) is expected to be generated. In this work, we showed that a natural Fad1p exists in yeast mitochondria and that, in its recombinant form, the protein is able, per se, to both enter mitochondria and to be destined to cytosol. Thus, we propose that *FAD1* generates two echoforms—that is, two identical proteins addressed to different subcellular compartments. To shed light on the mechanism underlying the subcellular destination of Fad1p, the 3′ region of *FAD1* mRNA was analyzed by 3′RACE experiments, which revealed the existence of (at least) two *FAD1* transcripts with different 3′UTRs, the short one being 128 bp and the long one being 759 bp. Bioinformatic analysis on these 3′UTRs allowed us to predict the existence of a *cis*-acting mitochondrial localization motif, present in both the transcripts and, presumably, involved in protein targeting based on the 3′UTR context. Here, we propose that the long *FAD1* transcript might be responsible for the generation of mitochondrial Fad1p echoform.

Keywords: FAD synthase; *FAD1*; mitochondria localization; *Saccharomyces cerevisiae*; mRNA; mitochondrial localization motif

Citation: Bruni, F.; Giancaspero, T.A.; Oreb, M.; Tolomeo, M.; Leone, P.; Boles, E.; Roberti, M.; Caselle, M.; Barile, M. Subcellular Localization of Fad1p in *Saccharomyces cerevisiae*: A Choice at Post-Transcriptional Level? *Life* **2021**, *11*, 967. https://doi.org/10.3390/life11090967

Academic Editor: Theodoros Rampias

Received: 29 July 2021
Accepted: 13 September 2021
Published: 14 September 2021

1. Introduction

Riboflavin (Rf or vitamin B2) deficiency in humans and experimental animals has been linked to several diseases, such as cancer, cardiovascular diseases, anemia, abnormal fetal development, different neuromuscular and neurological disorders, some of which are treatable with high doses of riboflavin. Among the latter, there are the multiple acyl-CoA dehydrogenase deficiency (MADD) and the Brown–Vialetto–Van Laere syndrome (BVVLS) (see [1] and refs therein). The role of this vitamin in cell metabolism depends on its conversion into flavin mononucleotide (FMN) and flavin adenine dinucleotide (FAD), which are the redox cofactors of a large number of dehydrogenases, reductases, and oxidases. Most of these flavoenzymes are compartmented in the cellular organelles, where they are involved in energy production and redox homeostasis as well as in different cellular regulatory events including protein folding, apoptosis, and chromatin remodeling [2–4]. The relevance of such processes merits further research aimed to better describe flavin homeostasis and flavoenzyme biogenesis, especially in those organisms that can be simple and suitable models for human pathologies. The conservation of the main biological processes within all eukaryotes, together with the possibility of simple and quick genetic manipulation, make the budding yeast, *Saccharomyces cerevisiae*, a suitable model to understand the molecular mechanisms underlying human diseases [5–8].

Yeasts, as well as other fungi, plants and bacteria, have the ability either to synthesize Rf de novo or to take it from outside, whereas mammals must obtain Rf from diet. The first eukaryotic gene coding for a cellular Rf transporter was identified in *S. cerevisiae* as the product of the *MCH5* gene [9]. More recently, three human Rf transporters have been cloned and characterized; they belong to a novel family of Rf transporters, namely, RFVT/SLC52, which exhibit no significant similarity to Mch5p [10,11]. Mutations in RFVTs have been recently linked to BVVLS, a neurodegenerative disorder characterized by cranial nerve deficits, bilateral sensorineural deafness, respiratory insufficiencies, and the degeneration of some spinal cord neurons [1,12].

Intracellular conversion of Rf to FAD is a ubiquitous pathway and occurs via the sequential actions of ATP:riboflavin 5′-phosphotransferase or riboflavin kinase (RFK, EC 2.7.1.26), which phosphorylates the vitamin into FMN, and of ATP:FMN adenylyl transferase or FAD synthase (FADS, EC 2.7.7.2) that adenylates FMN to FAD. Eukaryotes generally use two different enzymes for FAD production, whereas most prokaryotes depend on a single bifunctional enzyme [13,14]. The first eukaryotic genes encoding for RFK and FADS were identified in *S. cerevisiae* and named *FMN1* [15] and *FAD1* [16], respectively.

Fmn1p is a 24.5 kDa protein showing a sequence and structure similarity to the RKF-module of prokaryotic FADS and appears largely conserved through evolution. Immunoblotting analysis of subcellular fractions revealed that Fmn1p is localized in microsomes and in mitochondria [15]. Orthologs of *S. cerevisiae FMN1* have been cloned and the corresponding proteins purified from the yeast *Schizosaccharomyces pombe* and from *Homo sapiens*. The crystal structures of both these proteins have been solved, revealing a novel ATP and Rf-binding fold [17–19].

Fad1p, the sole known protein isoform generated by *S. cerevisiae FAD1* gene, is a 35.5-kDa soluble enzyme that is essential for yeast life, whose crystal structure has been solved in a complex with FAD in the active site [20]. Fad1p is a single-domain monofunctional enzyme belonging to the 3′-phosphoadenosine 5′-phosphosulfate (PAPS) reductase family and shows little or no sequence similarity to the prokaryotic FAD-forming enzymes. The human gene for FADS, named *FLAD1* [14,21], generates different alternatively spliced transcripts encoding different isoforms, most of them containing the PAPS reductase domain fused with an N-terminal molybdopterin-binding (MPTb) domain with FAD hydrolyzing activity [22].

Our group previously demonstrated that *S. cerevisiae* mitochondria are able to catalyze FAD hydrolysis via enzymatic activity, which is different from the already characterized NUDIX hydrolases and regulated by the mitochondrial NAD redox status [23].

The relevance of *FLAD1*, postulated in our previous papers [24,25], emerged in 2016 when *FLAD1* was identified as a novel Riboflavin-responsive MADD disease gene [26]. MADD defines a heterogeneous class of lipid storage myopathies associated with impaired fatty acid, amino acid and choline metabolisms [27]. In the frame of these studies, a novel isoform containing the sole PAPS domain, hFADS6, was identified and then characterized in some detail [28]. The homology modeling of the PAPS reductase domain of human FADS was performed using the orthologue from *Candida glabrata* as a template [28].

The simultaneous presence in the same cell of the different *FLAD1* transcripts is in line with a possible different subcellular localization for the different isoforms in mammals. Consistently, the existence of distinct cytosolic, nuclear and mitochondrial FADS was demonstrated in different rat models [29,30]. The human transcript variant 1 encodes a 65.3 kDa protein, hFADS1, which contains a predictable mitochondrial-targeting peptide [21]. Mitochondrial localization of hFADS1 was proven in vitro by mitochondrial import assay and confocal microscopy [29].

Mitochondrial supply of FAD is crucial also in yeast, where it is expected to be delivered to a number of nascent client apo-flavoenzymes [31]. This occurs in some cases, such as for the flavoprotein subunit of succinate dehydrogenase (Sdh1p), via an autocatalytic covalent flavinylation mechanism, which could be assisted by ancillary proteins [32–36].

The first eukaryotic gene encoding for a mitochondrial FAD translocator was identified in *S. cerevisiae* and named *FLX1*, whose human orthologue is *SLC25A32*. Interestingly, mutations in the *SLC25A32* gene have been identified as the cause for RR-MADD [37,38], thus confirming the essential role of flavin cofactor supply for mitochondrial proteome.

The existence of a mitochondrial FADS isoform in yeast is still controversial as well as the physiological role of the mitochondrial FAD translocator, Flx1p. Initially, it was reported that FAD is synthesized by Fad1p exclusively in the cytosol [16] and, consequently, imported into mitochondria via Flx1p [39]. However, results from our laboratory showed that, besides in the cytosol, FAD-forming activities can be specifically revealed in mitochondria, entailing FAD precursors uptake in mitochondria and mitochondrial FAD export to cytosol via Flx1p [40,41]. Moreover, FAD movement across the mitochondrial membrane catalyzed by Flx1p plays an additional regulatory role on apo-Sdh1p biogenesis at the post-transcriptional level [42].

As a matter of fact, the protein responsible for FAD synthesis in *S. cerevisiae* mitochondria remains to be identified and characterized. In this paper, we report the existence of *S. cerevisiae* Fad1p as two distinct echoforms localized to both cytosol and mitochondria, and the presence of two populations of *FAD1* mRNAs, which differ for their 3′UTRs. Whether and how the 3′UTRs play a role in the mechanism that destines Fad1p to either cytosol or mitochondria will be dealt with.

2. Materials and Methods

2.1. Materials

All reagents and enzymes were from Sigma-Aldrich (St. Louis, MO, USA), ThermoFisher Scientific (Waltham, MA, USA) and Merck KGaA (Darmstadt, Germany). Zymolyase was from MP Biomedicals (Aurora, OH, USA). Bacto Yeast Extract and Yeast Nitrogen base were from BD Difco (Franklin Lakes, NJ, USA). Monoclonal antibody against STREP-tag (α-STREP) was obtained from IBA Lifesciences (Göttingen, Germany), monoclonal antibody against actin (α-ACTIN) was from Abcam (Cambridge, MA, USA), secondary anti-rabbit or anti-mouse IgG antibodies conjugated with peroxidase were from ThermoFisher Scientific (Waltham, MA, USA). The RNeasy Midi Kit was from Qiagen (Hilden, Germany). The enhanced AMV Reverse transcriptase kit was from Sigma-Aldrich (St. Louis, MO, USA). The Dynabeads mRNA Purification kit was from Invitrogen (Waltham, MA, USA). The Wizard SV Gel and PCR Clean-up system was from Promega (Madison, WI, USA).

2.2. Yeast Strains and Recombinant Multicopy Plasmids

The wild-type *S. cerevisiae* strain CEN.PK113-13D (also named EBY157, WT, or K26; genotypes *MAT*α, *MAL2-8*^{c}, *SUC2* and *ura3-52*) was previously described in [40,42]. The CENPK2-1C, derived from the CEN.PK yeast series, was used as a recipient strain to express a recombinant Fad1p carrying a STREP tag at the N- or C-terminal end. To construct the multicopy plasmids p426HXT7-FAD1-STREP, p426-MET25-FAD1-STREP, and p426-MET25-STREP-FAD1, the *FAD1* gene was cloned by PCR with a selected pair of primers designed to amplify the complete *FAD1* open reading frame (ORF), skipping the contiguous 5′ and 3′ UTR regions. The amplified fragments were cloned into the multicopy plasmid p426-HXT7-STREP or p426-MET25-STREP using the standard gap-repair procedure. In these vectors, the recombinant *FAD1* construct, encoding Fad1p fused to the STREP tag (Trp-Ser-His-Pro-Gln-Phe-Glu-Lys) at the N- or C-terminal end, was placed between the *HXT7* or the *MET25* promoter and *CYC1* terminator. The transformation of the CEN.PK113-13D and CENPK2-1C strains with the recombinant or empty vectors was carried out according to the frozen competent cell procedure [43]; the transformed strains were selected for the presence of URA3 genetic marker, which confers the ability to grow in a minimal synthetic liquid medium (SM, 6.7 g/L yeast nitrogen base supplemented with 25 mM Histidine, 0.44 mM Leucin and 0.19 mM Tryptophan) without uracil. All the transformed strains used in this study are summarized in Table 1.

Table 1. *S. cerevisiae* strains used in this study.

Strain	Multicopy Plasmid
CEN.PK113-13D (EBY157, WT, K26)	/
K26 ^{CTR}	p426-HXT7-STREP
K26 $^{FAD1-STREP}$	p426-HXT7-FAD1-STREP
CENPK2-1C $^{FAD1-STREP}$	p426-MET25-FAD1-STREP
CENPK2-1C $^{STREP-FAD1}$	p426-MET25-STREP-FAD1

2.3. *Media and Growth Conditions*

The wild-type K26 cells were grown aerobically at 30 °C with constant shaking in rich liquid medium (YEP, 10 g/L yeast extract, 20 g/L Bacto Peptone). The K26-transformed strains were grown aerobically at 30 °C with constant shaking in SM medium. The CENPK2-1C strain and the novel derived strains (see Table 1) were grown aerobically at 30 °C with constant shaking in SM medium. Ethanol, glycerol, galactose or glucose (2% each) were used as carbon sources. The solid medium (YEP or SM) contained 18 g/L agar.

2.4. *Preparation of Cellular Extracts*

Cells grown up to the early exponential phase (5 h) were harvested by centrifugation (8000× *g* for 5 min), washed with sterile water, resuspended in 250 μL of lysis buffer (10 mM Tris-HCl, pH 7.6, 1 mM EDTA, 1 mM dithiothreitol, 0.2 mM phenylmethanesulfonylfluoride (PMSF), supplemented with one tablet of Roche protease inhibitor cocktail every 10 mL of lysis buffer), and vortexed with glass beads for 10 min at 4 °C. The supernatant was removed and centrifuged at 3000× *g* for 5 min to remove cell debris.

2.5. *Preparation of Spheroplasts, Cytoplasm, Mitochondria, and Submitochondrial Fractions*

Spheroplasts were prepared using Zymolyase. Mitochondria and cytoplasm were isolated from spheroplasts as described in [23,40]. Mitochondria, ruptured by osmotic shock, were centrifuged at 20,000× *g* for 30 min to separate the mitochondrial soluble fraction and the mitochondrial membrane-enriched fraction.

2.6. *Western Blotting*

Protein from subcellular fractions or from cellular extracts were separated by SDS-PAGE and transferred onto a PVDF membrane as in [42]. The immobilized proteins were incubated with either a monoclonal antibody against the STREP-tag raised in mouse (α-STREP) to detect the recombinant Fad1p or a polyclonal α-FADS antiserum (2000-fold dilution) raised in rabbit to detect Fad1p [29]. Immuno-reactive materials were visualized with the aid of a secondary anti-rabbit or anti-mouse IgG antibody conjugated with peroxidase. PVDF membranes were also probed with monoclonal α-ACTIN antibody (10,000-fold dilution, raised in mouse), to reveal yeast actin (Act1p).

2.7. *RNA Isolation*

The RNeasy Midi Kit was used to extract total RNA from WT cells grown up to early exponential phase. Total RNA concentration was quantified by monitoring absorbance to 260 nm ($OD_{260\ nm}$), RNA integrity was verified by formaldehyde agarose gel electrophoresis and its purity was evaluated by measuring the $OD_{260\ nm}/OD_{280\ nm}$ ratio.

2.8. *3′RACE Analysis*

3′RACE analysis was performed on poly(A)$^+$ RNA obtained from 75 μg of total RNA using the Dynabeads mRNA Purification kit, according to the manufacturer's instructions. The purified poly(A)$^+$ RNA was reverse-transcribed using the Enhanced AMV Reverse Transcriptase Kit and the oligo(dT)-anchor primer [oligo(dT)-AP], according to the manufacturer's instructions. The cDNA pool was diluted 10-fold and these were used as a template (2.5 µL) for PCR reactions using a gene-specific primer and the AP primer. The specificity of the first round PCR-amplified fragment was proved by performing a semi-nested PCR. The sequence of primers, used in this study, are reported in Table 2. The 3′RACE products were separated by 2% agarose gels. The fragments obtained with the first round PCR were purified from preparative agarose gel by using the Wizard SV Gel and PCR Clean-up system and then sequenced.

Table 2. Primers used in this study.

Gene	Primer	Sequence
FAD1	FAD1A.for	5′-ATCGGCGGAATTAACAACTCA-3′
	FAD1A.rev	5′-TTGCCAAATGCATGAATGATTT-3′
	FAD1B.for	5′-GCCTAGCGGC GTGATAGTTAA-3′
	FAD1B.rev	5′-TGCTGGCTTAGTAACGGAATTG-3′
	FAD1C.for	5′-CATTTGGCAAGGACGCAGAA-3′
SDH1	SDH1.for	5′-GCCAATTCCTTGTTGGATCTTG-3′
	SDH1.rev	5′-TGGCAACCCAGGCTGTAAAG-3′
ACT1	ACT1.for	5′-TTCCATCCAAGCCGTTTTGT-3′
	ACT1.rev	5′-GGCGTGAGGTAGAGAGAAACCA-3′
	Oligo(dT)-AP	5′-GACCACGCGTATCGATGTCGAC(T)$_{16}$V-3′
	AP.rev	5′-GACCACGCGTATCGATGTCGAC-3′

2.9. Semiquantitative RT-PCR Assay

Semiquantitative RT-PCR assay was performed on 200 ng of total RNA that was reverse-transcribed using the Enhanced AMV Reverse Transcriptase Kit. The cDNA pool was diluted 10-fold and 1 µL was used as a template in PCR-amplification using gene-specific primers (Table 2). Amplification conditions were as follows: 94 °C for 2 min, then 25 cycles of 94 °C for 15 s, 60 °C for 30 s, and 68 °C for 1 min. For each sample, *ACT1* mRNA was used as an internal control. *SDH1* mRNA was used as a control of inducible expression under nonfermentable carbon source growth conditions.

2.10. Other Assays

Fumarase (FUM), succinate dehydrogenase (SDH) and phosphoglucoisomerase (PGI) activities were spectrophotometrically measured as in [40,42,44]. Protein concentration was assayed according to Bradford [45], using bovine serum albumin as the standard.

3. Results

3.1. Mitochondrial Localization of Recombinant Fad1p

Our previous results based on functional approaches showed that, besides in the cytosol, FAD synthesizing activity can be specifically revealed in mitochondria from *S. cerevisiae* [40,46]. This result was somewhat surprising since a single gene, named *FAD1*, was reported as coding for a sole cytosolic FADS [16]. *FAD1* gene, as normally occurs in yeasts, has no introns; thus, different transcripts generated by alternative splicing are not expected. Consistently, a single transcript is reported in the Ensemble database (http://www.ensembl.org/index.html, accessed on 13 September 2021) from which a single translation product, Fad1p, is predictable with NCBI tool ORF finder (www.ncbi.nlm.nih.gov/orffinder, accessed on 13 September 2021). No mitochondrial-targeting peptide could be found in this product using several prediction programs, such as iPSORT, MitoProt, TargetP and PSORT II. Concerning these aspects, the yeast *FAD1* differs from the human *FLAD1* gene, which generates alternatively spliced isoforms with different subcellular localizations [29,30]. Nevertheless, one of these isoforms, hFADS6, strongly resembles the yeast protein Fad1p, showing a high level of conservation for the PAPS reductase domain (Figure 1).

```
Fad1p     ------------------------------------------------------------  0
hFADS6    MKGLFQNPAVQFHSKELYVAADEASIAPILAEAQAHFGRRLGLGSYPDWGSNYYQVKLTL  60

Fad1p     -----------------------------MQLSKAAEMCYEITNSYLHIDQKSQIIASTQ  31
hFADS6    DSEEEGPLEECLAYLTARLPQGSLVPYMPNAVEQASEAVYKLAESGSSLGKKV---A---  114
                                        :.:*:*  *::::*   :.:*    *

Fad1p     EAIRLTRKYLLSEIFVRWSPLNGEISFSYNGGKDCQVLLLLYLSCLWEYFFIKAQNSQFD  91
hFADS6    GALQT-----IETSLAQY--SLTQLCVGFNGGKDCTALLHLFHAAVQRKL----------  157
           *::      :.  :.::     ::...:****** .** *: :.: . :

Fad1p     FEFQSFPMQRLPTVFIDQEETFPTLENFVLETSERYCLSLYESQRQSGASVNMADAFRDF  151
hFADS6    ---PDV-PNPLQILYIRSISPFPELEQFLQDTIKRYNLQMLEAEG------SMKQALGEL  207
             ..  : *  ::* . . ** **:*: :* :** *.: *::       .* :*: ::

Fad1p     IKIYPETEAIVIGIRHTDPFGEALKPIQRTDSNWPDFMRLQPLLHWDLTNIWSFLLYSNE  211
hFADS6    QARHPQLEAVLMGTRRTDPYSCSLCPFSPTDPGWPAFMRINPLLDWTYRDIWDFLRQLFV  267
             :*: **:::* *:***:. :* *:. ** .** ***::***.*   :**.**

Fad1p     PICGLYGKGFTSIGGINNSLPNPHLRKDSNNPALHFEWEIIHAFGKDAEGERSSAINTSP  271
hFADS6    PYCILYDRGYTSLGSRENTVRNPALKCLSPGGHPTYRPAYL--LENE-EERNSRT----  320
          * * **.:*:**:*. :*:: ** *:  * .     :.   :  : :: * **.*

Fad1p     ISVVDKERFSKYHDNYYPGWYLVDDTLERAGRIKN  306
hFADS6    -----------------------------------  320
```

Figure 1. Alignment of FAD synthase sequences from yeast and humans. The alignment of yeast Fad1p with human FAD synthase isoform 6 (hFADS6) was performed by the *Pairwise Sequence Alignment* tool available at the EMBL-EBI website. The PAPS reductase domain (InterPro accession: IPR002500), present in both *S. cerevisiae* Fad1p (amino acids 55-231) and hFADS6 (amino acids 132-286), is underlined and colored in red. Asterisks depict identical residues in the same positions, whereas colons and periods indicate strongly and weakly similar residues, respectively.

To gain some insight into the cellular destination of Fad1p, we investigated the subcellular localization of the recombinant form of Fad1p carrying the STREP-tag at the C-terminus and overexpressed in the CENPK2-1C strain (Table 1). Therefore, subcellular fractions (spheroplasts, mitochondria and cytoplasm) were prepared from CENPK2-1C$^{FAD1\text{-}STREP}$ cells grown in ethanol up to early exponential growth phase and equal amounts of each fraction were analyzed by immunoblotting using the α-STREP antibody (Figure 2a). Even though the recombinant Fad1p–STREP was enriched in the cytoplasmic fraction, an evident immuno-reactive band was observed in the mitochondrial fraction, suggesting a mitochondrial localization for the recombinant Fad1p–STREP. Interestingly, the recombinant Fad1p–STREP was immunodecorated at the same size both in mitochondria and in cytoplasm. When the STREP-tag was moved from the C- to the N-terminal end, the ability of the recombinant STREP–Fad1p to enter mitochondria was completely lost, accompanied by a significant enrichment in the cytoplasmic fraction (Figure S1).

To further prove the ability of the recombinant Fad1p to enter mitochondria, we placed Fad1p–STREP expression under the control of HXT7 promoter in the multicopy plasmid p426-HXT7-FAD1-STREP (Table 1) and used a different genetic strain, namely, K26. Subcellular fractions were prepared from K26$^{FAD1\text{-}STREP}$ cells grown in ethanol at the early exponential phase and analyzed by immunoblotting carried out with the α-STREP (Figure 2b). The recombinant protein Fad1p–STREP was immunodecorated at the same size in both cytoplasm and mitochondria, where it was significantly enriched. When the K26$^{FAD1\text{-}STREP}$ cells were in the presence of the uncoupler carbonyl cyanide-p-trifluoromethoxy-phenylhydrazone (FCCP), the Fad1p–STREP import into mitochondria was prevented (Figure 2b), as expected for a process depending on mitochondrial membrane potential [47]. This finding clearly indicates that the recombinant Fad1p is imported and localized to mitochondria.

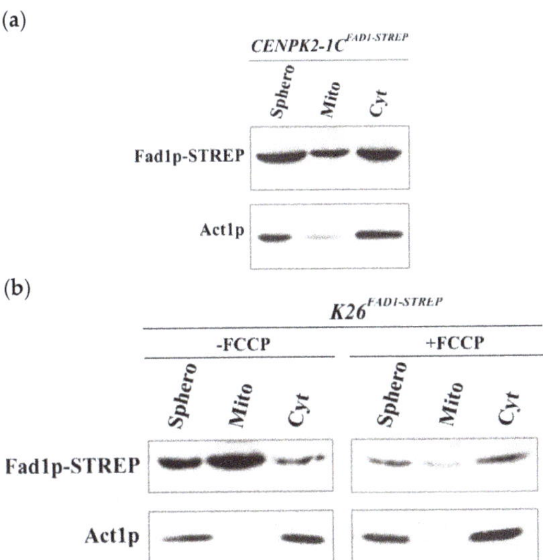

Figure 2. Mitochondrial localization of recombinant Fad1p. Spheroplasts (Sphero), mitochondria (Mito) and cytoplasm (Cyt) were prepared from CENPK2-1C (**a**) and K26 (**b**) cells transformed with the multicopy recombinant plasmid (Table 1). Both strains were grown up to the early exponential phase on ethanol. K26$^{FAD1\text{-}STREP}$ cells were incubated in the absence or presence of FCCP (25 μM). Proteins (0.1 mg) were separated by SDS-PAGE, transferred onto PVDF membrane and detected with either α-STREP or α-ACTIN antibodies.

3.2. Natural Fad1p Exists in Mitochondria

To ascertain whether the activity of FADS that we detected inside mitochondria is actually accountable to Fad1p, we analyzed the subcellular localization of natural Fad1p in cells grown either on nonfermentable (glycerol and ethanol) or fermentable (glucose and galactose) carbon sources up to the early exponential phase (5 h) (Figure 3a). A Fad1p band was detected in mitochondria prepared from glycerol-grown cells by using the α-FADS antiserum. It should be noted that even though actin (Act1p) was used as a cytosolic marker [48], a minimal amount of Act1p was also found to be associated to the mitochondrial fraction in agreement with [49]. In ethanol, an even more intense band was detected in the mitochondrial fraction, whereas in glucose-grown cells, the mitochondrial Fad1p level was very low. Thus, the Fad1p content in mitochondria seems to depend on the carbon source. As a control, the dependence of the mitochondrial SDH specific activity on different carbon sources was analyzed (Figure 3b), thus verifying the expected glucose repression due to both transcriptional and post-transcriptional events [50,51]. It is noteworthy that in the mitochondria prepared from galactose-grown cells, no band was detected (Figure 3a), even though we observed a high (induced) level of SDH activity (Figure 3b), in agreement with the presence of mitochondria in cells cultured in galactose as the carbon source [52]. Differently from what was observed for the mitochondrial Fad1p, the cytosolic echoform did not appear to depend on the carbon source (Figure 3a).

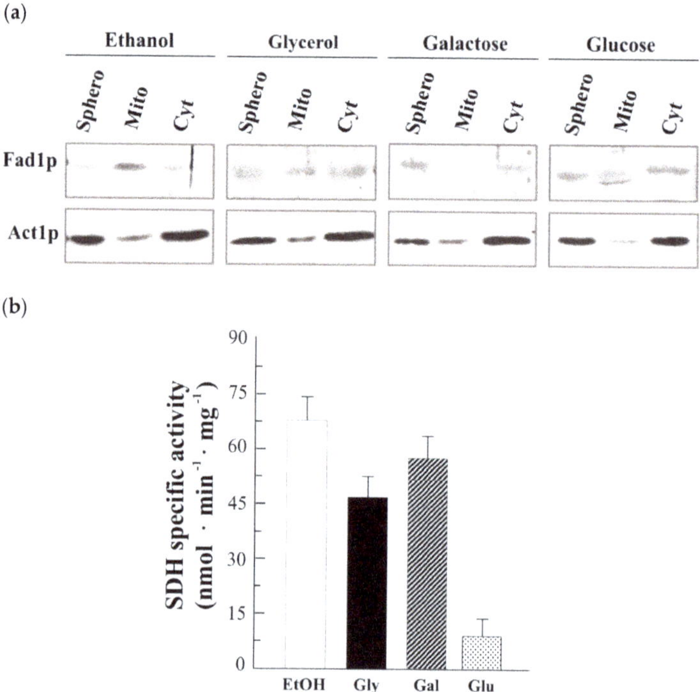

Figure 3. Subcellular distribution of Fad1p in different carbon sources. (**a**) Spheroplasts (Sphero), mitochondria (Mito) and cytoplasm (Cyt) were prepared from WT cells grown up to the early exponential phase at 30 °C in YEP liquid medium supplemented with the indicated carbon sources (2% each). Proteins (0.1 mg) were separated by SDS-PAGE, transferred onto PVDF membrane and detected with either α-FADS or α-ACTIN antibodies. (**b**) The carbon source dependence of the mitochondrial SDH-specific activity is reported as the means (±SD) of two independent measurements.

Cellular sublocalization of Fad1p was also evaluated in K26 cells grown in glycerol up to the stationary phase (24 h), as in [40]. Cytoplasmic and mitochondrial fractions were prepared from spheroplasts and equal amounts of each fraction were used to reveal Fad1p by immunoblotting analysis (Figure S2). A faint α-FADS immuno-reactive band was found in the cytoplasmic fraction. A much more evident band migrating at the same Mr was observed in total mitochondrial fraction, consistently enriched with respect to the spheroplast specific amount. A similar enrichment was observed for the fumarase (FUM) and succinate dehydrogenase (SDH), two mitochondrial markers (Figure S2, histogram). The absence of cytosolic contamination in mitochondria was demonstrated by the absence of phosphoglucoisomerase (PGI) activity, a marker of cytosolic fraction (Figure S2, histogram). Isolated mitochondria were further subfractionated in a membrane-enriched fraction (M_{fr}) and in a soluble fraction (S_{fr}). The Fad1p band was clearly detectable in both submitochondrial fractions, whose purity was confirmed by measuring the inner membrane marker SDH and the matrix marker FUM enzymatic activities (Figure S2, histogram).

Overall, on a molecular basis, these results support our previous functional data, proving that Fad1p exists in yeast mitochondria. Since no difference in size was observed between cytoplasmic and mitochondrial Fad1p, we postulated that the *FAD1* gene might generate two identical echoforms, destined to two distinct subcellular localizations.

3.3. Detection of Two FAD1 mRNAs with Different 3′UTR and Their Carbon Source Dependence Profile

In order to unravel the mechanism responsible for the subcellular destination of Fad1p, we considered a series of indications derived from the literature, which proposes a role for *cis*-acting 3′UTR elements in targeting transcripts to mitochondria [53,54], and from a genome-wide analysis, which suggests a high probability for *FAD1* mRNA to be located on mitochondria-bound polysomes [54]; hence, the protein would be imported while being translated in a co-translational process [55–57]. Therefore, we hypothesized that important information for Fad1p localization could reside in the 3′UTR of *FAD1* mRNA. Since no information was available, a combination of bioinformatic analyses and experimental procedures were carried out to gain knowledge about this untranslated region.

Firstly, we searched for putative polyadenylation sites (PASs), despite their complexity [58,59], and found two canonical signals (ATTAAA) in the 1 kbp region downstream of *FAD1* ORF using DNAFSMiner [60]. The proximal one was placed 50 nucleotides after the stop codon and the distal one was placed 339 nucleotides after the stop codon. The bioinformatic tool also predicted other different PASs with similar scores, namely, AATAAA and TATCAA (Figure S3). These predictions suggested that *FAD1* mRNA may be present in different forms, whose length depends on the alternative use of distinct PASs in the transcript 3′UTR.

To test this hypothesis, 3′RACE experiments were performed on a template cDNA obtained from purified poly(A)$^+$ RNA prepared from WT cells grown at the early exponential phase on glucose as a carbon source and the gene-specific primers (Table 2; Figure 4a). When the primers FAD1A.for and AP were used, a single amplified product of about 400 bp was obtained (Figure 4b, lane 1). The specificity of the 400 bp product was verified by a semi-nested PCR using the gene-specific primer FAD1C.for located 91 nucleotides downstream of FAD1A.for. The amplified product was obtained at the expected size, which is about 300 bp (Figure 4b, lane 2). The 400 bp fragment was sequenced, revealing the existence of *FAD1* transcript with a 3′UTR of 128 nucleotides (Figure S3). To search for the existence of a longer 3′UTR region of *FAD1* mRNA, a PCR was performed using the primers FAD1C.for and FAD1B.rev, the last being located downstream of the end of the 3′UTR previously identified (Figure 4b); in this case, a product of about 450 nucleotides was obtained (Figure 4b, lane 3), indicating the existence of a longer *FAD1* mRNA. To define the entire length of the 3′UTR of this transcript, a PCR reaction was performed using the primers FAD1B.for and AP, obtaining a product of 550 bp (Figure 4b, lanes 4 and 5). Its specificity was proved by a semi-nested PCR with the primers FAD1B.for and FAD1B.rev, which gave rise to a fragment of about 100 bp, as expected (Figure 4b, lane 6). The 550 bp product was sequenced, showing the existence of an additional *FAD1* mRNA with a longer 3′UTR region of 759 bp (Figure S2). Similar results were obtained when the analysis was performed using cDNA as a template, obtained starting from purified poly(A)$^+$ RNA prepared from WT cells grown on glycerol (Figure S4). Thus, *FAD1* generates (at least) two *FAD1* mRNAs with different 3′UTR lengths. Based on the different length of the 3′UTR, we named them 'short *FAD1* mRNA' and 'long *FAD1* mRNA', respectively.

To understand whether or not the length of 3′UTR is responsible for the fate of Fad1p echoforms, the carbon source dependence of the *FAD1* transcript levels was investigated. To this purpose, total RNA was extracted from WT cells grown at the early exponential phase on different carbon sources, and the amount of *FAD1* transcripts was evaluated by semiquantitative RT-PCR, with *ACT1* mRNA used as an internal control, essentially as in [42]. The long *FAD1* mRNA was amplified using the primer pair FAD1A.for and FAD1B.rev, whereas the total *FAD1* mRNA (short + long) was amplified using the primer pair FAD1A.for and FAD1A.rev (Figure S5). The RT-PCR product (540 bp) relative to the long *FAD1* mRNA complemented the 3′RACE data, confirming the presence of a longer transcript. The relative amount of total *FAD1* mRNA did not depend on the carbon source, whereas a carbon source dependence was observed for the long *FAD1* mRNA: its amount was higher in ethanol grown cells, reduced in cells grown on glucose and almost absent in galactose-grown cells. As a control, the expression pattern of *SDH1* mRNA was analyzed; it was repressed by glucose, but not by galactose (Figure S5), in agreement with SDH activity reported here (Figure 3, histogram) and in literature [50].

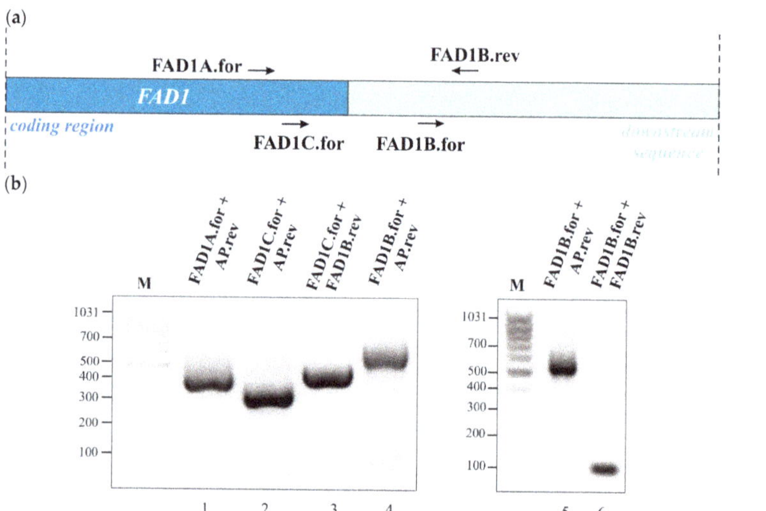

Figure 4. Characterization of *FAD1* mRNA 3′UTR by 3′RACE analysis. (**a**) *FAD1* coding region and 1 kbp downstream region are schematized. Arrows designate the position and direction of gene-specific primers used for 3′RACE. (**b**) The experiment was carried out on a poly(A)$^+$ RNA prepared from WT cells grown up to the early exponential phase at 30 °C on glucose. PCR products were separated by a 2% agarose gel. Fragments in lanes 1 and 4 were gel-extracted, purified and sequenced (Figure S3). On the left side of each gel, band sizes (bp) of the marker [M, *MassRuler Low Range DNA Ladder*, ThermoFisher Scientific (Waltham, MA, USA)] are indicated.

3.4. A Mitochondrial Localization Motif in FAD1 3′UTRs

To further confirm that Fad1p localization mechanism might be influenced at the mRNA level, we searched for *cis*-acting elements that could be responsible for targeting *FAD1* mRNA to the outer mitochondrial membrane. Sequence inspection revealed the existence of a putative *cis*-acting motif TGTATATACA containing the consensus mitochondrial localization motif M1 (TGTA(a/c/t/)ATA), as defined in [61], and resembling the motif 6 (WTATWTACADG) reported as a mitochondrial downstream motif [62]. This element was located about 90 nucleotides downstream the *FAD1* ORF stop codon; therefore, it was present in both *FAD1* transcripts. The same motif (or similar sequences) was not found in either the upstream *FAD1* ORF region or the *FAD1* ORF.

The 1 kbp genomic downstream region *FAD1* ORF contains, on the opposite strand the *MRP10* gene, encoding a mitochondrial ribosomal protein of 10.7 kDa [63,64]. Mrp10p lacks a mitochondrial-targeting sequence (as revealed by bioinformatic analysis); however, the identified M1 mitochondrial localization motif is present in its 3′UTR, as revealed by the Shalgi's database [61]. This motif is placed 56 nucleotides downstream of the *MRP10* ORF stop codon. It should be noted that the M1 motif in 3′UTR of *FAD1* and *MRP10* is a palindromic sequence that could be read on both DNA strands (Figure 5). This property implies a rather nontrivial extension of the M1 motif and allows the same region to play the role of mitochondrial localization motif for both the *MRP10* and the *FAD1* transcripts.

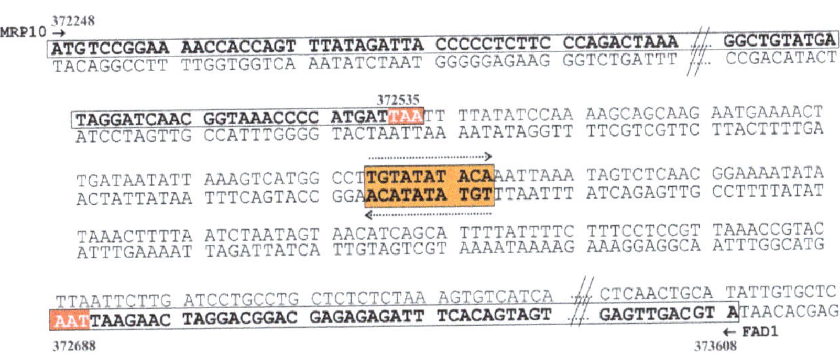

Figure 5. The *cis*-acting motif M1 in the 3′UTR of *FAD1* transcripts. Schematic representation of the partial sequence of the *S. cerevisiae* genomic region comprised between nucleotides 372248 and 373617 (chromosome IV). The white boxes indicate *FAD1* ORF on the Crick strand and *MRP10* ORF on the *Watson* strand; the orange box indicates the mitochondrial localization motif M1, placed about 87 and 55 nucleotides downstream of the stop codon (TAA, highlighted in red) of *FAD1* and *MRP10* ORFs, respectively. The dotted arrows highlight the bidirectionality of the M1 palindromic sequence.

4. Discussion

Experiments reported in this paper point out the issue of the origin of mitochondrial FAD in *S. cerevisiae* mitochondria, strictly related with the problem of the identity and of the subcellular localization of the FAD forming enzyme, FADS. Immunoblotting experiments reported here reveal the presence of an α-FADS immuno-reactive band in mitochondria, showing the same Mr of that found in the spheroplasts and cytoplasm. This is in line with the absence of a putative cleavable presequence in Fad1p, the sole peptide generated by *FAD1* gene. Thus, these results prove, on a molecular basis, Fad1p localization to yeast mitochondria and support the absence of a cleavable mitochondrial signal peptide in Fad1p

It is well-documented that in eukaryotic cells, a protein can be located to two different subcellular compartments; this dual localization results in different echoforms—that is, proteins with identical or nearly identical amino acid sequences distinctly placed in the cell [65–67]. This is an important and frequent phenomenon; in fact, more than one third of the yeast mitochondrial proteome seems to be dual localized [68,69]. It has also become evident that dual localization can be regulated, induced and rebalanced in response to either cellular signaling or changing extracellular conditions. Different dual-targeting mechanisms of mitochondrial proteins have been described or suggested, and, in some case, they seem to be tightly regulated in time, location and function [65]. Our results demonstrate that *FAD1* generates two echoforms.

Analyzing the dependence of the Fad1p levels on fermentable and nonfermentable carbon sources (Figure 3), we verified that the amount of Fad1p in mitochondria, but not in the cytoplasm, depends on the carbon source being higher in ethanol and glycerol than in the fermentable glucose. Interestingly, mitochondrial Fad1p almost disappeared in galactose-grown cells. This finding may settle the discrepancy between results from our (see Introduction) and other laboratories [32,39] about the existence of mitochondrial FADS and, consequently, about the direction of FAD transport via the inner mitochondrial membrane translocator Flx1p. The absence of mitochondrial Fad1p, observed in galactose-grown cells, implies that in this condition FAD must be transported from cytosol to mitochondria via Flx1p (Figure 3 [32,39]). Vice versa, in a nonfermentable carbon source and in glucose, in the presence of mitochondrial Fad1p, Flx1p is expected to mediate FAD efflux from mitochondria to cytosol acting as a regulator of apo-flavoprotein biogenesis [40,42].

Using yeast strains overexpressing a recombinant form of Fad1p, the dual location of natural Fad1p was confirmed. Our data demonstrate that Fad1p is, per se, able to enter mitochondria, despite the absence of a canonical mitochondrial-targeting sequence. A "mitochondrial destination" message should be localized at the N-terminal end, since Fad1p loses its ability when the STREP-tag is moved from the C- to the N-terminal end (Figure S1). Presumably, the α1-helix in the first twenty amino acids at the N-terminal end, as well as the contiguous α2-helix, indeed have amphipathic characteristics, with the basic residues exposed to the solvent from the same side of the central axis, as revealed by the crystal structure solved at 1.9 Å resolution (PDB entry: 2wsi) [20]. This would be in agreement with the involvement of this protein moiety in the import process. Alternatively, another putative noncanonical/weak mitochondrial-targeting sequence [70] could be implicated in the mitochondrial destination of Fad1p.

We also observed that in galactose-grown cells, the recombinant Fad1p-STREP entered mitochondria per se [71], behaving differently from the natural Fad1p that was unable to enter mitochondria. All together, these data suggest the existence of a fine mechanism regulating the localization of the natural Fad1p, presumably at the transcript level and dependent on the carbon source. It is well-established that intracellular localization of mRNA has a significant impact on the efficiency of its translation and, presumably, determines its response to changing metabolic conditions or cellular stress [72].

Our attention was focused on the 3′UTR region of *FAD1* mRNA. Since no information was available on this region, we carried out a 3′RACE analysis and found the existence of (at least) two *FAD1* mRNAs with different 3′UTR lengths (Figure 4). This finding could be relevant for Fad1p cell biology, since it is well-known that 3′UTR regulates multiple aspects of mRNA metabolism, including subcellular localization, translation efficiency and stability, in cooperation with different RNA binding proteins [73]. A mechanism that could have a role in generating different transcripts with specific subcellular localizations and/or functions might be alternative polyadenylation [74,75]. We predicted different canonical and noncanonical putative PASs localized in the first 1 kbp downstream of *FAD1* ORF. Whether or not these PASs are actually active and responsible for the generation of the two *FAD1* transcripts are matter of ongoing investigation in our laboratory.

Searching for possible *cis*-acting motifs responsible for the peripheral mitochondrial localization of *FAD1* transcripts, we revealed the existence of the mitochondrial localization motif M1, placed 87 nucleotides downstream the *FAD1* ORF stop codon (Figure 5). The same motif has been identified within a set of genes encoding mitochondrial proteins, whose mRNAs are translated by polyribosomes attached to the cytosolic side of the outer mitochondrial membrane [76]. This motif is present in both *FAD1* transcripts, which might localize close to the mitochondrial outer membrane. Intriguingly, this finding is in agreement with a value of mitochondrial localization of mRNA (MLR) for *FAD1* equal to 87 [54]. This value is similar to the MLR values measured for other mRNAs, whose products have been demonstrated to be internalized into mitochondria (i.e., 87.8 for *SDH1* transcript), thus indicating that Fad1p translation has a high probability to occur on the mitochondrial outer membrane. Recently, it has been reported that mRNAs that initiate translation away from mitochondria experience a significant drop in mobility toward the outer mitochondrial membrane and tend to remain there [77]. The M1 motif is a candidate binding site of the RNA binding protein Puf3p [78], localized on the cytosolic side of the outer mitochondrial membrane, where it interacts with some components of the mitochondrial protein import machinery [57,79]. Besides its role in mRNA targeting to mitochondria, Puf3p controls mRNA stability and translation efficiency according to growth conditions and in response to oxidative stress [80–82]. The identified palindromic M1 motif is shared by the 3′UTR of *FAD1* transcripts and the 3′ downstream region of *MRP10* gene, which is transcribed in the opposite direction with respect to *FAD1* gene. Since an interaction between the M1 of *MRP10* mRNA and Puf3p has been proven [83,84], an involvement of Puf3p in mitochondrial localization of Fad1p could be expected. By analyzing the carbon source dependence profile of *FAD1* transcripts (Figure S3), we found that the long transcript is expressed in ethanol and in glycerol at a higher level than in glucose; it is almost absent in galactose. The comparison between this profile and mitochondrial Fad1p levels (Figure 3) shows that the carbon source dependence of the long

FAD1 mRNA level strictly parallels the carbon source dependence of the mitochondrial Fad1p level. This observation suggests that the long *FAD1* mRNA might be responsible for the generation of Fad1p mitochondrial echoform. Therefore, although the M1 motif is present in both the short and long *FAD1* transcripts, we propose that the different length of the 3′UTRs could be responsible for the distinct subcellular destinations of Fad1p. This diverse destination could be explained by a differential Puf3p–M1 interaction caused by the exposure of M1 motif in different RNA contexts. In this scenario, the secondary structure of the two 3′UTRs may be causative of the selective interaction with Puf3p. Moreover, the recruitment of additional RNA binding proteins, presumably involved in mRNA decay or Puf3p interaction with mitochondrial import machinery [85], could also be affected by 3′UTR length and structure. Therefore, Puf3p might regulate the dynamic balance between translation efficiency, protein folding and import rate, thus establishing whether Fad1p may be released in the cytosol or imported into mitochondria [81]. Additionally, it has been shown that cytosolic ribosomes have the ability to 'sense' features relative to nascent polypeptide chains; between those features, the sequence and structure of 3′UTR mRNAs can influence the polypeptide synthesis rate, modulating translation accordingly [76,86]. It is noteworthy that the M1 motif was not included in the constructs used for the expression of recombinant Fad1p. Nonetheless, the protein was capable, per se, of localizing to mitochondria in the presence of the mitochondrial membrane potential (Figure 2; [71]). Therefore, the M1 sequence would serve to fine-tune the cellular localization of Fad1p echoforms in various metabolic conditions, such as different carbon sources.

Our hypothesis is summarized in Figure 6. We speculate that, when Puf3p is bound to the M1 motif present in the long 3′UTR context (panel a), Fad1p synthesis efficiency and/or folding rate are slowed down, allowing the co-translational import to proceed and generate the Fad1p mitochondrial echoform. Vice versa, when Puf3p interacts with the M1 in the short 3′UTR (panel b), a higher translation rate might cause a premature folding, thus favoring the cytosolic destination of Fad1p.

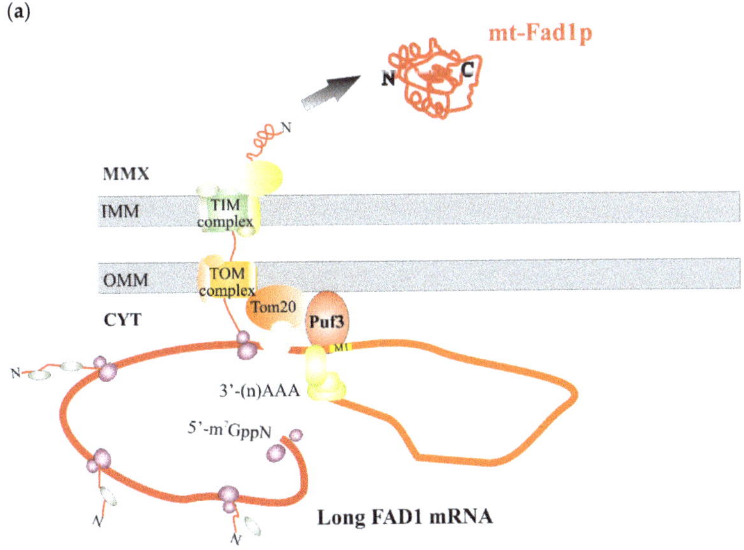

Figure 6. *Cont.*

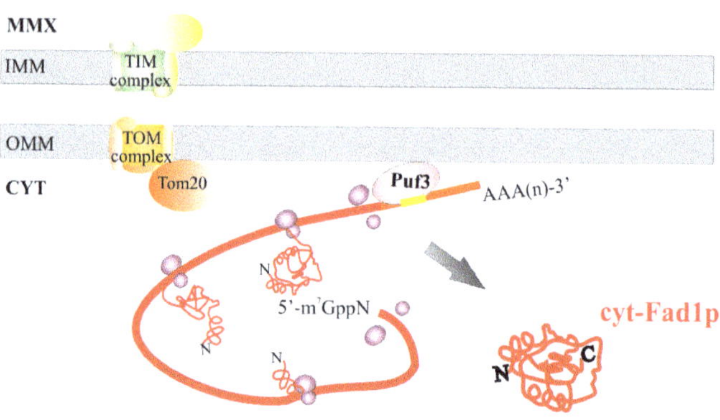

Figure 6. Model depicting the dual location of Fad1p. The cartoon shows a possible role of Puf3p in the different destination of Fad1p echoforms. (**a**) Puf3p binds to the long *FAD1* mRNA and drives the nascent polypeptide in the mitochondrial matrix (MMX) via TOM–TIM complexes before the fully translated protein starts to fold. (**b**) Puf3p binds to the short *FAD1* mRNA. As translation is presumably not sufficiently slowed down, premature and rapid folding of the nascent polypeptide in the cytosol (CYT) prevents Fad1p import across mitochondrial membranes (OMM—outer mitochondrial membrane; IMM—inner mitochondrial membrane).

Supplementary Materials: The following are available online at https://www.mdpi.com/article/10.3390/life11090967/s1. Figure S1: Sublocalization of recombinant STREP-Fad1p. Figure S2: Endogenous Fad1p mitochondrial localization in WT cells grown up to the stationary phase (24 h) on glycerol. Figure S3: *FAD1* ORF and its 1 kbp downstream region. Figure S4: 3′RACE analysis carried out on cells grown on glycerol (early exponential phase, 30 °C). Figure S5: Carbon source dependence of *FAD1* transcripts.

Author Contributions: Conceptualization, M.B., M.R., M.C., and E.B.; methodology, E.B. and M.R.; formal analysis, M.C., P.L., and F.B.; investigation, F.B., M.O., T.A.G., M.T., and P.L.; resources, E.B., M.R., and M.B.; writing—original draft preparation, M.B. and T.A.G.; writing—review and editing, M.R., F.B., and M.T.; supervision, M.C. and M.R.; project administration, M.B., M.O., M.R., and M.C.; funding acquisition, M.B. All authors have read and agreed to the published version of the manuscript.

Funding: This work was supported by grants from MIUR PON-ricerca e competitività 2007–2013 (project PON01_00937: "Modelli sperimentali biotecnologici integrati per lo sviluppo e la selezione di molecole di interesse per la salute dell'uomo") and from Università degli Studi di Bari "Aldo Moro" (Progetti Competitivi "Effetto di mutazioni di FLAD1 e di alterazioni dell'omeostasi delle flavine sullo stato redox e sulla biogenesi mitocondriale: uno studio integrato su fibroblasti umani") to M.B.

Institutional Review Board Statement: Not applicable.

Informed Consent Statement: Not applicable.

Data Availability Statement: Not applicable.

Acknowledgments: The helpful collaboration of Emilia Dipalo, who participated as a student in the early stages of this work, is gratefully acknowledged.

Conflicts of Interest: The authors declare no conflict of interest.

References

1. Tolomeo, M.; Nisco, A.; Leone, P.; Barile, M. Development of Novel Experimental Models to Study Flavoproteome Alterations in Human Neuromuscular Diseases: The Effect of Rf Therapy. *Int. J. Mol. Sci.* **2020**, *21*, 5310. [CrossRef]
2. Joosten, V.; Van Berkel, W.J. Flavoenzymes. *Curr. Opin. Chem. Biol.* **2007**, *11*, 195–202. [CrossRef] [PubMed]
3. Macheroux, P.; Kappes, B.; Ealick, S.E. Flavogenomics—A genomic and structural view of flavin-dependent proteins. *FEBS J.* **2011**, *278*, 2625–2634. [CrossRef] [PubMed]
4. Barile, M.; Giancaspero, T.A.; Leone, P.; Galluccio, M.; Indiveri, C. Riboflavin transport and metabolism in humans. *J. Inherit. Metab. Dis.* **2016**, *39*, 545–557. [CrossRef]
5. Pimentel, C.; Batista-Nascimento, L.; Rodrigues-Pousada, C.; Menezes, R.A. Oxidative stress in Alzheimer's and Parkinson's diseases: Insights from the yeast Saccharomyces cerevisiae. *Oxidative Med. Cell. Longev.* **2012**, *2012*, 132146. [CrossRef]
6. Botstein, D.; Fink, G.R. Yeast: An experimental organism for 21st Century biology. *Genetics* **2011**, *189*, 695–704. [CrossRef]
7. Tenreiro, S.; Outeiro, T.F. Simple is good: Yeast models of neurodegeneration. *FEMS Yeast Res.* **2010**, *10*, 970–979. [CrossRef] [PubMed]
8. Bastow, E.L.; Gourlay, C.W.; Tuite, M.F. Using yeast models to probe the molecular basis of amyotrophic lateral sclerosis. *Biochem. Soc. Trans.* **2012**, *39*, 1482–1487. [CrossRef] [PubMed]
9. Reihl, P.; Stolz, J. The monocarboxylate transporter homolog Mch5p catalyzes riboflavin (vitamin B2) uptake in Saccharomyces cerevisiae. *J. Biol. Chem.* **2005**, *280*, 39809–39817. [CrossRef] [PubMed]
10. Moriyama, Y. Riboflavin transporter is finally identified. *J. Biochem.* **2011**, *150*, 341–343. [CrossRef]
11. Yonezawa, A.; Inui, K. Novel riboflavin transporter family RFVT/SLC52: Identification, nomenclature, functional characterization and genetic diseases of RFVT/SLC52. *Mol. Asp. Med.* **2013**, *34*, 693–701. [CrossRef]
12. O'Callaghan, B.; Bosch, A.M.; Houlden, H. An update on the genetics, clinical presentation, and pathomechanisms of human riboflavin transporter deficiency. *J. Inherit. Metab. Dis.* **2019**, *42*, 598–607. [CrossRef]
13. Yruela, I.; Arilla-Luna, S.; Medina, M.; Contreras-Moreira, B. Evolutionary divergence of chloroplast FAD synthetase proteins. *BMC Evol. Biol.* **2010**, *10*, 311. [CrossRef]
14. Giancaspero, T.A.; Colella, M.; Brizio, C.; Difonzo, G.; Fiorino, G.M.; Leone, P.; Brandsch, R.; Bonomi, F.; Iametti, S.; Barile, M. Remaining challenges in cellular flavin cofactor homeostasis and flavoprotein biogenesis. *Front. Chem.* **2015**, *3*, 30. [CrossRef]
15. Santos, M.A.; Jimenez, A.; Revuelta, J.L. Molecular characterization of FMN1, the structural gene for the monofunctional flavokinase of Saccharomyces cerevisiae. *J. Biol. Chem.* **2000**, *275*, 28618–28624. [CrossRef]
16. Wu, M.; Repetto, B.; Glerum, D.M.; Tzagoloff, A. Cloning and characterization of FAD1, the structural gene for flavin adenine dinucleotide synthetase of Saccharomyces cerevisiae. *Mol. Cell. Biol.* **1995**, *15*, 264–271. [CrossRef]
17. Bauer, S.; Kemter, K.; Bacher, A.; Huber, R.; Fischer, M.; Steinbacher, S. Crystal structure of Schizosaccharomyces pombe riboflavin kinase reveals a novel ATP and riboflavin-binding fold. *J. Mol. Biol.* **2003**, *326*, 1463–1473. [CrossRef]
18. Karthikeyan, S.; Zhou, Q.; Mseeh, F.; Grishin, N.V.; Osterman, A.L.; Zhang, H. Crystal structure of human riboflavin kinase reveals a beta barrel fold and a novel active site arch. *Structure* **2003**, *11*, 265–273. [CrossRef]
19. Karthikeyan, S.; Zhou, Q.; Osterman, A.L.; Zhang, H. Ligand binding-induced conformational changes in riboflavin kinase: Structural basis for the ordered mechanism. *Biochemistry* **2003**, *42*, 12532–12538. [CrossRef] [PubMed]
20. Leulliot, N.; Blondeau, K.; Keller, J.; Ulryck, N.; Quevillon-Cheruel, S.; Van Tilbeurgh, H. Crystal structure of yeast FAD synthetase (Fad1) in complex with FAD. *J. Mol. Biol.* **2010**, *398*, 641–646. [CrossRef] [PubMed]
21. Brizio, C.; Galluccio, M.; Wait, R.; Torchetti, E.M.; Bafunno, V.; Accardi, R.; Gianazza, E.; Indiveri, C.; Barile, M. Over-expression in Escherichia coli and characterization of two recombinant isoforms of human FAD synthetase. *Biochem. Biophys. Res. Commun.* **2006**, *344*, 1008–1016. [CrossRef] [PubMed]
22. Leone, P.; Galluccio, M.; Brizio, C.; Barbiroli, A.; Iametti, S.; Indiveri, C.; Barile, M. The hidden side of the human FAD synthase 2. *Int. J. Biol. Macromol.* **2019**, *138*, 986–995. [CrossRef]
23. Giancaspero, T.A.; Locato, V.; Barile, M. A regulatory role of NAD redox status on flavin cofactor homeostasis in S. cerevisiae mitochondria. *Oxidative Med. Cell. Longev.* **2013**, *2013*, 612784. [CrossRef]
24. Vergani, L.; Barile, M.; Angelini, C.; Burlina, A.B.; Nijtmans, L.; Freda, M.P.; Brizio, C.; Zerbetto, E.; Dabbeni-Sala, F. Riboflavin therapy. Biochemical heterogeneity in two adult lipid storage myopathies. *Brain A J. Neurol.* **1999**, *122 Pt 12*, 2401–2411. [CrossRef]
25. Gianazza, E.; Vergani, L.; Wait, R.; Brizio, C.; Brambilla, D.; Begum, S.; Giancaspero, T.A.; Conserva, F.; Eberini, I.; Bufano, D.; et al. Coordinated and reversible reduction of enzymes involved in terminal oxidative metabolism in skeletal muscle mitochondria from a riboflavin-responsive, multiple acyl-CoA dehydrogenase deficiency patient. *Electrophoresis* **2006**, *27*, 1182–1198. [CrossRef] [PubMed]
26. Olsen, R.K.J.; Konarikova, E.; Giancaspero, T.A.; Mosegaard, S.; Boczonadi, V.; Matakovic, L.; Veauville-Merllie, A.; Terrile, C.; Schwarzmayr, T.; Haack, T.B.; et al. Riboflavin-Responsive and -Non-responsive Mutations in FAD Synthase Cause Multiple Acyl-CoA Dehydrogenase and Combined Respiratory-Chain Deficiency. *Am. J. Hum. Genet.* **2016**, *98*, 1130–1145. [CrossRef]
27. Gregersen, N.; Andresen, B.S.; Pedersen, C.B.; Olsen, R.K.; Corydon, T.J.; Bross, P. Mitochondrial fatty acid oxidation defects–remaining challenges. *J. Inherit. Metab. Dis.* **2008**, *31*, 643–657. [CrossRef]
28. Leone, P.; Galluccio, M.; Barbiroli, A.; Eberini, I.; Tolomeo, M.; Vrenna, F.; Gianazza, E.; Iametti, S.; Bonomi, F.; Indiveri, C.; et al. Bacterial Production, Characterization and Protein Modeling of a Novel Monofuctional Isoform of FAD Synthase in Humans: An Emergency Protein? *Molecules* **2018**, *23*, 116. [CrossRef]

29. Torchetti, E.M.; Brizio, C.; Colella, M.; Galluccio, M.; Giancaspero, T.A.; Indiveri, C.; Roberti, M.; Barile, M. Mitochondrial localization of human FAD synthetase isoform 1. *Mitochondrion* **2010**, *10*, 263–273. [CrossRef] [PubMed]
30. Giancaspero, T.A.; Busco, G.; Panebianco, C.; Carmone, C.; Miccolis, A.; Liuzzi, G.M.; Colella, M.; Barile, M. FAD synthesis and degradation in the nucleus create a local flavin cofactor pool. *J. Biol. Chem.* **2013**, *288*, 29069–29080. [CrossRef]
31. Gudipati, V.; Koch, K.; Lienhart, W.D.; Macheroux, P. The flavoproteome of the yeast Saccharomyces cerevisiae. *Biochim. Et Biophys. Acta* **2014**, *1844*, 535–544. [CrossRef]
32. Hao, H.X.; Khalimonchuk, O.; Schraders, M.; Dephoure, N.; Bayley, J.P.; Kunst, H.; Devilee, P.; Cremers, C.W.; Schiffman, J.D.; Bentz, B.G.; et al. SDH5, a gene required for flavination of succinate dehydrogenase, is mutated in paraganglioma. *Science* **2009**, *325*, 1139–1142. [CrossRef] [PubMed]
33. Ghezzi, D.; Goffrini, P.; Uziel, G.; Horvath, R.; Klopstock, T.; Lochmuller, H.; D'Adamo, P.; Gasparini, P.; Strom, T.M.; Prokisch, H.; et al. SDHAF1, encoding a LYR complex-II specific assembly factor, is mutated in SDH-defective infantile leukoencephalopathy. *Nat. Genet.* **2009**, *41*, 654–656. [CrossRef]
34. Robinson, K.M.; Lemire, B.D. Covalent attachment of FAD to the yeast succinate dehydrogenase flavoprotein requires import into mitochondria, presequence removal, and folding. *J. Biol. Chem.* **1996**, *271*, 4055–4060. [CrossRef] [PubMed]
35. Kim, H.J.; Winge, D.R. Emerging concepts in the flavinylation of succinate dehydrogenase. *Biochim. Biophys. Acta* **2013**, *1827*, 627–636. [CrossRef] [PubMed]
36. Rutter, J.; Winge, D.R.; Schiffman, J.D. Succinate dehydrogenase-Assembly, regulation and role in human disease. *Mitochondrion* **2010**, *10*, 393–401. [CrossRef]
37. Schiff, M.; Veauville-Merllie, A.; Su, C.H.; Tzagoloff, A.; Rak, M.; Ogier de Baulny, H.; Boutron, A.; Smedts-Walters, H.; Romero, N.B.; Rigal, O.; et al. SLC25A32 Mutations and Riboflavin-Responsive Exercise Intolerance. *N. Engl. J. Med.* **2016**, *374*, 795–797. [CrossRef]
38. Hellebrekers, D.; Sallevelt, S.; Theunissen, T.E.J.; Hendrickx, A.T.M.; Gottschalk, R.W.; Hoeijmakers, J.G.J.; Habets, D.D.; Bierau, J.; Schoonderwoerd, K.G.; Smeets, H.J.M. Novel SLC25A32 mutation in a patient with a severe neuromuscular phenotype. *Eur. J. Hum. Genet.* **2017**, *25*, 886–888. [CrossRef]
39. Tzagoloff, A.; Jang, J.; Glerum, D.M.; Wu, M. FLX1 codes for a carrier protein involved in maintaining a proper balance of flavin nucleotides in yeast mitochondria. *J. Biol. Chem.* **1996**, *271*, 7392–7397. [CrossRef]
40. Bafunno, V.; Giancaspero, T.A.; Brizio, C.; Bufano, D.; Passarella, S.; Boles, E.; Barile, M. Riboflavin uptake and FAD synthesis in Saccharomyces cerevisiae mitochondria: Involvement of the Flx1p carrier in FAD export. *J. Biol. Chem.* **2004**, *279*, 95–102. [CrossRef]
41. Barile, M.; Passarella, S.; Bertoldi, A.; Quagliariello, E. Flavin adenine dinucleotide synthesis in isolated rat liver mitochondria caused by imported flavin mononucleotide. *Arch. Biochem. Biophys.* **1993**, *305*, 442–447. [CrossRef]
42. Giancaspero, T.A.; Wait, R.; Boles, E.; Barile, M. Succinate dehydrogenase flavoprotein subunit expression in Saccharomyces cerevisiae–involvement of the mitochondrial FAD transporter, Flx1p. *FEBS J.* **2008**, *275*, 1103–1117. [CrossRef]
43. Gietz, R.D.; Schiestl, R.H. Frozen competent yeast cells that can be transformed with high efficiency using the LiAc/SS carrier DNA/PEG method. *Nat. Protoc.* **2007**, *2*, 1–4. [CrossRef]
44. Giancaspero, T.A.; Locato, V.; De Pinto, M.C.; De Gara, L.; Barile, M. The occurrence of riboflavin kinase and FAD synthetase ensures FAD synthesis in tobacco mitochondria and maintenance of cellular redox status. *FEBS J.* **2009**, *276*, 219–231. [CrossRef]
45. Bradford, M.M. A rapid and sensitive method for the quantitation of microgram quantities of protein utilizing the principle of protein-dye binding. *Anal. Biochem.* **1976**, *72*, 248–254. [CrossRef]
46. Pallotta, M.L.; Brizio, C.; Fratianni, A.; De Virgilio, C.; Barile, M.; Passarella, S. Saccharomyces cerevisiae mitochondria can synthesise FMN and FAD from externally added riboflavin and export them to the extramitochondrial phase. *FEBS Lett.* **1998**, *428*, 245–249. [CrossRef]
47. Martin, J.; Mahlke, K.; Pfanner, N. Role of an energized inner membrane in mitochondrial protein import. Delta psi drives the movement of presequences. *J. Biol. Chem.* **1991**, *266*, 18051–18057. [CrossRef]
48. Traglia, H.M.; O'Connor, J.P.; Tung, K.S.; Dallabrida, S.; Shen, W.C.; Hopper, A.K. Nucleus-associated pools of Rna1p, the Saccharomyces cerevisiae Ran/TC4 GTPse activating protein involved in nucleus/cytosol transit. *Proc. Natl. Acad. Sci. USA* **1996**, *93*, 7667–7672. [CrossRef] [PubMed]
49. Leadsham, J.E.; Gourlay, C.W. Cytoskeletal induced apoptosis in yeast. *Biochim. Biophys. Acta* **2008**, *1783*, 1406–1412. [CrossRef] [PubMed]
50. Lombardo, A.; Cereghino, G.P.; Scheffler, I.E. Control of mRNA turnover as a mechanism of glucose repression in Saccharomyces cerevisiae. *Mol. Cell. Biol.* **1992**, *12*, 2941–2948. [CrossRef]
51. Prieto, S.; De la Cruz, B.J.; Scheffler, I.E. Glucose-regulated turnover of mRNA and the influence of poly(A) tail length on half-life. *J. Biol. Chem.* **2000**, *275*, 14155–14166. [CrossRef]
52. Polakis, E.S.; Bartley, W. Changes in the enzyme activities of Saccharomyces cerevisiae during aerobic growth on different carbon sources. *Biochem. J.* **1965**, *97*, 284–297. [CrossRef]
53. Zahedi, R.P.; Sickmann, A.; Boehm, A.M.; Winkler, C.; Zufall, N.; Schonfisch, B.; Guiard, B.; Pfanner, N.; Meisinger, C. Proteomic analysis of the yeast mitochondrial outer membrane reveals accumulation of a subclass of preproteins. *Mol. Biol. Cell* **2006**, *17*, 1436–1450. [CrossRef] [PubMed]

54. Marc, P.; Margeot, A.; Devaux, F.; Blugeon, C.; Corral-Debrinski, M.; Jacq, C. Genome-wide analysis of mRNAs targeted to yeast mitochondria. *EMBO Rep.* **2002**, *3*, 159–164. [CrossRef] [PubMed]
55. Ahmed, A.U.; Fisher, P.R. Import of nuclear-encoded mitochondrial proteins: A cotranslational perspective. *Int. Rev. Cell Mol. Biol.* **2009**, *273*, 49–68.
56. Kellems, R.E.; Allison, V.F.; Butow, R.A. Cytoplasmic type 80S ribosomes associated with yeast mitochondria. IV. Attachment of ribosomes to the outer membrane of isolated mitochondria. *J. Cell Biol.* **1975**, *65*, 1–14. [CrossRef]
57. Eliyahu, E.; Lesnik, C.; Arava, Y. The protein chaperone Ssa1 affects mRNA localization to the mitochondria. *FEBS Lett.* **2012**, *586*, 64–69. [CrossRef]
58. Dichtl, B.; Keller, W. Recognition of polyadenylation sites in yeast pre-mRNAs by cleavage and polyadenylation factor. *The EMBO J.* **2001**, *20*, 3197–3209. [CrossRef]
59. Zhao, J.; Hyman, L.; Moore, C. Formation of mRNA 3′ ends in eukaryotes: Mechanism, regulation, and interrelationships with other steps in mRNA synthesis. *Microbiol. Mol. Biol. Rev.* **1999**, *63*, 405–445. [CrossRef]
60. Liu, H.; Han, H.; Li, J.; Wong, L. DNAFSMiner: A web-based software toolbox to recognize two types of functional sites in DNA sequences. *Bioinformatics* **2005**, *21*, 671–673. [CrossRef] [PubMed]
61. Shalgi, R.; Lapidot, M.; Shamir, R.; Pilpel, Y. A catalog of stability-associated sequence elements in 3′ UTRs of yeast mRNAs. *Genome Biol.* **2005**, *6*, R86. [CrossRef] [PubMed]
62. Kellis, M.; Patterson, N.; Endrizzi, M.; Birren, B.; Lander, E.S. Sequencing and comparison of yeast species to identify genes and regulatory elements. *Nature* **2003**, *423*, 241–254. [CrossRef]
63. Jin, C.; Myers, A.M.; Tzagoloff, A. Cloning and characterization of MRP10, a yeast gene coding for a mitochondrial ribosomal protein. *Curr. Genet.* **1997**, *31*, 228–234. [CrossRef]
64. Kitakawa, M.; Graack, H.R.; Grohmann, L.; Goldschmidt-Reisin, S.; Herfurth, E.; Wittmann-Liebold, B.; Nishimura, T.; Isono, K. Identification and characterization of the genes for mitochondrial ribosomal proteins of Saccharomyces cerevisiae. *Eur. J. Biochem.* **1997**, *245*, 449–456. [CrossRef]
65. Yogev, O.; Pines, O. Dual targeting of mitochondrial proteins: Mechanism, regulation and function. *Biochim. Et Biophys. Acta* **2011**, *1808*, 1012–1020. [CrossRef]
66. Bader, G.; Enkler, L.; Araiso, Y.; Hemmerle, M.; Binko, K.; Baranowska, E.; De Craene, J.O.; Ruer-Laventie, J.; Pieters, J.; Tribouillard-Tanvier, D.; et al. Assigning mitochondrial localization of dual localized proteins using a yeast Bi-Genomic Mitochondrial-Split-GFP. *eLife* **2020**, *9*, e56649. [CrossRef] [PubMed]
67. Ben-Menachem, R.; Pines, O. Detection of Dual Targeting and Dual Function of Mitochondrial Proteins in Yeast. *Methods Mol. Biol.* **2017**, *1567*, 179–195. [CrossRef] [PubMed]
68. Ben-Menachem, R.; Tal, M.; Shadur, T.; Pines, O. A third of the yeast mitochondrial proteome is dual localized: A question of evolution. *Proteomics* **2011**, *11*, 4468–4476. [CrossRef]
69. Kisslov, I.; Naamati, A.; Shakarchy, N.; Pines, O. Dual-targeted proteins tend to be more evolutionarily conserved. *Mol. Biol. Evol.* **2014**, *31*, 2770–2779. [CrossRef]
70. Dinur-Mills, M.; Tal, M.; Pines, O. Dual targeted mitochondrial proteins are characterized by lower MTS parameters and total net charge. *PLoS ONE* **2008**, *3*, e2161. [CrossRef] [PubMed]
71. Barile, M.; University of Bari 'Aldo Moro', Bari, Italy. Personal communication, 2021.
72. Lashkevich, K.A.; Dmitriev, S.E. mRNA Targeting, Transport and Local Translation in Eukaryotic Cells: From the Classical View to a Diversity of New Concepts. *Mol. Biol.* **2021**, 1–31. [CrossRef]
73. Andreassi, C.; Riccio, A. To localize or not to localize: mRNA fate is in 3′UTR ends. *Trends Cell Biol.* **2009**, *19*, 465–474. [CrossRef]
74. Sparks, K.A.; Dieckmann, C.L. Regulation of poly(A) site choice of several yeast mRNAs. *Nucleic Acids Res.* **1998**, *26*, 4676–4687. [CrossRef]
75. Turner, R.E.; Pattison, A.D.; Beilharz, T.H. Alternative polyadenylation in the regulation and dysregulation of gene expression. *Semin. Cell Dev. Biol.* **2018**, *75*, 61–69. [CrossRef]
76. Sylvestre, J.; Vialette, S.; Corral Debrinski, M.; Jacq, C. Long mRNAs coding for yeast mitochondrial proteins of prokaryotic origin preferentially localize to the vicinity of mitochondria. *Genome Biol.* **2003**, *4*, R44. [CrossRef] [PubMed]
77. Horton, P. A simulation study of mRNA diffusion and mitochondrial localization. *bioRxiv* **2019**. [CrossRef]
78. Gerber, A.P.; Herschlag, D.; Brown, P.O. Extensive association of functionally and cytotopically related mRNAs with Puf family RNA-binding proteins in yeast. *PLoS Biol.* **2004**, *2*, e79. [CrossRef] [PubMed]
79. Garcia-Rodriguez, L.J.; Gay, A.C.; Pon, L.A. Puf3p, a Pumilio family RNA binding protein, localizes to mitochondria and regulates mitochondrial biogenesis and motility in budding yeast. *J. Cell Biol.* **2007**, *176*, 197–207. [CrossRef] [PubMed]
80. Rowe, W.; Kershaw, C.J.; Castelli, L.M.; Costello, J.L.; Ashe, M.P.; Grant, C.M.; Sims, P.F.; Pavitt, G.D.; Hubbard, S.J. Puf3p induces translational repression of genes linked to oxidative stress. *Nucleic Acids Res.* **2014**, *42*, 1026–1041. [CrossRef] [PubMed]
81. Quenault, T.; Lithgow, T.; Traven, A. PUF proteins: Repression, activation and mRNA localization. *Trends Cell Biol.* **2011**, *21*, 104–112. [CrossRef]
82. Saint-Georges, Y.; Garcia, M.; Delaveau, T.; Jourdren, L.; Le Crom, S.; Lemoine, S.; Tanty, V.; Devaux, F.; Jacq, C. Yeast mitochondrial biogenesis: A role for the PUF RNA-binding protein Puf3p in mRNA localization. *PLoS ONE* **2008**, *3*, e2293. [CrossRef]
83. Freeberg, M.A.; Han, T.; Moresco, J.J.; Kong, A.; Yang, Y.C.; Lu, Z.J.; Yates, J.R.; Kim, J.K. Pervasive and dynamic protein binding sites of the mRNA transcriptome in Saccharomyces cerevisiae. *Genome Biol.* **2013**, *14*, R13. [CrossRef] [PubMed]

84. Lapointe, C.P.; Stefely, J.A.; Jochem, A.; Hutchins, P.D.; Wilson, G.M.; Kwiecien, N.W.; Coon, J.J.; Wickens, M.; Pagliarini, D.J. Multi-omics Reveal Specific Targets of the RNA-Binding Protein Puf3p and Its Orchestration of Mitochondrial Biogenesis. *Cell Syst.* **2018**, *6*, 125–135.e6. [CrossRef] [PubMed]
85. Chacinska, A.; Koehler, C.M.; Milenkovic, D.; Lithgow, T.; Pfanner, N. Importing mitochondrial proteins: Machineries and mechanisms. *Cell* **2009**, *138*, 628–644. [CrossRef] [PubMed]
86. Gong, F.; Yanofsky, C. Instruction of translating ribosome by nascent peptide. *Science* **2002**, *297*, 1864–1867. [CrossRef] [PubMed]

MDPI
St. Alban-Anlage 66
4052 Basel
Switzerland
Tel. +41 61 683 77 34
Fax +41 61 302 89 18
www.mdpi.com

Life Editorial Office
E-mail: life@mdpi.com
www.mdpi.com/journal/life

www.ingramcontent.com/pod-product-compliance
Lightning Source LLC
LaVergne TN
LVHW070952170726
843515LV00005B/973